BEHAVIOURAL RESEARCH FOR TRANSPORT POLICY

Of related interest ...

New Survey Methods in Transport
2nd International Conference,
Hungerford Hill, Australia 1983.

Proceedings of the Ninth International
Symposium on Transportation and
Traffic Theory, Delft, 1984.

BEHAVIOURAL RESEARCH FOR TRANSPORT POLICY

The 1985 International Conference on Travel Behaviour

Noordwijk, The Netherlands
16-19 April 1985

VNU SCIENCE PRESS

Utrecht, The Netherlands
1986

VNU Science Press BV
P.O. Box 2073
3500 GB Utrecht
The Netherlands

First published 1986

CIP-DATA KONINKLIJKE BIBLIOTHEEK, DEN HAAG

Behavioural research for transport policy: The 1985 International Conference on Travel Behaviour: 16-19 April 1985, Noordwijk, the Netherlands - Utrecht: VNU Science Press. - ILL.
ISBN 90-6764-060-3 bound.
SISO 377.4 UDC 656.1.001.5:351.81.
Subject headings: travel behaviour; research/transport policy.

Printed in The Netherlands by ICG Printing, Dordrecht

Contents

Preface

Transport policy has traditionally dealt with technical matters; construction of waterways, ports, roads and railways, all require techinical skills of different sorts. Also, the construction of networks was often required for strategic reasons: roads constructed by the Roman empire 2000 years ago and under Napoleon 200 years ago still influence the road network of Europe.

People responsible for transport policy decisions therefore consulted people with technical and/or strategic expertise, or in some cases they possessed this expertise themselves. In several countries, the civil engineers (ingénieurs des ponts et chaussées) formed one of the most prestigious groups of the civil service.

Gradually economics played an increasingly important role as policy makers were confronted with scarce resources and pricing problems: the considerable sums needed for investing in the road system had to be found from public funds as there was no appropriate market for infrastructure use. In many cases this was arranged by levying special taxes from road users. The nonexistent, or at least imperfect, market for road use gave rise to difficult situations for other transport modes, and the normal rules of the market system could no longer apply.

Although economics is a behavioural science by definition - it deals with people handling scarce resources to satisfy their needs - it is not always recognized as such. Often people consider it to be restricted to finance and efficiency. More recently, methodologies such as social cost-benefit calculations have been used and developed to give a wider application of economic science, but critics considered that still too limited, since it did not consider effects unquantifiable in monetary terms.

Even the restricted form of economics was placed secondary to engineering: some 50 years ago, it was very strange for an economist to be chairman of a large railway company, and even now it is uncommon for the head of the road directorate of a national government not to be trained as an engineer.

Supported by the technical sciences and economics many governments and transport companies realized infrastructure, vehicles and operating systems of very high quality indeed. The very bad state of repair of many roads and bridges in the United States is clearly not due to lack of technical skills in the field of maintenance, but to insufficient consideration of the economics of road infrastructure.

Nevertheless, even in countries where the transport system could obtain enough resources for its operation and extension the results are far from ideal:

- traffic safety on the road system still leaves much to be desired;
- public transport systems suffer from high deficits, insufficient levels of service, or both;
- road congestion is now normal in and around large cities;
- spatial organization of activities on the basis of generalized car use makes life very difficult for those not using this mode of travel;
- longer distances and dangerous crossings restrict the travel possibilities of unaccompanied children, obliging older people to travel with them.

These shortcomings have not arisen out of ill-will or negligence, but are due to insufficient consideration of the real purpose of the transport system: that is to serve people. The needs of these individuals, and the spatial interactions that accompany them, have often not been recognized, and therefore transport policies have, while satisfying some needs, indirectly created others.

Until now it has not been generally recognized that travel is an important aspect of human behaviour, and that effective policy in the field should start from the idea that it is not public authorities or transport enterprises that decide how and when people will travel, but the individuals themselves. Behavioural research should provide information on people's response to policy measures. It is of no use to build roads with the objective of relieving congestion if this creates congestion elsewhere. It is of no use to run a high-quality train service if it does not provide the amenities that potential users require.

Those whose task it is to actually take policy decisions or give advice to politicians have been accused of disregarding important aspects of their field, and aiming at results that seem beautiful, but are not always relevant. We should realize, however, that nowadays many policy makers work under circumstances that prevent them from thinking in a relaxed way about the subject matter they are responsible for, and from communicating with others working on the same problems. However, those on the research side may develop some hope that one day policy makers will discover that they really need their work.

The main problem with behavioural research on transport is not so much that it is insufficiently advanced, but that is split into many different approaches, in some cases without intercommunication. Some approaches are dictated by mathematical considerations, others by types of data collection, and others again by the context in which they see people's travel. A policy maker wanting to obtain information from behavioural research would have considerable difficulty on deciding which stream to address and would obtain very different answers depending on which expert he or she contacts.

Another problem for the policy maker is that research results are not always easy to understand: conclusions are often vague and subject to conditions, and many reports are written in jargon containing words unknown or having quite a different meaning to that in common use. In a way, the more mathematical approaches are at an advantage: people tend to consider formulae incomprehensible, but by definition right, and they accept the results without discussion, whereas a written text that cannot be understood causes doubt in the reader, which is added to the doubts already expressed in the text itself.

Publications on behavioural research might be better placed in *Le Monde, The Times, Washington Post* or *NRC Handelsblad*, making them more accessible and understandable to the general public, and those responsible for policy decisions.

Another area in which research tends to fall short is the timing of contributions. A government installed for a period of four years does not want to wait three of them until questions to researchers have been answered. Of course, there may be good arguments to ask for a sufficient period to complete appropriate research, but this should not prevent researchers from telling immediately what they can contribute now. The choice is often either between results of high academic value and low practical value, or the converse. So, behavioural researchers should alter their own behaviour first and become far more cooperative, responsive, understandable and active. They should actually develop answers to questions before these have been tabled, and discuss important policy issues publicly. Perhaps they can take as an example political economists, who regularly discuss issues of general economic policy, addressing themselves to the public in an understandable way, criticizing government or opposition alike. In fact, all people travel, most people use cars, why cannot a learned exposé on the background of travel draw enough interest to be acceptable for the opinion page of a large newspaper?

What is needed in the first place is communication: policy makers should take the time to explain to researchers what the issues are that they are faced with; researchers should tell policy makers their conclusions in plain words. Neither of them should hesitate to communicate with those that are most influenced by their work: the users of the transport system.

We now come to the question of why we need international contacts. Large countries like the United States or France, that have spent considerable resources on behavioural research in transport, might be able to go their own way. Nevertheless, much can be learned by mutually comparing methods and results. However, for a small country, even if it traditionally spends important efforts on transport and research, and therefore also on transport research, it is essential to draw on work done in other countries. We should realize that, notwithstanding important differences in economic and social situations and political approaches, our problems and solutions are far more similar then one would expect at first sight.

Therefore, this and similar conferences are important as a stimulus and meeting place, forming the starting point for new work and further contacts.

Aad Rühl
Chairman of the
International Association
for Travel Behaviour

(Those wanting information on activities in the field of travel behaviour can write to: The International Association for Travel Behaviour, c/o PTRC, 110 Strand, London WC2R 0AA; attention Ms Elizabeth Smith.)

List of participants

Australia

D. Hensher
Macquarie University
School of Economics
North Ryde 2113

Denmark

P. Andersen
Danish State Railways
Passagert jenesten
Sølvgade 40
1349 København

J. Keldborg
Ministry of the Environment
Miljøministeriet Udviklingskontoret
Slotholmsgade 12
DK-1215 København

France

P. Bonnel
Ecole Nationale des Travaux
Publics de l'Etat
Rue M-Audin
69120 Vaux-en-velin

P. Fargier
CETUR
8 Avenue Aristide Briand
Bagneux 92220

M. Fenois
Transexel/Transport en Commun Lyonnais
50 Cours Lafayette
69423 Lyon

M. Houée
Ministère de l'Urbanisme
du Logement et des Transports SAEP
55-57 Rue Brillat Savarin
75013 Paris

A. Lasserre
SNCF, Direction Commerciale
Voyageurs Département Marketing
10 Place du Budapest
75436 Paris Cedex 09

F. Loiseau-Van Baerle
Union des Transports Publics
5 Rue d'Aumale
75009 Paris

R. Marche
Institut de Recherche des Transports CESA
2 Avenue Général Malleret
Joinville
94114 Arcueil Cedex

F. Moïsi
Ministère de l'Urbanisme du Logement et des Transports
Direction Terrestres/S.A.E.S.T.
244 Boulevard Saint-Germain
75007 Paris

F. Potier
Institut de Recherche des Transports
2 Avenue Général Malleret
Joinville
94114 Arcueil Cedex

E. Quinet
Ecole nationale des Ponts et Chaussees
Dept Economie et Sciences Sociales
28 Rue des Saints-Pères
75007 Paris

C. Raux
Laboratoire d'Economie des Transports
16 Quai Claude Bernard
69007 Lyon

Finland

S. Parkko
Roads and Waterways Administration
P.O. Box 33
00152 Helsinki

Germany

U. Barth
Institut für Regionalwissenschaft der Universität Karlsruhe
Postfach 6380
7500 Karlsruhe

U. Kunert
D.I.W.
Königin-Luise strasse 5
D-1000 Berlin 33

H. Nüsser
Deutsche Forschungs- und Versuchsanstalt für Luft- und Raumfahrt
Linder höhe
D-5000 Köln 90

A. Schwipps
Technische Universität
Institut für Verkehrsplanung und Verkehrswegebau
Str. d. 17 Juni 135
1000 Berlin 12

H. Verron
Technische Universität
Str. d. 17 Juni 135
1000 Berlin 12

D. Wilken
DVFLR German Aerospace Research Establishment
Linder höhe
D-5000 Köln 90

W. Young
c/o Monash University, Australia
Institüt für Verkerswesen
Universität Karlsruhe
Postfach 6380
D-7500 Karlsruhe

Hungary

E. Nagy
Institute for Transport Sciences
Thán K.U. 3/5
1119 Budapest

The Netherlands

A. Baanders
Ministerie van Verkeer en Waterstaat
Postbus 20901
2500 EX Den Haag

L. van den Berg
Erasmus Universiteit
Postbus 1738
3000 DR Rotterdam

S. Bexelius
Ministerie van Verkeer en Waterstaat
Postbus 20901
2500 EX Den Haag

P. Bovy
Institute for Town Planning
Delft University of Technology
Berlageweg 1
2600 GA Delft

A. van den Broecke
Social Research
Van Nijenrodeweg 355
1982 HD Amsterdam

F. le Clerq
Physical Planning Department
Room 8020
Wibautstraat 3
1091 GH Amsterdam

A. Daly
Cambridge Systematics Europe
Laan van Meerdervoort 32
2517 AL Den Haag

F. van Dam
Ministerie van Verkeer en Waterstaat
Postbus 20901
2500 EX Den Haag

H. van Eijs
Emancipatieraad
Lutherse Burgwal 10
2512 CB Den Haag

M. van Es
Gemeente Amsterdam
Afd. Verkeer en Vervoer
O.Z. Voorburgwal 282
Amsterdam

J. Gantvoort
Institute for Town Planning Research
Delft University of Technology
P.O. Box 5043
2600 GA Delft

J. Golob
Ministerie van Verkeer en Waterstaat
Postbus 20901
2500 EX Den Haag

Th. Golob
Bureau Goudappel en Coffeng
Parkweg 4
Postbus 161
7400 AD Deventer

V. de Groot
Stuurgroep Verkeer en Vervoer
Ministerie van Verkeer en Waterstaat
Postbus 20901
2500 EX Den Haag

C. Gundy
Stichting Wetenschappelijk Onderzoek Verkeersveiligheid
Postbus 170
2260 AD Leidschendam

H. Gunn
Cambridge Systematics Europe
Laan van Meerdervoort 32
2517 AL Den Haag

D. Hatch
N.V. Nederlandse Spoorwegen
Postbus 2025
3500 HA Utrecht

H. Hoestra
Vereniging Reizigers Openbaar Vervoer
Herenstraat 64
3911 JG Rhenen

G. Jansen
Institute for Town Planning
Delft University of Technology
Berlageweg 1
2628 CR Delft

A. van der Hoorn
Ministerie van Verkeer en Waterstaat
Postbus 20901
2500 EX Den Haag

H. Katteler
Institute for Applied Sociology
Graafseweg 274
6532 ZV Nijmegen

L. Klaassen
Nederlands Economisch Instituut
Burgemeester Oudlaan 50
3063 PA Rotterdam

M. Klok
Ministerie van Verkeer en Waterstaat
Postbus 20901
2500 EX Den Haag

C. van Knippenberg
Verkeerskundig Studiecentrum
Rijksuniversiteit Groningen
Rijksstraatweg 76
9752 AH Haren

E. Kroes
Nederlands Vervoerswetenschappelijk
Instituut
Polakweg 12
2288 GG Rijswijk

W. van Lierop
Economic and Social Institute
Free University
P.O. Box 7161
1007 MC Amsterdam

M. van Maarseveen
Traffic and Transportation Group TNO
P.O. Box 45
2600 AA Delft

H. Pol
Ministerie van Verkeer en Waterstaat
Postbus 20901
2500 EX Den Haag

J. van Rens
Nederlands Vervoerswetenschappelijk
Instituut
Polakweg 13
2288 GG Rijswijk

A. Smyth
Cambridge Systematics Europe
Laan van Meerdervoort 32
2517 AL Den Haag

L. Schreurs
Instituut voor Longitudinaal Beleids-
onderzoek
Singel 399
1912 WN Amsterdam

F. Timmerman
Rijksplanologische Dienst
Ministerie VROM
Willem Witsenplein 6
2596 BK Den Haag

J. Vogelaar
Nederlands Economisch Instituut
Burgemeester Oudlaan 50
3062 PA Rotterdam

M. van Wamelen
Ministerie van Verkeer en Waterstaat
Postbus 20901
2500 EX Den Haag

L. van Wissen
Bureau Goudappel en Coffeng
Parkweg 4
Postbus 161
7400 AD Deventer

Portugal

L. Countinho
Companhia Carris de Ferro de Lisboa
Rua Primeiro de Maio 101
1399 Lisboa

South Africa

R. Sweet
National Institute for Transport and Road
Research
Council for Scientific and Industrial Research
Navorspad, Box 395
Pretoria 0001

Sweden

A. Lindeberg
Statistics
S-11581 Stockholm

S. Algers
Stockholm County Council
Public Transport Board
Sergels Torg, 12 tr
S-11157 Stockholm

R. Waara
TRUM
University of Umea
S-90187 Umea

S. Widlert
Stockholm County Council
Public Transport Board
Sergels Torg 12, 12tr
S-11157 Stockholm

Switzerland

H. Berger
Eidgenössiches Verkehrs- und Energie-
wirtschaftsdepartem.
Effingerstrasse 14
3003 Bern

M. Keller
Institut für Verkehrsplanung
Eidg. Technische Hochschule
ETH-Hönggerberg
CH-8093 Zürich

United Kingdom

D. Ashley
The MVA Consultancy
112 Strand
London WC2R 0AA

J. Bates
John Bates Services
The Old Coatch House
Wyke, Gillingham
Dorset

M. Bradley
Transport Studies Unit
Oxford University
11 Bevington Road
Oxford OX2 6NB

M. Clarke
Transport Studies Unit
Oxford University
11 Bevington Road
Oxford OX2 6NB

E. Forsyth
Transport Studies Unit
Oxford University
11 Bevington Road
Oxford OX2 6NB

P. Goodwin
Transport Studies Unit
11 Bevington Road
Oxford OX2 6NB

Y. Hayashi
Department of Civil Engineering
Nagoya University
Furo-cho, chikusa-ku
Nagoya 464
Japan
c/o (in 1985)
Institute for Transport Studies
University of Leeds
Leeds LS2 9JT

P. Hills
University of Newcastle upon Tyne
Division of Transport Engineering
Claremont Tower
Newcastle upon Tyne NE1 7RU

A. Hitchcock
Transport and Road Research Laboratory
Crowthorne RG11 6AU

A. Layzell
Transport Studies Unit
11 Bevington Road
Oxford OX2 6NB

L. Pickup
Transport and Road Research Laboratory
Crowthorne RG11 6AU

J. Segal
British Rail
Rail House
Euston Square, P.O. Box 100
London NW1 2DZ

R. Sheldon
Steer Davis & Gleave
11 Worple Road
Richmond TW10 6DG

N. Wrigley
University of Bristol
Department of Geography
Bristol BS8 1SS

USA

M. Ben-Akiva
Massachusetts Institute of Technology
77 Mass. Avenue
Cambridge, MA 02139

J. Benjamin
Economics Department of Transportation
Institute
North Carolina Agrocultural and Technical
State University
Greensboro, North Carolina

D. Hartgen
New York State Department of
Transportation
1220 Washington Avenue
State Campus
Albany, NY 12232

C. McKnight
Urban Transportation Center
University of Illinois at Chicago
P.O. Box 4348
Chicago, IL 60680

R. Kitamura
University of California at Davis
Department of Civil Engineering
Davis, CA 95616

L. Kostyniuk
Civil Engineering Department
Michigan State University
East Lansing, MI 48824

V. Kozel
The World Bank
1818 H-Street, NW
Washington, DC 20433

J. Louvière
Dept of Marketing
University of Iowa
P.H.B.A.
Iowa City, IA 52242
from 15 August 1985:
Dept of Marketing and Economic Analysis
Faculty of Business
University of Alberta
Edmonton, Alberta T69 2GI
Canada

A. Meyburg
Cornell University
Dept of Environmental Engineering
305 Hollister Hall
Ithaca, NY 14853

E. Pas
Duke University
Dept of Civil Engineering
Durham, NC 27706

W. Recker
Institute of the Transportation Studies
University of California
Irvine, CA 92717

E. Weiner
Office of the Secretary
Department of Tranportation
400 7th Street SW
Washington, DC 20590

1
Activity analysis and trip chaining

Behavioural Research for Transport Policy
pp. 3–15. © 1986 VNU Science Press.

ACTIVITY MODELLING - A RESEARCH TOOL OR A PRACTICAL PLANNING TECHNIQUE?

Mike Clarke
Oxford University Transport Studies Unit

Models of how people spend their time have been under development for at least ten years, and they have been considered for applications in transport planning for at least eight. After that length of time, one would expect that the approach should be showing some practical returns for the effort expended on its development. This paper attempts to take stock of how far that expectation has been achieved, and discusses how far the approach has lived up to the claims made for its potential utility. In doing so, some alternative approaches to activity modelling are compared and the essential features of the approach are discussed. The points which are raised are illustrated by describing the model which has been built at Oxford University, and its application in a case study of rural bus services.

The phrase 'activity modelling' does not have a unique interpretation; study of the literature reveals that different people have sometimes very different views of what it entails. It may therefore be worth commencing this paper by restating the rationale behind the approach as it has been applied at Oxford.

The essential theme of the activity approach to transport modelling is the role of time in every-day travel behaviour. Travel inevitably involves movement through space and this has always been the central focus of transport planning models. In conventional models time has usually been considered merely as one of the 'costs' incurred in making a trip; suggesting that time is a resource like money which is consumed by travel. However we know that time is in fact a very special commodity which unlike money cannot be saved for some future eventuality; it 'spends' itself inexorably - all we can do is choose how (within certain limits) we occupy each unit of time as it passes. Activity models set out to simulate the way in which people make those choices, subject to all the constraints which are imposed by the physical and social environment in which they live.

Not only is individual travel behaviour greatly influenced by the way in which time is organised; many important transport policy questions are also dependent on detailed aspects of the temporal dimension. The scale of facilities needed for transport whether by private or public modes depends not on the total number of people travelling between two points, but rather on the number travelling between those points at any one time. As an alternative to the costly and often socially and environmentally undesirable policy of providing for peak demand, planners have available the alternative of investigating, and possibly modifying, the behaviour which causes the peaks to exist (see for example South Australian Energy Research Associates, 1983).

THE ROLE OF ACTIVITY MODELS

The models are based on an attempt to go beyond the notion that if we simply observe the trips made by a sample of individuals we are in a position both to explain their behaviour and to predict how they would react to changes to their personal or external circumstances. In an attempt to get at a more basic motivation for travel behaviour we look at the activities which a person undertakes, and how these are organised into a schedule describing the way in which that person spends his/her time. Travel is relegated to a supporting role - as a mechanism whereby the individual can move through space to complete a desired sequence of activities.

It is important to realise that, while activity approaches are potentially able to deal with some situations which could not be handled by a trip-based formulation, there are very many circumstances in which trip-based aproaches are entirely adequate, and the added sophistication, complexity and expense of an activity-based approach would in such cases be a positive disadvantage. What the activity approach can potentially do is to provide added realism to predictions of changes in behaviour when the policy under study is one which directly concerns questions of timing. Amongst these one would include public transport timetabling, transport policies which apply differentialy to peak and off-peak travel and 'secondary' effects such as flexible work hours or alterations to the timing of the availability of other activities. In addition to these policy measures there are a number of important topics of behaviour which have a strong time component. The whole question of trip-chaining for example is difficult to consider unless activities are taken into account, and yet the amalgamation of several activities into a single home to home circuit may be an important response to some policy measures. Similarly, the use made of a family car by non-working members of the family may give rise to far-reaching and unexpected effects from a policy aimed at reducing car use for journeys to work.

A STATEMENT OF THE ACTIVITY APPROACH

The basic premiss of the activity model, then, (and all authors seem to be agreed on this point) is that travel should be viewed as a derived demand resulting from the need to participate in activities at different locations. However there are detailed differences between alternative statements of the activity approach, and not all models which claim to be based on activity methods fall into the realm considered in this paper. The salient points of what this author means by 'the activity approach' are as follows:-

(a) The primary 'need' is to take part in a set of activities which depend on one's circumstances preferences and obligations.

(b) The 'problem' then is to develop a schedule of these activities which can be performed subject to the various constraints which limit our behaviour.

(c) These constraints stem from various sources....

- Obligations We must attend work (or school, or hospital appointments etc) at particular times and at particular locations.

- Institutional considerations Shops, libraries, pubs etc. are only open at particular times of day.

- Transport considerations If successive activities in a schedule take place at different locations then we must allow time to travel from one to another. Public transport is only available at particular times between particular locations; there may be periods of time during which private transport is not available (when a family's only car is in use by someone else).

- Inter-personal effects Most people react with other people while carrying out an activity programme - of these interactions those which actually shape the activity schedule are mostly within the family.

(d) In general there will be a number of alternative activity schedules which are feasible according to the above criteria. Each will have advantages and disadvantages for the subject, and he/she will choose to adopt that schedule which on balance suits him/her best.

SOME DETAILED CONSIDERATIONS

In a very perceptive article Holly (1978) emphasises the problem of scale in time-space research. At the outset it is important to establish the degree of resolution at which we expect any activity model to apply. All the evidence suggests that we should be studying discrete events in time (activities) rather than some aggregate measure of behaviour. This implies a detailed (minute-by-minute) representation of the temporal dimension so that subtleties of the interaction between time spent in different ways can be accommodated. If this level of detail is somewhat daunting, some recompense can be gained from the fact that a corresponding simplification in the level of spatial detail may often be admissable.

The facets of behaviour in which we are interested tend to be particular to a single person in a given situation. Although some aspects of spatio-temporal constraints are common to everyone (shop opening hours for example), most of the factors influencing the form of someone's activity schedule are properties of that person's circumstances. His home address, family structure, work obligations - not to mention his personal preferences - are all likely to have a profound effect on his activity behaviour. It therefore really only makes sense to talk about activity schedules of a particular person, and so our model has to be truly disagregate - making predictions of the behaviour of a single person. At the same time we must recognise the great importance of interactions with other members of that person's household. The unit for modelling should therefore generally be the individual in the context of the household. Although we must inevitably consider the detailed behaviour of individuals, in most cases it will be desirable to be able to predict the behaviour of populations or sub-groups of the population. We are therefore faced with the need to be able to generalize our individual predictions into a more useful aggregate prediction.

It is important to be clear what time-scale applies to the reaction we are setting out to model. There is in fact a continuum of decisions at different levels - some detailed and short-term, others major and long-term - which a person is constantly making. And these decisons are all interrelated, since the result of one is likely to affect the making of others. This system may never actually be in equilibrium, neither internally with respect to itself nor with respect to external stimuli. What we observe as human behaviour is some sort of instantaneous picture of a complex unstable dynamic process. The complete system is extrordinarily difficult to define, but for our purposes it is convenient to think of a hierarchy of responses ranging from long- to short- term. Such a hierarchy (adapted from Salomon, 1983) is presented in Figure 1. The precise form of the hierarchy is of course open to argument. However consideration of this hierarchy does allow us to define the limits of application of any particular model, be it based on activities or otherwise. In general, activity models will be addressed at questions involving the lower levels of figure 1, but given the interrelation between different levels of the hierarchy they are also likely to have something to offer to the modelling of

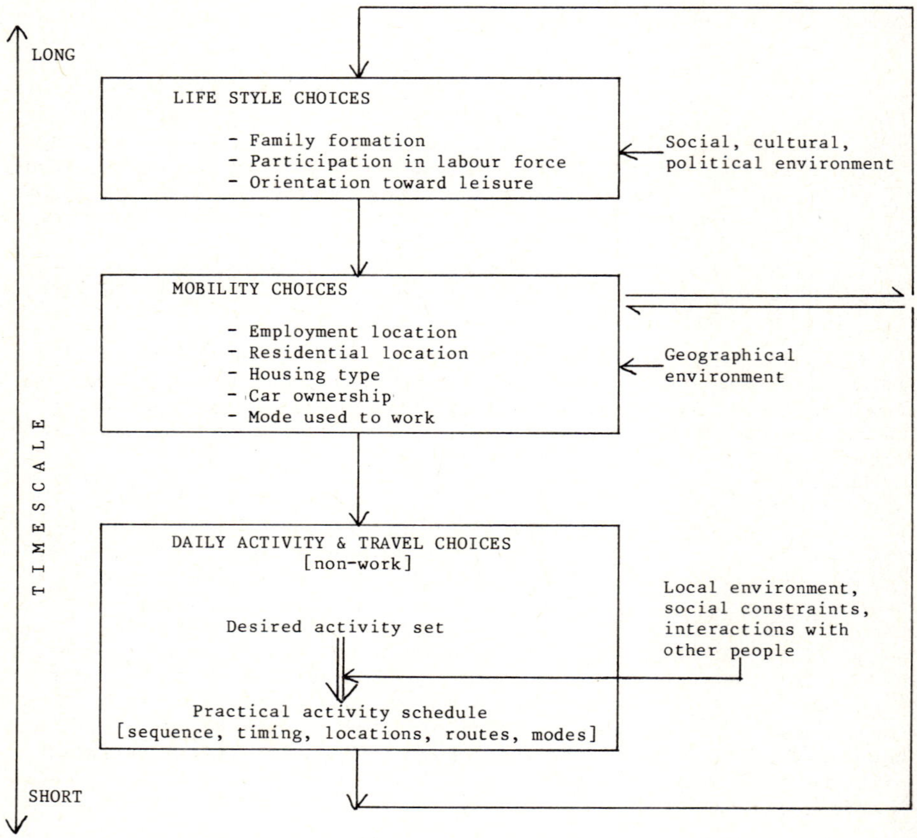

FIGURE 1 - An hierarchy of behavioural choices (based on Salomon, 1983)

longer term decisions.

The idea of an activity schedule for an individual carries with it no specific time interval over which that schedule should be defined. Any self-contained time interval could be used ranging from 'the lunch-hour' to a week or even a month. The time interval which most naturally suggests itself is one day since it is conveniently bounded by periods of sleep. Nevertheless activity schedules can be, and often are, planned over longer time periods; and one week might be more appropriate for some applications.

Finally we have to ask ourselves: 'What exactly are we modelling?' There are philosophical problems caused by the very detailed level at which this kind of model works. In an aggregate model we can gloss-over the question of the typicality of the day we are modelling since such questions can be averaged out over a large number of members of the sample. In a model of a single day's behaviour of a single person however we really should address the question explicitly. Are we interested in a 'typical' day (whatever that means), or a specific day which we happened to sample in the survey on which the model is based. It is clear that the variation between the behaviour of the same person on different days can be greater than the variation of behaviour between different individuals observed on the same day. In this case, how should we choose what days to consider? A random day (or days) for respondent? each person's major shopping day?

Those points which are raised above and which have not yet been resolved need consideration before activity modelling can be of very much practical use to the policy-maker.

THE RANGE OF APPROACHES TO ACTIVITY MODELLING

What follows is not meant to be a comprehensive review of the activity-modelling literature, but rather a short description of some typical examples of various approaches to the subject. For a more complete description of both the development and the state of the art of activity modelling and analysis, the reader is referred to Damm (1983) and Jones (1983).

There is a class of models which can be described as activity-based but which we will not consider in detail here. These models look only at time budgets - the aggregate way in which time is spent - and while such models reveal much of interest about activity behaviour, and they are likely eventually to be used to provide an input to activity scheduling models - they lie outside the scope of the current paper.

Of more direct relevance are those models which take account of people's activity patterns; i.e. not only what activities they take part in, but how those activities are arranged in time. In considering alternative formulations which have been proposed or developed it is convenient to adopt the classification of Stephens (1976) who recognizes two general approaches: one which looks at the choices made by people in arriving at a pattern of behaviour, and a contrasting approach which concentrates on the constraints under which those choices are made.

The choice-based models tend to aim to explain the observed activity schedules by calibrating individual-choice models relating some description of the schedule to various independent variables. Damm (1979), for example, used models which predict whether or not a person participates in a particular activity during a particular time interval, and given participation, a second model of activity duration.

The models he developed include terms which stand proxy for interactions between behaviour in different periods of the day.

Van der Hoorn (1983) uses equations which predict location choice for a particular activity depending on the previous location and (separately) activity choice as a function of the previous location and the period of the day. Only 'discretionary' activities are included in the model; 'mandatory' activities act as 'pegs' around which the others are modelled. Some degree of constraint is applied so that some activities are available only at certain times and at certain locations and time must be allowed for participation in 'mandatory' activities. Variables are included which represent linkages between each activity and the preceding and succeeding ones.

This model has been used to investigate the effect of hypothetical scenarios such as reductions in working hours and in real income.

Perhaps the most complete specification of this type of model is that developed by Hirsh et al (1984). Although described as a model of 'weekly' behaviour, like most other activity models this one divides a period of time into a number of sub-periods and predicts probabilities of activity participation in each. The system of equations in this model is very rigorously developed but has to be simplified somewhat before the model can be applied in practice. This model too has been used to investigate the effect of reduced working hours.

The second group of models comprises those which consider detailed paths through time and space taking particular note of the constraints which circumscribe those paths. This is what we shall refer to as an activity scheduling model. Stephens (1976) presents a Monte-Carlo simulation which traces the space-time paths between fixed 'pegs' of mandatory activities, also taking into account constraints which are largely self-imposed (so called 'subjective constraints'). Of particular interest is the fact that the survey on which this model is based attempted to gather measures of subjective constraints from respondents themselves. Stephens states that the aim was to use the simulation as a descriptive technique rather than a predictive one, and his findings include the fact that the subjective constraints were rather less influentual in determining behaviour than he had expected them to be. Despite the lack of expanatory power provided by the technique he concludes that simulation provides the most appropriate approach to activity scheduling since purely anaytical solutions are precluded by the complexity of the processes involved.

The Lund School's contribution to activity model also takes the form of a simulation, known as PESASP (Lenntorp 1978). This model traces people's paths through space and time with particular emphasis on the opportunities open to an individual as a result of the situation in which he finds himself. It would typically be used for example to enumerate the number of people who could have access to a particular facility in a particular window of time.

The last model mentioned in this brief summary is 'STARCHILD' developed by Recker and his colleagues at the University of California at Irvine (Recker et al 1983). This is a complex system of models which combines simulation, pattern recognition, optimization and individual choice techniques. The model is designed to generate alternative activity schedules in a combinatorial manner from a planned set of activities to be undertaken by an individual, and treats this as a choice set from which a particular course of action is chosen. This system has been developed to the stage where the choice sub-model has been calibrated on a travel diary survey data set.

The model which is described in the next section of this paper is also of the constraint-based simulation type, similar in concept to STARCHILD, but with a greater emphasis on the choice-set generation sub-model.

It is interesting to note that the above classification into choice- and constraint-based models does not define two alternative approaches to solving the same problem. It turns out that the two types of model also tackle reactions at quite different time-scales and levels of detail. The choice models tend to have been used to examine long-term strategic questions involving perhaps changes in the structure of social constraints (typically changes in working hours, availability of leisure time, female participation in the workforce etc.). The constraint-based models on the other hand tend to be more suitable for looking at detailed short-term behaviour. The emergance of two levels of activity modelling - an association of short-term with constraints and long-term with choice - is entirely in keeping with Thrift (1976):-

> In the short term people have very little choice at all their lives are day-to-day routine. ..... It is only on a larger time scale that people may be able to exercise any significant choice. They can deliberate whether to change their workplace, house etc. These decisions .... are taken through the filter of the relevant culture and it is at this scale that attitudes become important per se. Once a decision is taken it is usually only a matter of time before a new day-to-day routine is fixed, which will of necessity be a function of the physical environment.

ACTIVITY MODELLING AT OXFORD

The activity approach to the study of travel behaviour was the subject of research at Oxford University Transport Studies Unit at various times between 1974 and 1983. One product of this work has been a simple activity model (CARLA), and this model is now discussed in terms of the points which were introduced above. For a detailed description of CARLA see Clarke (1984).

The model's design was based on the following principles:-

(a) it should enable us to model the detailed time-space path of an individual over any defined period of time (but typically one day).

(b) it should represent the effects of constraints on behaviour resulting from obligation, institutional constraints and interactions within the family, while allowing a realistic degree of choice in both timing and location of activities.

(c) it should allow for travel between the locations of successive activities.

(d) it should impose the logic of the activity system - by preventing activities from overlapping etc.

In order to be useful the model should be sensitive to the kinds of policy questions likely to be of interest and should produce results in a form which is of practical use. In addition it was felt strongly that the model should be designed as far as possible so as to be both explicit in its assumptions and clear in its analogy with actual behaviour.

These principles themselves decide some parameters of the model design. They suggest that we should be considering a detailed simulation of the behaviour of an individual in a particular environmental setting. The primary unit of account will be activities, with travel representing the ability to move between successive activity locations.

The model is aimed at the lowest levels of Figure 1; at the detailed organization of day-to-day life. It predicts how people can arrange the activities they wish to take part in, given the constraints imposed by the external environment. At least for the time being we ignore the possibility of altering the set of activities to be undertaken. Ideally we would work with the set of activities which our subject would like to undertake. But given the difficulty, even for the subject himself, of defining that set we must usually work instead with the set of activities which the subject is observed to undertake at present. It must be remembered that this set is likely to be limited by constraints, some of which might be relaxed in the 'after' situation we are modelling. In this case the model will not be able to predict the emergence of a previously supressed demand.

The starting point then is the 'current' activity schedule of a particular person. This schedule might have been collected in a diary survey, or it could be generated as the output of a sub-model at the next higher level in Figure 1 (that level which deals with changes in an activity set). Either way we start with a description of how our subject behaves at present. We then wish to be able to predict how that behaviour would change given an imposed change to the circumstances of our individual. We introduce here the concept of a choice-set of activity schedules. In any particular set of circumstances we may think of a set of alternative schedules which the subject could have chosen to perform. These schedules are all equally feasible in as much as they do not imply our subject breaking any imposed constraints on the timing or location of activities and they all involve participation in the same set of activities. They differ only in the ordering, timing and location of those activities, and perhaps in the travel arrangements made to move between locations.

The set of feasible schedules under current conditions will include the observed schedule (which was the starting-point for the model), but will also include several other alternatives which are (presumably) less attractive than the one which the subject chose to adopt. We can now vary the conditions in some way to reflect the policy measure which we wish to examine. In a typical application of this model the imposed change might be a variation in the allowed timing of certain activities, or a change to the transport provision. We can now generate combinatorially the set of alternative activity schedules using the same activity set as before, but taking into account the new circumstances. The resulting set of alternative schedules may include the one which the subject originally carried out, in which case 'no-change' is a valid reaction to the policy measure. In many cases however that will not be the case and the subject would have to select his/her reaction from amongst this new choice set of schedules.

Representation of 'the Environment'

Our modelled actors must be able to carry out their activity patterns in some kind of description of the environment in which their real-life counterparts live. In principle the model should perhaps contain a description of the complete land-use and transport networks of the locality, with the added complication that extra temporal detail is required over and above that usually collected for a more conventional modelling exercise. However in most circumstances something rather simpler will suffice.

While in one respect the model is inevitably working at a very detailed level - in as much as it must apply to an individual person -

there is in fact considerable choice as to the level of detail at which the activity patterns may be modelled. In contrast to more traditional transport models, there tends to be an emphasis on the time dimension in activity models and a corresponding lack of emphasis on the spatial dimension. In the Dutch work reported below, for example, the geographical description of the area was collapsed into a single dimension along one bus route. In many cases the only detail needed from the spatial side will be a matrix of travel times between the locations likely to be visited. However it will usually be necessary to supply some kind of description of which activities can take place at which locations, and to define the times between which certain activities can take place - shop opening hours for example. Lastly if public transport trips are to be considered suitable timetables must be supplied.

In very many respects the choice of how detailed a description of the environment is required must depend on the particular application of the model.

The Activity Diaries

The same sort of pragmatism must apply to the preparation of suitable activity data. There is no single answer to how detailed an activity diary should be, and in a single survey different respondents will have different ideas as to at what level of detail to report their behaviour. What for some people might be a single activity such as 'housework' might for others be a series of discrete activities 'clear table, wash-up, vacuum floor'. In this trivial example we are clearly not very interested in the detailed version of events from a transport modelling point of view, but one can imagine other cases in which the respondent may omit useful information (for example by failing to report intermediate stops to drop passengers during a car journey). All this implies that the survey data may need some careful massaging before it is in a suitable form for use in the model.

Although the aim in collecting the data is to account for every minute of the subject's day, we do not aim to model all this time. It is surely the case that some activities are merely 'time-fillers' and these should not be included in the simulation of how people wish to spend their time. It is assumed that the subject fits such activities into the spare time in the predicted schedules in a similar manner to the observed schedule.

However we must ensure that all relevant detail is preserved. Included here are all activities outside the home since these by definition give rise to travel. In addition some in-home activities have an important influence on how the day can be scheduled and so they must be included in the simulation. Meal times are of particular importance here (see below for their influence on intra-family linkages) as do the chores that 'must' be included in the daily schedule.

Linkages within the Family

Members of a household do not act as individuals in scheduling their activities. Some activities are undertaken jointly through choice, some through convenience (travelling together) and others through obligation (child care). The degree of interaction clearly varies according to the household structure and particularly according to the ages of any children. The simplest way to implement this effect seems to be to insist that any activities which take place jointly in the observed diary must also be capable of taking place jointly in the modelled schedules. We model each individual separately and then compare the schedules to see which combinations of individual schedules obey the linkage constraints, giving rise to valid household schedules.

A PRACTICAL APPLICATION OF THE MODEL

Whereas most applications of activity models which have been reported tend to have been hypothetical desk studies, CARLA has been tested in a live policy study of rural bus services in the Netherlands. The study was undertaken by the University of Groningen, and is reported briefly in van Knippenberg and Clarke (1984), and in detail in van Knippenberg and Splinter (1984). In many ways this study was an ideal application for a simple activity model. The policy under examination directly concerned the timing of bus services, and the spatial dimension was simplified by considering only a single bus route connecting a small village to its regional centre.

Four-day activity diaries were collected from all people aged over 11 in some 95 households who lived in the village and who used the current good quality bus service to the city. The activity model was then used to simulate the effect on those households of implementing various reduced bus services. The model determines whether each household can continue to carry out their current set of activities given the newly inposed bus timings, taking account of timing constraints on their activities, and linkages between family members. In those cases where the household could make use of the bus service, an attempt was made to assess the degree of inconvenience to which they would be subject - in terms of disruption to their current routine, and in terms of excess waiting time associated with their bus journey. Note that, although CARLA does not contain a choice sub-model, and so does not predict a particular course of action for each respondent, its ability to generate a choice-set of alternative activity schedules is sufficient to allow it to be used to assess this policy question.

The model was run in parallel with a programme of interviews with the households themselves. This enabled the assumptions about temporal flexibility which are used in the model to be checked against respondents' own views about their likely behaviour. In addition the model's predictions of behaviour were shown to some households, and an encouraging degree of agreement was obtained between model output and respondents' opinions.

The model proved itself not only able to evaluate timetables of different overall levels of service, but it could also distinguish between the convenience of pairs of timetables which differ only in the detailed timings of the services. A particularly valuable aspect of the model is that, once it has been set up, the marginal cost of invetigating a further timetable is comparatively low.

THE CURRENT STATE OF ACTIVITY MODELS

This particular model has now been developed from a simple demonstration program into a practical planning tool. The model was put to the test in the Dutch rural bus study, and this showed that even a comparatively simple model is capable of producing useful practical results.

The model does not yet exist as a self-contained package which transport planners could use for activity analysis. Indeed it is doubtful if such a product could or should be produced at present. It is still something of a large step for most planners to accept the radical recasting of concepts which the activity approach implies. And it seems from experience gained so far that each application of activity models requires subtly different facilities and so a general model may not yet be possible.

Although CARLA is a fairly large piece of software which is designed to run on powerful mainframe computers, a simplified version has also been developed to run on a micro-computer. The limited experience which it has been possible to gain with this version suggests that activity scheduling on a micro-computer is both feasible and useful. Because of its very direct interface with the user, and the use made of graphical displays, this model is particularly suitable for demonstration purposes, in addition to its role as an investigative tool for planners and researchers.

The lesson to be learned, I believe, from the rural bus study and the use of the micro-based version of the model, is that activity models do not deserve their reputation of being large, cumbersome, expensive and impractical. The development of a completely specified activity-modelling system is indeed a major task, and one which requires further effort before a practical tool can be produced. But the state of the art of activity modelling is now sufficiently advanced that simple models can be used to provide answers to policy questions. These simple models contain all the essential features which make the activity approach useful for specific problems, and it is neither necessary nor wise to think of activity models as being limited to being research tools until the complete models have been developed.

However, one drawback to the technique as applied in the rural bus study is that activity data are undoubtedly expensive to collect and process. In this comparatively small-scale research project the extra effort needed was considered to be well-justified, but before this kind of technique can find wider application in transport planning, procedures will have to be adopted which are less data-hungry. It is suggested that the general approach used here can be adapted so that its data requirements are no greater than those of much simpler models. One way in which this might be achieved is to use simple travel survey data as a basis for <u>generating</u> activity diaries. Several researchers have used travel data as an input to activity-based techniques, but the problem is that some aspects of in-home behaviour, which cannot be inferred directly from travel data, are important in shaping family behaviour. However, we have learned sufficient about the ways in which different types of household organize their behaviour to be able to build models which would flesh-out bare travel data with generated in-home activities.

CONCLUDING REMARKS

There is still much scope for the further development of activity scheduling simulation models, but even in their current stage of development they are no longer merely research tools, but are available for use in the analysis of real policy questions.

It is important to be clear about how these models fit into the battery of techniques available to the planner. They should be seen as a particular class of model which can provide very powerful, but outwardly simple, analysis of a particular class of problems. It would be particularly unfortunate if the models were neglected because of their apparent simplicity. As soon as one starts to examine the processes involved in activity scheduling in sufficient detail to code them as computer simulations it becomes clear that what is apparently a simple arrangement of periods of time actually involves some very subtle effects. That the structure of this kind of model makes these effects explicit and visible is one of the strengths of activity scheduling.

Transport models have traditionally involved mathematical equations which over the years have become more complex and less easily understood by non-experts. This is perfectly justifiable if the additional complexity leads to more accurate predictions or a more comprehensive description of behaviour. But in the field of activity scheduling we are dealing with highly complex multi-dimensional systems of activities and persons moving in time and space, and it is inevitable that a statistical representation of such a system will have to be much more complex than the equivalent descriptions of travel behaviour. Even if such models are soluble their complexity is likely to cloud the essentially elegant processes involved in activity scheduling.

It seems appropriate to end by quoting Thrift (1976) once again:

> Time-space budget data are ..... very difficult to analyse not only for logistic reasons .... I would suggest that the way ahead lies in the use of statistics only as summary measures and for inferential purposes it might be wise to construct some new edifice based on combinatorics.

If it does eventually prove possible to develop complex mathematical models of activity behaviour which out-perform the simpler activity scheduling models then let us not underestimate the importance that we understand the processes involved. It is suggested that scheduling models have a particularly useful role to play in achieving that understanding.

Acknowledgements

The author is indebted to the U.K. Economic and Social Research Council and the Directoraat Generaal van het Verkeer for their financial support of the development and application of CARLA. Thanks are also due to his colleagues in Oxford and Groningen who contributed to the work.

REFERENCES

Carpenter, S.M. & Jones, P.M. (eds) (1983) Recent Advances in Travel Demand Analysis Gower, Aldershot.

Clarke. M.I. (1984) An Activity Scheduling Model of Travel Behaviour TSU publication 263, Oxford University.

Damm D. (1983) "Theory and Empirical Results: A comparison of recent activity-based research" in: Carpenter & Jones (1983)

Damm, D. (1979) Toward a Model of Activity Scheduling Behavior Unpublished PhD. Thesis, MIT, Cambridge Mass.

Hirsh, M., Prashker, J.N. & Ben-Akiva, M. (1984) Theoretical Model of Weekly Activity Pattern Paper presented at the Transportation Research Board Annual Meeting, Washington D.C., January 1984.

Holly, B.P. (ed) (1976) Time-Space Budgets and Urban Research: A Symposium Discussion paper no 1, Department of Geography, Kent State University.

Holly, B.P. (1978) "The Problem of Scale in Time-Space Research" in: Carlstein et al. (1978), volume 3.

Hoorn, T. van de (1983) Experiments with an Activity-Based Travel Model Transportation vol 12, no 1.

Jones, P.M. (1983) A Review of Human Activity Literature. TSU Publication 206, Oxford University.

Knippenberg, C. van & Splinter, J. (1984) De Relatie Tussen Openbaar Vervoer en Mobilitie in Kleine Kernen. Traffic Reserach Centre, University of Groningen, The Netherlands.

Knippenberg, C. van & Clarke, M.I. (1984) Taking Account of When Passengers Want to Travel. Traffic Engineering & Control, Dec. 1984.

Lenntorp, B. (1978) "A Time-Geographic Simulation Model of Individual Activity Programmes" in: Carlstein et al. (eds), volume 2.

Recker, W.W., McNally, M.G., & Root, G.S. (1983) Chaining Behavior in Urban Tripmaking - Interim Report Institute for Transportation Studies, University of California, Irvine.

Salomon, I. (1983) "Life Styles - A Broader Perspective on Travel." in: Carpenter & Jones (1983)

South Australian Energy Research Associates (1983) Temporal Flexibility as a Peaking Strategy Report to the Director-General of transport, South Australia.

Stephens, J.D. (1976) "Daily Activity Sequences and Time-Space Constraints" in: Holly (ed) (1976)

Thrift, N. (1976) "Some More Problems and a Utopian Prospectus" in: Holly (ed) (1976)

Behavioural Research for Transport Policy
pp. 17–30.

RELATIONS BETWEEN THE QUALITY OF THE TRANSPORT CHAIN AND THE CHOICE OF MODES IN FRANCE

Michel HOUEE - French Ministry of Town Planning, Housing and Transport - S.A.E.P.

Perhaps more than elsewhere amongst the countries endowed with a well-developed public transport network, the way in which the various modes of transport are linked with one another is very important in France : as regards the heavily built-up areas, the density of the urban networks contrasts with the poorness of the non-radial suburban services; in a more general way, outlying urban areas, where the private car is a common feature, suffer from poor access to the town centre and servicing within them; moreover, the hyper-centralized nature of the interregional transport network all too often necessitates connecting up with radial services for cross-country inter-regional links; that is one of the roles capable of being played by the regional and departmental links which therefore have to meet a double requirement of connection to inter-regional links and with each other.

However, concern about the inter-relations between the modes is of relatively recent origin. There are many reasons for this : institutional - for example, responsibility for transport eludes the urban organisational authority outside the perimeter of urban transport; economic - those in charge of networks being inclined to argue more in terms of competition than of complementarity; financial - as the harmonization of fares poses delicate problems of compensation; and finally sociological - the demand for use of a public transport "chain" being perceived as captive in origin and its satisfaction therefore being assumed to be more a matter of social than commercial logic.

If the law of orientation of interior transport and likewise the recommendations of the Ninth Plan's transport policy group do indeed give inter-modal complementarity as one of the major objectives to be reached, the ways which enable this to be done are sufficiently varied to merit close examination of those which appear to be the most effective in a given situation : for their pendular journeys, do suburbanites think more in terms of the comfort of their terminal journeys or of the existence of an attractive fare zone? Does access to the inter-regional network for the inhabitants of enclosed regions suffer primarily due to the regional network's inadaptation or to a lack of information on the possibilities? The stakes are high : firstly, financial constraints will doubtless not permit, even on an experimental basis, the introduction of all the possible improvements; secondly, only identification of the strategic aspects will allow public transport, penalised as it is by the nuisance of having to change within a journey, to rival the private car which provides a door-to-door service.

Hence the reason for undertaking the analysis of the behaviour, criticism and suggestions of the users confronted with various possibilities of linking up the networks which form the public transport possibilities of their zone. Ten locations were thus selected for their diversity vis-à-vis the following criteria : the relative performance of the principal and terminal links in the chain as regards speed, frequency and comfort; the quality of the connections between modes of transport in terms of the compatibility of time-tables, the connection route and the availability of parking for cars and cycles at stations, etc.; the availability of information for the potential users of the chain; the existence of special fare facilities such as passes valid within a zone. Furthermore, one area may offer good chain possibilities for inter-regional travel and poor possibilities for regional travel, whereas the opposite may be true in another area.

Each of these ten areas may be characterized as a whole in the following way :

. The Massy-Palaiseau-Verrières area, in the relatively near suburbs south of Paris, is characterized by extremely rich possibilities of both public and private transport for short journeys with many possibilities of chains both in the direction of several places in Paris and of several suburban zones : access to Denfert, Châtelet, the gare du Nord and Roissy in the North, the Chevreuse valley to the South on line B of the RER; access to the stations Austerlitz, St. Michel, Invalides as well as Versailles Rive Gauche, St. Quentin en Yvelines and the outer suburbs in the South on line C of the RER : servicing the Bièvre valley and the South-East suburbs by a SNCF line (Massy-Palaiseau's 2 RER and SNCF stations and to a lesser degree the Massy-Verrières RER station each enjoy excellent connections through the RATP and APTR surface urban public transport networks as well as a large capacity for connecting parking); the immediate proximity of a junction onto the A6 motorway giving access to all directions. On the other hand, Massy-Palaiseau's station is away from the inter-city network and necessitates connecting in Paris (or Versailles for the Western network) for inter-regional journeys; however the discussions about the possibility of the Atlantic high speed train stopping at Massy station should be noted. Finally, Orly airport is 5 km away by the N 186 road and the A6 motorway.

. The Evry-Ville Nouvelle area, further from Paris on the South-Eastern side (zone 5 as opposed to zone 4 for Massy for the "carte-orange" pass) enjoys good accessibility to Paris from 3 stations, 2 of which have good link-ups through the urban bus network and one(Evry-Courcouronnes) has a large area for parking but is distinguished from the Massy area by a clearer unidirectionality of transport possibilities (SNCF line to the gare de Lyon and the motorway to the South) and greater competition from the motorway infrastructure (less than 1 km to the A6 motorway). On the other hand, it boasts a network of cycle paths which, in conjunction with the parking facilities for cycles at the stations, facilitates access to them by this mode. Finally, Evry's situation is comparable to that of Massy from the point of view of access to the SNCF inter-city network.

. The Villeurbanne and Bron areas in the Eastern suburbs of Lyon enable a comparison to be made of the "chain" practice in a large provincial built-up area, endowed with an initial section of an urban public transport network within it, with the Paris conurbation where the image of a transport network within the area concerned itself is being completed with the RATP/SNCF level B inter-connection projects. The Villeurbanne zone, in the immediate proximity to the North of the A line of the Lyon underground, enjoys excellent possibilities of providing connections by bus or by private car (the availability of free car parking facilities) with the terminus of this line (Laurent Bonnevay station is very well equipped from the point of view of connecting with the service); further away from this axis to the South, the Bron zone, although linked to Laurent Bonnevay by the urban bus network, has, in competition, access to the centre of Lyon by classic bus lines which could enable the extent of the underground's attractiveness to be tested. Fares to encourage a chain of travel are the urban time-table ticket permitting up to three connections excluding return and the monthly ticket for the sole zone within the Courly (Lyon Urban Community) perimeter. Finally, the introduction of the South-East high speed train is obviously the most consequential recent event where transport is concerned (considerable increase in connecting facilities with Paris, significant change in air/rail competition, significant impact on the regional transport network, constitution of "La Part-Dieu" as a centre for connecting), following a long period of very noticeable efforts in favour of the motorway network.

. The Lormont and Pessac areas in the suburbs of Bordeaux provide the oppurtunity of testing the impact of specific efforts made as regards the capacity for a chain of the surface public transport network in a large built-up area which does not have public transport on site. Indeed, while the Pessac zone to the North-West has classic access to the centre by a radial omnibus bus line, that of Lormont to the South-East has been the subject of a series of experiments : guaranteed free flow for the access route to the centre thanks to a system of bus lanes and traffic control in favour of buses, the introduction of express buses during rush hours, the establishment of a private car/bus exchange car park, numerous connections onto this main line from secondary radial lines, bypass lines and even shuttles with a smaller capacity serving the more distant outlying areas, some of which have a boundaries call system. Moreover, Bordeaux enjoys a good classic motorway and rail link with Paris, Spain and Toulouse, links with the other axes being less good at the inter-regional level. At the regional level, there is a satisfactory frequency of service for the towns situated on the inter-city rail network as well as for some towns serviced by coach on the heavy traffic axes (such as Libourne). The link between the coach and rail stations is not always very well guaranteed in the latter case.

The main reason for choosing the Chartres area is that, situated some 80 km to the West of the capital and linked to it both by frequent and fast trains and by the A10 motorway, it is one of the privileged places forming a wide circle round Paris which emits great daily alternating migrations, notably for reasons of going to and from work; it was therefore particularly interesting to study the choices of people living in the built-up area or on the outskirts of Chartres between using the motorway and linking up with an available efficient form of public transport (quite good connections with the urban and suburban buses, the problem of enough parking facilities at the stations). Furthermore, Chartres is one of the poles of one of the first successful experiments of a regional express road link, the Transbeauce, linking Chartres to Dreux to the North and Orléans to the South-East.

. Unlike that of Chartres, the Montereau area, situated hardly any closer to Paris, has since 1.7.83 been able to enjoy the extension of the "carte orange" pass to SNCF users, providing an encouragement to use public transport, of interest to test, particularly for access to Paris, which is a little less efficient on the whole in nature than in the case of Chartres, whether it be by train (less fast) or by car (no direct motorway access). The incentive effect was tested both with users having easy access to the station (on foot or by a single urban line) or more difficult access (by an infrequent inter-urban line, or by car in view of insufficient parking facilities). For access to the inter-regional network, a detour by Paris is almost obligatory in practice.

. The Creusot area mainly enables the impact to be tested of the considerable reduction in the time taken to reach Lyon (40 minutes) and Paris (1 hour 28 minutes) on the behaviour of transport users in a historically hemmed in zone, greatly affected by the economic crisis. This impact was tested with inhabitants living nearby or away from the new station, depending on whether they live in Creusot (9 km from the high speed train station) or out towards Autun or Montchanin (linked by road shuttle services), a huge free car park being available at the high speed train station, moreover. In addition, the Creusot area belongs to a multipolar group including Montchanin, Montceau-les-Mines and even Chalon/Saone, served by inter-urban lines for which information given to the high speed train users is very inadequate.

. The Hagondange and Rombas areas in the Orne valley in Lorraine offer the interest of being served and linked respectively with the rhythmic rail service of the Moselle valley, christened Metrolor, one of the first experiments to really improve the regional rail service. They make it possible to test, therefore, both, in the case of Hagondange, the influence of an efficient regional service on the use of trains both for access to the regional poles (Nancy and especially Metz and Thionville) and for the linking up with the inter-regional network in the direction of Paris, Brussels, Strasbourg or Dijon and, in the case of Rombas, the attractiveness of the regional service in view of rail and road linking services. In both cases, the motorway network (A31 giving access to the Paris-Strasbourg A40) exerts keen competition.

. The Valence area is uncommon from the point of view of the capacity of offering a transport chain in that it presents a very great contrast between a North-South inter-regional rail availablility which is considerable for a town of its size (including, notably, 20 high speed trains in the direction of Montpellier and excluding classic trains) and a non-existent regional rail availability to the West (Ardèche being the only French department to be totally lacking in any) and limited merely to the directions of Grenoble and Gap to the East. It follows that Valence represents a typical case for the need of an inter-relation between inter-urban road transport, ensuring that the whole of Ardèche and a large part of Drome are linked up, and rail transport. However, difficulties exist in view of the distance of the rail and coach stations and the fact that some urban lines do not serve the station as the project for setting up a connection centre within the larger framework of the renovation of a district shows.

. Finally, the Auch area, at the other extreme of the scale of chain capacities in relation to the Paris conurbation, is characteristic of a zone situated away from the inter-regional network and with no easy access to this network by the expedient of a developed regional network since the only regional rail link departing from Auch concerns the town of Toulouse. Thus, as per the time-table schedule, access to Paris could be made possible with a single link to the inter-regional network (the same with the coach providing a connection in the evening with the night train from Montauban), or require first a connection to the regional network (via Agen on the Bordeaux-Toulouse line for example). Not is the motorway network any more accessible, as Agen also represents the nearest entry point. For regional travel, 8 inter-urban lines link Auch to the principal neighbouring poles.

In each area, about 30 users whose homes and activities require them to link up several modes of transport were selected using a more widely circulated pre-questionnaire. This selection first focussed on satisfying quotas relating to the types of chain used: captive users using a chain for their travel to and from work; those with a choice as to whether they use a chain or not for their journeys to and from work, divided in both cases into sub-categories according to whether they said they also had recourse to chain transport for their holiday travel or for other personal reasons or for both reasons or for neither of the two reasons. The weighting of these quotas was different in each zone : Thus the representation of the cases of a transport chain between home and work is higher in the areas on the outskirts of large built-up areas, and that of chains for long distance journeys is greater in the areas of medium-sized towns served by high speed trains. In addition, setting up this fish-tank enabled the representativeness of the sample to be controlled in relation to such socio-economic parameters as age, sex, size of household, its income, its degree of motorization, etc. Each person selected for the survey was then asked in detail about four types of journey : to and from work, principal journeys for other personal reasons, last holiday trips and possible professional trips. Not only was the mobility behaviour described in detail and the reasons for it analysed, but the alternative chain possibilities were explored. When the persons questioned were not aware of them themselves, the researcher was able to suggest them by using cards, tables or any other relevant aids. Special attention was also paid to the incidence of the choice of mode having to satisfy several motives for one and the same journey. Finally, criticisms and suggestions about the quality of the chain were most welcome.

1 - A chain for journeys to and from work : the influence of the quality of public transport supply

The primary objective of this study being, as is said in the introduction, to identify the influence of the link-up capacity of the transport network on users' choice of mode, we began by seeking to link up the practices and opinions regarding travel to and from work to the nature of the choice of public transport available for making the journey from the point of view of forming a chain. We chose to characterize this using three indicators : the distance separating origin and destination, the number of links in the chain and the type of combination it implies. We added to these indicators from the description of the journey made if it implies a chain or, if not, from the chain considered to be a possible alternative by the inquiry or, if not this either, from the chain used if the vehicle should break down or, as a last resort, from the potential chain suggested by the researcher. Pending construction of a synthetic indicator of the nature of the choice from these three basic elements, we considered each basic indicator distinctly.

Firstly, distance seems to lead, in correlation to its length, to increasing use of the chain practice (at least if the independance of the distance criterion vis-à-vis the quotas per-zone is accepted). In relation to the proportion of users of the chain system for less than 10 km, the proportion is multiplied by two from 10 to 50 km and 2.4 for over 50 km - the probable consequence of the dual effect of having less and less recourse to the private car with the distance and the increasing probability with the same distance of not being able to link origin and destination by a single public transport mode. Moreover, this is confirmed by the increase in the number of links in the chain with the distance : from 2.4 links on average for less than 10 km, to 3 from 10 to 50 km (equivalent to the average for the whole of the sample) and to 4.1 for over 50 km.

The number of links in itself does not seem to be a very important factor in general in dissuading people from using chains; indeed, habitual users of chains use a significantly larger number of links than non-habitual users would have to use (3.4 as opposed to 2.4). Amongst the habitual users of chains, those obliged to use public transport have a slightly less complicated chain than those who can use private cars : 2.9 links as opposed to 3.5; this would seem to suggest in this sample that captive users are to be found in more direct proximity to their work (half of them work less than 10 km away as opposed to 40% for the rest of the sample). Finally, amongst those who are habitual users of a chain to and from work, those who also use a chain for their holidays are separated from their work by a smaller number of links : 3.3 as opposed to 3.7; from this it may be possible cautiously to advance the hypothesis that the habitual use of "easy" daily chains encourages the use of chains for longer and less frequent journeys.

The relation between the criterion of the number of links and that of the distance confirms the previous observations for each section of distance : at the very most it may be observed that it is particularly where there is less than 10 km that the captive users use noticeably fewer links than other users of chains (2.2 as opposed to 2.8) and that the rare users who do not use chains for over 50 km seem to be the only ones who are sensitive to the dissuasive effect of the number of links (2 of them would have to use 5 links in sucession).

The two most widespread combinations of modes at the level of the potential possibilities of getting to and from work in our sample are walking plus surface public transport, with or without the addition of public transport in its exclusive location, each covering more than 20% of the sample. Afterwards, in decreasing order, came having recourse to only surface public transport, walking associated with surface public transport and a combination of private car, public transport in its exclusive location and walking (from 15 to 11% each in the sample). The only remaining combination to represent more than 5% of the sample is access by car to public transport in its exclusive location without walking at the terminal. The following lessons may be drawn from these

Observations : the only currently practicable public mode without addition of private mode is surface public transport; in any place where there is public transport in its exclusive location, be it in association or not with surface public transport, walking and the private car appear (cycles only account for 2.3% of the cases in the combinations of modes despite the cycle paths of the Evry zone). Of note is that the private car is involved in more than a quarter of the combinations. On the other hand, despite the presence of the car park at La Butinière in Bordeaux, the possibility of access by car to surface public transport is found very rarely. Finally, two complicated combinations of mode involving a number of dissuasive links are rarely mentioned : there is rarely a chain involving private car, public transport in its exclusive location and surface public transport, with or without an additional walk.

Faced with this structure of possibilities the choice of using a chain or not is far from being uniformly spread: considerable recourse to the use of a chain can be noted whenever access to collective transport in its exclusive location is possible; 94% of cases when access by car is involved, 92% when there is a chain with walking and surface public transport, 84% where there is access by car completed by a terminal journey on foot. On the other hand, surface public transport in only integrated into a chain in 40% of the cases when it requires at least a walking journey, and in 20% of the cases when it is combined with another form of surface public transport (even so this is the case of captive users in almost half of each of these two cases). The decisive role for the choice of mode of public transport in its exclusive location within the chain can thus clearly be seen.

2 - The characteristics of habitual users of chains

As a second stage it is interesting to find out if users who have recourse to using chains differ from others. Amongst the criteria under which these users may be characterized, we first chose the level of motorization at the level of all journeys, sex and the socio-professional category for travel to and from work.

If, by definition, the members of households with no motorized transport come within the category of captive users and therefore make use of chains on a much more frequent basis, the members of households with one or several motor vehicles do not noticeably differ from this point of view. However, it is interesting to note that the availability of several cars does not seem to constitute an obstacle to the practice of using chains, since this is the case of 50% of the members of multi-motorized households, as opposed to 46% within mono-motorized households for travel to and from work and that, in addition, the proportion of individuals who never use a chain is only 15% in the first case as compared to 21% in the second. Multi-possession even seems to be an incentive to using chains for journeys for other personal reasons since it is the case of 29% of the users in this category, as opposed to only 20% within the one-vehicule households: perhaps a sign that the second car is not only used as the exclusive mode of travel but also as a mode of linking up with another.

The analysis based on sex shows firstly that, if the women questioned make journeys involving at least two collective transport links in more or less the same proportion as men (40% of the women as compared to 44% of the men), they are, however, more likely to use chains involving a single public transport link (21% as against 8%) to the detriment of their use of the car for reasons of getting to and from work (38% of them have recourse to it as against 47% of the men). The habitual use of chains differs according to sex in particular in the Lyon/Bordeaux (41% of women use chains with a single public transport link as opposed to 21% of men) and the Massy-Evry-Rombas zones (27% of female car users as opposed to 47% male). Moreover, as it has been seen that the chains become more complicated the longer the distance, this seems to indicate that women tend to use chains more than men for short distances, the latter being more often able to use their cars in this case.

The analysis according to socio-professional category shows that recourse to complex use of chains, involving at least two public transport links, is principally the case of two opposite categories from this point of view: executives and higher intellectual professions on the one hand and persons without a permanent professional activity[1] on the other, with 50% and 56% respectively of the journeys made with complicated chains as opposed to an average of 43% in the sample. But whereas in the first case this is done to the detriment of simple chains (5% of their journeys as against 14% on average), it corresponds, in the second case, to lesser use of the private car (22% of their journeys as opposed to an average of 43%). Complex chains may therefore play both the role of competitor to the private car on long distances for comfortably well-of categories and of privileged mode for all journeys for socially disadvantaged categories. An interesting observation, moreover, concerns manual workers who have less recourse than average to complex chains (33% as opposed to an average of 43% of their journeys to and from work) and more recourse to the private car as the sole mode of transport (55% as opposed to an average of 45%).

Another interesting aspect to explore is the comparison of the practices for the outward and return journeys according to whether the users make use or not of a chain on their outward journey for travelling to work. A different return from the outward journey may indeed indicate the need to make an additional journey when either going or coming back (taking a child to school, shopping before returning home, etc.).

It has been noted in this regard, that it is mainly the users of a "simple" chain (involving a single public transport

1 especially students

link) whose behaviour alters for the homeward journey (24% of the cases): two thirds of them return home from work by car, the last third uses a chain involving several public transport links (which in both cases may make more intermediate stops possible). On the other hand, habitual users of chains involving several public transport links on the outward journey, or users of the private car, rarely return by a different method than they go (in 7% and 4% of the cases respectively).

As an alternative solution to making the journey to and from work, using a chain is considered by car users in 26% of the cases when the chain is simple (a sole public transport link) and in 35% of the cases when it is complicated; use of the car (the family's second car, or as a passenger) being the most frequently cited alternative. In the symmetrical configuration, the car emerges as the dominant alternative; in 90% of the cases if there is a complicated chain and in 69% of the cases if the chain is simple (it being thus possible to replace the simple chain by a complex one in 12% of the cases).

3 - Advantages and problems felt in habitual travel to and from work

At a later stage, we turned our attention to the arguments put forward in favour of or against the commonly adopted modal solution for journeying to and from work, as a distinction being made according to whether a chain was used or not. To do this, we seperated the advantages mentioned into 10 entries and the problem into 34 entries using as inspiration the items of the mnemonic index for re-inquirement.

The frequency with which problems encountered was raised seems to grow with the distance: it is 1.25 times greater per person questioned in the 10 to 50 km bracket than for less than 10km, and 2.25 times greater for over 50 km. However, this concerns solely the habitual users of chains, responsible for two thirds of the problems raised and who raise 60% of these problems in the case of more than 50 km; on the other hand, for users of a single public transport link or individual modes, the problems seem to be linked for the most part to short journeys: 60% of the problems in the first case and two thirds in the second are raised with regard to journeys of less than 10 km.

On the whole, the most frequent problems mentioned are the physical discomfort of public transport (13% of the cases), then the frequency of the return service, traffic problems, the frequency of the outward service, delays, the length of time waiting for connections (from 8 to 7% of cases for each reason). But this distinction varies according to whether a chain is used or not: thus physical discomfort is mentioned in 20% of cases when a single public transport link is used, when we know that a method of surface public transport is most often involved; likewise, traffic problems account for a third of the problems referred to by the users of individual modes; on the other hand, for users of chains, there are, in addition to the afore-mentioned problems, those of time-table and psychological discomfort.

Likewise, distance has an influence on the nature of the problems raised: thus problems of frequency, delays, duration of connection and total duration of the journey occupy a relatively larger share of the problems at over 50 km; on the other hand, there are no traffic problems in this distance bracket.

If the zones are now regrouped by types of distance brought into play by journeys to and from work (zone 1 : Montereau-Chartres-Le Creusot characterized by extreme distances, either short or long; zone 2 : Massy-Evry-Rombas Hagondange presenting medium distances; zone 3 : Lyon-Bordeaux where short distances predominate; zone 4 : Valence-Auch predominantly medium and long distances) a very high concentration of problems is to be noted in zone 1 (44% of all problems). Here too, this is ascribable mainly to the users of chains who account for 85% of the references to problems in this zone. It is particularly in this zone that problems concerning the duration of the journey are concentrated. Zone 2 comes second with regard to problems with 30% of the references. Problems relating to connections are of particular importance to the users of the chains. Zone 3 only occupies 20% of the total of problems in general, but 48% amongst the users of a single public transport link (with 30% of the problems relating to physical discomfort) and 35% amongst the users of individual modes (with 30% of the problems connected with traffic problems). The zone 4 users raise few problems (7% of the total, taken as a whole) and these mainly involve traffic problems encountered by the users of individual modes.

Analysis of the problems raised according to the combination of modes used shows that the problems are mentioned more frequently by the users of chains when the chain includes a given public transport in its exclusive location associated with surface public transport and a walk (about one third of the problems raised), car and a walk (one fifth of the problems), and car or a walk (15% in each of the two cases). On the other hand, when the chain comprises only a single public transport link, the surface public transport/walk combination accounts for nearly half of the problems.

The advantages most frequently mentioned by the users of public transport are psychological comfort, speed and cost, in decreasing order. These are mentioned for the most part by the users of chains associating public transport in its exclusive location plus walking or private car. Car users, for their part, place importance on the speed and freedom of the travel schedule which the car provides. Special mention should also be made of the attitude of the users of complex chains in the Montereau-Chartres-Le Creusot zones, the advantages linked to the cost and fare structure on the one hand, and to the use of time on the other, are particularly significant to them (24% and 20% respectively of the advantages mentioned, whereas the relative weight of these arguments is negligible with habitual users of complex chains in the other zones).

4 - Improvements noted in the availability and reasons given for the choice of mode

Improvements were noted by the users in their journey to and from work in about two thirds of the cases. However, this proportion comes to almost three quarters with users of complex chains, whereas it is only about 50% with car drivers.

The improvements most frequently cited by the users of public transport relate for a third of them to physical comfort, for a seventh to frequency, and in slightly smaller proportions to the duration of the journey and to the fare structure. Obviously, for their part, car drivers refer in the main to better traffic flow and its repercussion on the duration of the journey.

The view points of the users differ a great deal, however, according to the zones. In Montereau-Chartres-Le Creusot, improvement of frequencies was felt more strongly than elsewhere (22% of the improvements mentioned) by a population partially composed of considerable distance travellers and satisfaction is higher than average (only 28% of users did not notice any improvement). In Massy-Evry-Rombas Hagondange, 40% of the improvements mentioned concern physical comfort. In Lyon-Bordeaux, it is the shortening of the duration of the journey which is especially appreciated (29% of the improvements cited) in particular for the users of complex chains probably influenced by the introduction of the underground. Finally, only a third of the Valence-Auch users consider they have benefited from improvements.

To finish, it is of note that nowhere were the organization of connections and the quality of information perceived to a significant extent to have been subject to improvements.

The main reasons for their choice as given by the users of public transport in a chain, are, in decreasing order, cost (nearly 30% of the replies), speed (more than 15%), safety and occupation of time (almost 10% each). In this overall picture, captive users occupy a place apart with the reason given being, for the most part, the absence of any other possibility. The reasons given by the users of individual modes are first speed (20% of the replies) and the freedom of journeys, stops and time (18% of the replies).

Cost is an argument which is advanced in particular by those whose chain is made up of a mode of public transport in its exclusive location, while speed represents more than one fifth of the reasons for the choice of a combination of private car and a given public transport in its exclusive location.

Users of individual modes refer to speed in the main

when recourse to public transport would imply using surface public transport whether or not associated with a walk. The same applies for mentioning the freedom of use, while the argument for using the same means of transport for other reasons when making the journey is only given by those for whom the alternative is just surface public transport.

Furthermore, the reasons given are not insensitive to distance. Thus cost is more frequently mentioned to justify the choice made for long distances (over 50 km) not only where a chain is used involving one or several public transport links, but also where a car is used: this doubtless shows where there are efforts to be made with regard to fares for long distance public transport to and from work in cases where its high cost leads to the private car being preferred.

On the other hand, speed is most frequently mentioned on the whole where the distance is short (less than 10 km). This is almost certainly due to a great extent to the users of private cars, but this phenomenon can also be seen with the habitual users of chains comprising a single public transport link: in the latter case, it may reflect a problem of choice between public transport and going on foot for very short distances.

Safety takes on great importance for long distances, especially amongst the users of chains with several public transport links (it represents a plus for chains over such distances in relation to individual transport), while in the medium distance bracket, comfort is mentioned in particular by the habitual users of chains with several links, as well as the absence of an alternative possibility in the mediumdistance bracket by the habitual users of public transport chains with a single link.

Sex is another factor distinguishing the reasons given. Thus amongst the habitual users of chains with a single public transport link, males are twice as sensitive to cost as females, and three times more sensitive amongst car users. The same is also true for the time taken in long chains comprising several public transport links (8 % of the reasons given by men in this situation against 3 % for women).

On the other hand, the absence of another possibility amongst the users of public transport (with one or more links) is almost exclusively given by the women, which provides due confirmation of them having the family car less at their disposal. Furthermore, the reliability of time-tables is mentioned three times more often by women than by men amongst the habitual users of chains comprising a single public transport link, a fact which doubtless reflects the weight of family constraints on the women concerned. Men also give more importance to the freedom of journeys, stops and time, which is related to their more frequent use of private car.

In conclusion, it would appear that the use of public transport by way of a chain of journeys is very much affected by the quality of the choice, less, it would seem, in relation to the number of links forming the chain than to the presence within this chain of efficient public transport within the area itself. If the number of cars at the disposal of the household does not seem to have a considerable influence on the practice of using a chain, such a practice appears to be more current with women, executives and with students. The habitual users of complex chains are, moreover, those who experience the most problems in their journeys to and from work, especially when the distance is considerable : their complaints mainly relate to physical discomfort, frequencies, delays and the duration of waiting for connections; the presence of public transport with its exclusive locations, the incentive character of which has been seen, would appear on the other hand to be behind a large proportion of problems. We may deduct from this that users' requirements increase with the quality of servicing provided, and that reversely, under a minimum level, public transport, being not recognized as a possible alternative, stops being talked about. The habitual users of complex chains are also those who have noticed the greatest improvements recently in the service regard to physical comfort and frequency, although the most poorly serviced zones still come in for a high proportion of negative opinions; nowhere are the organization of connections and the quality of information perceived to have been subject to sigificant improvements. Finally, the main reasons for the mode chosen by the users of chains are cost, speed, safety and occupation of time in decreasing order, while car drivers are proportionally more sensitive to speed associated with the freedom of journeys, stops and time schedules; the relative importance of these reasons varies according to the distance of the journey and the sex of the user, thus reflecting the greater constraint which the journey to and from work represents for women and long-distance users.

Behavioural Research for Transport Policy
pp. 31–52.

An Activity-Based Modeling Framework for Transportation Policy Evaluation

W.W. Recker
School of Engineering and
Institute of Transportation Studies
University of California, Irvine
Irvine, California 92717 USA

M.G. McNally
Department of Civil Engineering and
School of Urban and Regional Planning
University Southern California
Los Angeles, CA 90089 USA

1. Introduction

Assessment of the impacts of transportation policies on complex travel behavior necessarily involves consideration of constraints, absent in conventional techniques, yet fundamental to the activity approach. Policies may be generally viewed as those which either relax or tighten constraints, or alternatively, those which expand or reduce the choice set of individual decision-makers. Whereas existing research has focused on the analysis of relevant constraints in activity/travel behavior, few extensions have been made to examine choice as central to policy assessment, due to both the strength of accepted relationships between constraints and feasible activity patterns, and limited results relative to choice set formation.

The STARCHILD model system (Recker et al, 1983b) generates feasible activity patterns utilizing a constrained, combinatoric scheduling algorithm. This pattern set is subsequently subject to alternate techniques of choice set formation, resulting in a calibrated choice model. The impact of a range of policies (e.g., budget restrictions and activity availability, both temporally and spatially) is estimated within the model structure by examining the resultant changes in individual choice sets.

A brief summary of pertinent research related to the model's development and a brief examination of the data base utilized in the policy application precedes a comprehensive treatment of model estimation results and the ensuing evaluation for a range of policy scenarios.

2. Activity-Based Approach

The activity-based approach to complex travel behavior analysis, a field spawned through the inadequacies of first aggregate, then disaggregate and attitudinal-based modeling systems often has been characterized as fragmented and lacking a sound methodological foundation. In a classification of activity-based approaches, Golob and Golob (1983) suggest this lack of cohesive theory is "compensated for by a profusion of concepts and methods." This plethora of research directions, while contributing strongly to a growing empirical base (see, for example,

Damm, 1983) and further illustrating the failures of existing analysis techniques and the promise of the activity approach, has made only marginal contributions toward a new theoretical basis.

Surprisingly, many of the major contributions to the state of the art were made early in the field's development. Perhaps the first comprehensive model of activity patterns and corresponding policy applications was developed by Lenntorp (1976), whose model represented the first operational application of the theoretical framework advanced in the well-known work of Hagerstrand that allowed for meaningful policy evaluation. The major limitation of his approach was the inability to assess probable responses to various policies, his model only producing the range of potential actions.

The most comprehensive work on activity-based approaches has been completed at the Transport Studies Unit at Oxford, and is summarized by Jones, et al (1983). Both the HATS gaming approach and the CARLA scheduling simulation model formed the basis for many full pattern analysis approaches which followed. The most promising aspect of the TSU approach was the immediate application in alternative policy scenarios.

Futher comprehensive, joint modeling and policy analysis was conducted by Recker et al (1983). An explicit quantification of the activity pattern led to a more easily interpretable policy evaluation, but the approach still lacked a behavioral basis which would lead toward explicit choice behavior.

Many researchers have used activity-based advances jointly with more conventional approaches with success (see, for example, Landau et al, 1980, or Hirsh et al, 1984). Specific policy application has supported much of this work, and Jones (1983) suggests continued research and application in this direction. There is still the need for a comprehensive theory and operationalization of the activity approach which is capable of policy analysis across all domains of behavior response (Heggie, and Jones, 1978); the STARCHILD modeling system of Recker et al (1983) has been developed in response to this need. The policy applications under consideration draw from initial work by Burns (1979) who first considered the spatial and temporal constraints on transportation patterns using an accessibility approach. This paper attempts to introduce a choice element capable of estimating potential behavioral response to changes in constraints facing transportation decision makers.

3. The Data

Data used in this policy application of the STARCHILD modeling system were drawn from a 1980 home interview survey of over 600 households in the Windham, Connecticut Planning Region. The survey includes a comprehensive, single day, travel activity diary for each household in addition to a basic socio-economic profile and transportation supply inventory (Davis et al, 1981).

The survey questions focus on the temporal (and spatial) flexibility of activity sojourns and tours which provides in addition to conventional purpose, location and mode variables, information concerning earliest departure times, allowable waiting times, fixed constraint times, and latest return times. Further information was provided regarding the importance, temporal and spatial flexibility, and degree of advanced planning associated with each activity. This information eliminates many of the restrictions associated with single day diaries as it is possible to establish the nature of the travel day reported as being typical of the actual travel patterns of the households.

Travel diaries were collected for all members of the household. Features of the Windham Data source allowed the simulation model to include the effects of household interaction rather than simply individual pattern formation in absence of explicit household constraints.

Base travel times utilized in this study are conventional network times provided by the Connecticut Department of Transportation which correspond to the locational coding used in the original Windham regional study (Davis et al, 1981). The 10-town Windham Planning Region road network was coded using a total of 276 nodes. The full travel time matrix was constructed from free flow automobile travel times specified for adjacent zone pairs using a modified Moore shortest path algorithm (Moore, 1963). A statistical procedure was used to adjust these network times to produce values which were compatible with reported (perceived) travel times.

In the survey, activities were categorized into 17 types, by purpose. Activity frequency was divided into four categories: daily, weekly, monthly and occurring less often than once a month. The vast majority (approximately 85%) of the activities reported in the Windham survey occur on a daily basis, while another 12% fall in the weekly category. The majority of the activities that occur daily are work activities (54%), which accounted for 45% of all activities reported in the survey.

Mode choice was heavily dominated by automobile use. The automobile was used for 87% of all trips (77% - personal auto, 7% - other private auto, 3% - other auto). Only school activities made significant use of any other mode.

A significant portion of the sample (greater than 10%) indicated a willingness to wait at least 30 minutes both before and after all activities in which they engaged. Many others (25%) were selectively willing to wait as long as 5 minutes for certain activity types, such as: church, medical and dental appointments, while 10% were willing to wait that long for: work, movies and theaters, other recreation, shopping, and after-school activities.

To help clarify the data regarding activity participation, the households surveyed were separated into groups on the basis of characteristics and relationships within the household (Table 1). The average durations for activities based on the classifications of Fried et al (1979) are shown in Fig. 1 together with the total daily travel time, DT, for travelers in each group, and for the total sample, group O. (It is noted that the daily travel is constantly slightly greater than one hour.) Because much of the college student group (9) live on campus, their school-related activities are not reported in the travel survey. Thus, their computed activity durations are probably underestimated. By far, the WC category predominates the average travelers' day.

These durations are further illustrated in Figure 2 with travelers who made a work trip on the left bar versus travelers who made no work trips on the right bar for each life cycle group. Because the WC category includes only work and school activities, the WC duration on a left bar may include both work and school activities while the WC duration on the right bar is the average school activity time period. The numbers appearing at the top of each bar indicate how many people belong to that segment of analysis.

The distributions over activity types of the average percentage of household time spent away from home for each group and total sample are shown in Figure 3.

Finally, a series of summary statistics on a broad range of characteristics of the travel/activity behavior of the sample are displayed in Table 2.

TABLE 1
HOUSEHOLD STRATIFICATION

Group	Characteristics	N
1	Households with only one individual (individuals less than 65 years of age and non-students)	36
2	Households with at least one child 6 years of age or less and only one worker	67
3	Households with at least one child 6 years of age or less and two or more workers	38
4	Households with one or more children 7-17 years of age and only one worker (no children 6 years of age or less)	26
5	Households with one or more children 7-17 years of age and two or more workers (no children 6 years of age or less)	70
6	Households with no children (youngest person at least 18 years of age) and only one worker	56
7	Households with no children (youngest person at least 18 years of age) and two or more workers	65
8	Elderly households (either all people 65 years of age or older; or at least one person over 65 and no workers	48
9	Student households (each person in household a student)	153
10	Unrelated individuals (each person in household unrelated to the others)	25
	Entire Sample	600

4. The Model

The travel/activity characteristics of a subsample of 77 respondents to the Windham survey, selected on the basis of completeness of data for all members of the respondent's household, were analyzed relative to the expected impacts of several transportation policy options and scenarios. This analysis was performed as part of the continuing development of the STARCHILD modeling system (Recker et al., 1983b).

The STARCHILD model which has been developed to examine the formation of household travel/activity patterns utilizes a simulation approach comprised of six stages:

(1) Specification of individual activity programs from an examination of household activity programs and constraints, and the interactions between the household members given the existing supply environment.

(2) Generation of the set of feasible, individual travel/activity patterns through a constrained, combinatoric scheduling algorithm.

(3) Identification of distinct members of the set of feasible travel/activity patterns by means of pattern recognition techniques.

(4) Identification of non-inferior (perceived) pattern set for individual choice utilizing a multi-objective programming approach.

(5) Formulation of a pattern choice model, which specified individual travel/activity pattern choice probabilities.

The output of this modeling process is a calibrated model of the individual's choice of activity/travel pattern. The results of the calibration of the choice model (logit) for the selected subsample of individuals are displayed in Table 3. All coefficients are correctly signed and test as significant at the 0.05 level. The model is able to correctly classify approximately 82% of the activity/travel pattern choices.

TABLE 3

ESTIMATION RESULTS FOR CHOICE OF ACTIVITY/TRAVEL PATTERN

VARIABLE	COEFFICIENT	STD. ERROR	T
1. TRAVEL TIME:VI	-0.23806E+01	0.85895E+00	-2.772
2. TRAVEL TIME:HM	-0.21382E+01	0.83780E+00	-2.552
6. HOME TIME:ALL	0.75132E+00	0.38921E+00	1.930
7. POTENTIAL:ACT	0.21786E+01	0.63477E+00	3.432
9. RISK:VI+I	-0.17727E+01	0.36977E+00	-4.794

PERCENT OF CHOICES PREDICTED CORRECTLY = 82%

PSEUDO R^2 = 0.34

Two characteristics directly related to travel time were found to significantly impact the choice of activity/travel pattern: the travel time to activities that are considered very important to the well-being of the household, TRAVEL TIME:VI; and the travel time associated with the return trip to home from activities of all types, TRAVEL TIME:HM. In initial testing of the model activities were categorized into four importance levels: (1) very important, (2) important, (3) relatively unimportant, and (4) unimportant, consistent with information contained in the data set used for estimation of the prototype model. Travel times associated with accessing activities of each type were also categorized accordingly. In addition, a separate category for the return trip home was also established.

Over two thirds of the respondents characterized the activity types: work, church, school, and medical, dental, legal appointments as "very important" to the well-being of the household. Approximately two thirds of those surveyed categorized theaters and movies, spectator sports, grocery and small item shopping, and after-school activities as either "very important" or "important." Trips associated with these two groups of activity types comprise 75% of the total trips in the sample. All other activity types were either considered unimportant or lacked a general consensus on the importance of the activity in question.

Although the estimated coefficients for the two travel time variables included in the model are comparable, their respective elasticities are noticeably dissimilar (Table 4).

TABLE 4

ESTIMATED CHOICE ELASTICITIES

VARIABLE	ELASTICITY
TRAVEL TIME:VI	-0.98
TRAVEL TIME:HM	-0.37
HOME TIME:ALL	4.55
POTENTIAL:ACT	1.00
RISK: VI+I	-0.92

The choice of activity/travel pattern is marginally elastic with respect to the travel time associated with activities considered very important to the well-being of the household and relatively inelastic with respect to travel time home from activities.

Travel time associated with activities in an individual's program that are judged as being less than very important to the well-being of the household was found to be insignificant in the choice of activity/travel patterns. The explanation of this result is rooted in an understanding of the nature of the types of activities which typically fall within this category in the sample. Such activities typically were of the nonrepetitive variety (e.g. spectator sports, movies and theater, restaurant, etc.). The implication is that, because these are "rare" events, not much attention is devoted to "fine tuning" the repetitive portion of the activity/travel pattern to minimize travel to these activities. A second feature typical of these activities is that they tend to involved more than one member of the household. Virtually all potential traveltime savings associated with the activity/travel pattern choice alternatives involved complex travel behavior (i.e., trip chaining) of one form or another. The implication is (expectedly) that trip chaining is not conducive to activities involving coordination among several individuals.

Conversely, travel time associated with very important activities tends to be repetitive and involving only the traveler.

The variable TRAVEL TIME:HM measures the time required to return home following and out-of-home activity rather than continuing onto the out-of-home activity scheduled next in the activity program. As such, it reflects the additional travel time associated with nonoptimal (in the travel sense) travel behavior.

That the home is an activity location of special significance need not be argued. It is the location that simultaneously offers the maximum amount of privacy from non-household members and the maximum potential for interaction among household members. More importantly, it represents the base for staging an individual's activity pattern. Furthermore, the value of time spent at this location is as much determined by the activity schedules of other members of the household as by the inherent characteristics of the location itself.

The results of the model estimation indicated a tendency among individuals to choose activity/travel patterns which allow them to be home at times when all other members of the household are there, HOME TIME: ALL. The individual's choice of specific pattern is highly elastic relative to his characteristic of the activity/travel pattern (Table 4).

It has been hypothesized that individuals consider their potential to participate in unplanned activities (i.e. activities that were not explicitly planned at the beginning of the day) when selecting an activity schedule (Recker et al., 1983b). The utility of the total potential to participate in unplanned activities as derived by Recker et al. involves both temporal and spatial characteristics of potential activity participation as well as the probability of such participation arising as a function of the time since the last participation. The procedure used to estimate this utility is based on identification of the set of locations at which any given activity type can be performed within the space-time constraints (the locations of the planned activities and the times than an individual must arrive and is free to leave these locations) that permit sufficient time for an individual to travel to the specific location, spend the desired amount of time participating in the activity and then reach the next planned activity prior to the time when he/she must participate in it.

For computational efficiency, the individual activity locations were aggregated into zones and the set of feasible activity locations (i.e., the number of zones in which an individual could perform an unplanned activity) were then determined using a zonal travel time matrix. In addition, the set of activity types that were evaluated as potential unplanned activities, were aggregated into the following five categories:

(1) grocery shopping
(2) clothes/small appliance shopping
(3) shopping other than (1) or (2)
(4) restaurant
(5) other (banking, post office, visit friend, etc.)

The probability of participating in some unplanned activity of a specific type in a particular zone was calculated by summing the number of trips made to all locations in that zone for that specific type of activity and dividing this number by the total number of trips made for activities of that type. The mean duration of each of the five activity types listed above was calculated and used as the required duration of the unplanned activities. Only the durations of activities that were not planned at least twenty-four hours in advance were included in the calculation of the mean durations. Finally, the probability of the need arising for an individual to participate in an unplanned activity of specific type was set equal to the inverse of its frequency since it may be shown that this probability is equal to the mean time interval between occurances of activity.

These values were then used in constructing the utility measure (as proposed by Recker et al) for the potential that each alternate activity/travel pattern offered for the individual's participation in unplanned activities. The estimated coefficient associated with this portion of the total utility of the activity travel pattern (as shown in Table 3) is highly significant.

The final variable employed in the model is related to the risk associated with not being able to participate in a planned activity due to stochastic variations in travel time (and/or activity duration, although not included in this study). In this model, risk was calculated as the sum of the probabilities that sufficient time would not be available to complete each activity, categorized according to the importance of the activity. In this calculation it was assumed that the random component of travel time to any particular activity was uniformly distributed around the minimum and maximum reported travel times between the two zones corresponding to the origin and destination of the trip. It was further assumed that the duration of any discretionary activity at the home location was indeterminant and flexible; accumulated "risk" could not be carried forward to succeeding tours.

Calibration results indicate that individuals are sensitive to the risk involved in not being able to participate in activities deemed important to the household, RISK:VI+I. These results indicated that the additional travel time to home while between activities, which biases choice toward patterns which involved trip chaining, may be counterbalanced by the risk involved in chaining activities together. That such risk is insignificant where unimportant activities are involved suggest that the preferred positioning of activities in complex tours is one reflective of decreasing activity importance.

5. Policy Application

From a policy perspective, the STARCHILD modeling system provides a potential methodology whereby the impact of various transportation-related policy options on the travel/activity behavior of individuals can be assessed. Consistent with the theoretical premise of the model, travel behavior is seen as resulting from activity scheduling behavior. This activity scheduling behavior is subject to constraints imposed by the specific characteristics of the transportation, activity and household systems (i.e., the spatial/temporal connectivity of activity locations by travel modes and the interaction between household members). Any policy that changes the characteristics of the transportation, activity or household system will therefore change the nature of the constraints imposed on the individual, which in turn, will alter the individual's set of alternatives. Policies that may be investigated based on this framework include:

(1) Changes in operating hours of activity locations.
(2) Flexible work hours (flex time).
(3) Restrictions on total daily personal travel time budget.
(4) Changes in the spatial distribution of activity locations.
(5) Changes in the performance of the transportation system.

To estimate the impact that these various policies have on activity scheduling behavior (and hence, on travel behavior) the following procedure is employed:

(1) The new set of constraints imposed on the individual by the proposed policy is specified and input to the simulation model, as part of the individual actvty programs or as a system-wide parameter,
(2) The set of feasible activity patterns resulting from the new constraints is calculated,
(3) The new feasible activity patterns are classified to construct the new choice set, and
(4) Using the choice model parameters estimated previously, choice probabilities for the new pattern alternatives are obtained.

Any policy involving a change in the operating hours of a specific activity type can be incorporated by simply changing the temporal availability parameter associated with that activity type. For example, flextime can be introduced into the model by increasing the temporal availability of the "work" activity and allowing the start time of the "work" activity to occur over some period of time (as opposed to being constrained to occur at a particular point in time). The duration of the work activity, however, would remain unchanged. To estimate the impacts of a restriction on personal travel time budget, the total daily travel time associated with each feasible activity pattern could be calculated and all those patterns involving travel time in excess of the limit are eliminated from the individual's opportunity set prior to the implementation of the classification and choice models. The impacts of changes in the spatial distribution of activities could be estimated by changing the observed locations of the activities contained in the individual's activity program. The impact of changes in the performance characteristics of the transportation system can be assessed through adjustments to the various travel time matrices used in the generation of the feasible pattern set.

5.1. Policy Constraint Specifications.

Three policy areas have been selected for detailed analysis to illustrate the actual application of the STARCHILD modeling system to policy analysis. Alternate formulations of these areas produce policies which may be considered either expansive or restrictive, based on the resulting changes in individual choice space. An expansive policy (such as flextime) produces, an increased choice space from which feasible patterns and thus choice sets are drawn. Minimally, the base choice set is drawn from the expanded choice space. In general, new alternatives are introduced to the choice set. Conversely, restrictive policies reduce the size of the choice space, and similarly shrink the set of potential patterns which form the set of actual distinct alternatives.

The three policy areas selected in this study are: (1) Variable work hours (ie. Flextime), (2) Changes to Network travel speed (resulting from any of a variety of transportation planning strategies), and (3) Changes to individual travel time budgets (which may, for example, result from shortened work hours). More specifically, these policies may be stated as:

Expansive

1. A 10% increase in overall network travel speed.
2. A 20% increase in overall network travel speed.
3. A 25% increase in individual travel time budgets.
4. A 1 hour flexibility in work star and end time.

Restrictive

1. A 10% decrease in overall network travel speed.
2. A 20% decrease in overall network travel speed.
3. A 25% decrease in individual travel time budget.

To simplify the present analysis, the effect of each policy is analyzed independently, although the impact of multiple policies acting simultaneously or choice behavior is easily examined within the STARCHILD system.

5.2. Generation of Feasible Activity Patterns

The STARCHILD model generates the complete set of feasible activity patterns for each individual, reflecting the impact of the proposed policy. As expected, expansive policies create larger feasible pattern sets, and the more substantial the policy change is, the greater the increase in the pattern set. The converse is true of restrictive policies.

Figure 4 depicts the sensitivity of network speed on pattern generation. A ten-percent change in speed, either an increase or a decrease, will affect only 32.5 and 37.7 percent of the sample, respectively, although some of the individuals do face significant expansion or reductions in their feasible pattern sets. Twenty-percent changes in speed produce greater changes across the board, with 41.6% of the sample benefiting from an increase in feasible choices with increase speeds, and over half (50.7%) facing reductions in the number of feasible patterns based on decreased speeds. Only one individual faced an empty feasible pattern set when this restrictive policy was applied.

A second policy scenario involves changes to individual travel time budgets. Impacts were estimated for a 25% increase and a 25% decrease in the observed individual budget, and also for a budget policy in which the individual is restricted to the time budget implied by the observed activity pattern. The sensitivity of the size of the feasible pattern set relative to the base case (which had no exogeneous budget enforced) is presented in Figure 5. The results indicate that policies involving travel budgets can have significant effects on the feasible pattern set. When the observed budget is enforced, only 35% of the sample remains unaffected. In fact, 54.5% of the sample have at least a 40-60% reduction in the feasible pattern set. The other two policy scenarios illustrate quite opposite impacts, with almost half of the sample unable to complete any activity pattern when faced with a 25% decrease in the travel budget, whereas 55% of the sample were unaffected by the 25% budget increase. In all cases, of course, a restrictive policy reduces, or at best maintains the individual feasible pattern set. Conversely, an expansive policy increases, or at worst maintains, the individual pattern set.

The last policy tested was flextime. Since 27 percent of the sample did not have work activities, they were not impacted by flextime policies. The outstanding majority of the sample enjoyed a significant increase in the size of the feasible pattern set resulting from a one hour flextime policy. Over 53% of the individuals investigated had at least a 200% increase in the feasible pattern set. (see Figure 6).

5.3. Classification of New Choice Sets

The feasible pattern sets generated for each individual under each policy scenario are input to the classification of a distinct choice set. As with impacts expected of alternate policies on feasible pattern sets, expansive policies, at worst, maintain existing choice sets, while the converse holds true for restrictive policies. An examination of changes within the actual choice set is of greater importance than corresponding analysis at the feasible pattern set since even a substantial increase or decrease in the latter set could result in an insignificant change in choice space dimensionality, due to the inherent redundancy in the feasible pattern generation procedure.

Results of changes in network speeds on choice set dimensionality support those based on examination of the feasible pattern set - a large proportion of the sample is unaffected (between 43% and 54% for all policy scenarios). Figure 7 illustrates the increases in choice set dimensionality by the number of representative pattern alternatives added due to increased speed. For example, with a 20% increase in network speed, over a quarter of the sample (26%) had their choice sets increased by an additional alternative, distinct from those in the existing set. Increasing the choice set, of course, provides the added potential for alternatives whose overall utility exceeds that of the existing selected alternative. Conversely, restrictive policies such as an overall decrease in network speeds (see Figure 7) reduce the dimensionality of the new choice set, and potentially decrease the expected utility of choice. The decreases in dimensionality closely parallel the increases evident with corresponding expansive policies, i.e., a 20% change in speed produces similar changes in magnitude of dimensionality, with the nature of the policy determining the direction of impact.

Figure 8 presents similar results for the temporal strategies investigated. A 25% increase in personal travel budgets produces changes in choice set dimensionality similar to those associated with increase in network speed. The majority of the sample is unaffected (55%) and only a small proportion (7%) have an increase in the choice set of at least 3 alternatives.

The remaining temporal policies produce the most significant changes in dimensionality, and in opposite directions. The majority of those unaffected by flextime policies are non-workers. Of workers in the sample, 82 percent enjoyed an increase in dimensionality, with 29% of the overall sample (39% of the workers) gaining at least 3 additive distinct pattern choices. On the other hand, introduction of a policy enforcing a 25% reduction in personal travel time budgets severely reduces choice set dimensionality. In fact, one half the sample faced a total reduction in dimensionality leaving an empty choice set. Figure 7 depicts only those individuals who have at least one alternative remaining (e.g. 33% of those with patterns remaining, have the same number of alternatives as the base case). Over 46% of this subsample have reductions in dimensionality of at least three choice alternatives.

The greatest sensitivity of choice set dimensionality to the various policies considered is associated with the expansive flextime policy and the restrictive travel budget reduction policy. Choice set size was somewhat insensitive to network speed changes for the range of adjustments analyzed.

5.4. Model Application to Assess Probable Response

Using the choice model parameters estimated previously, choice probabilities are obtained for the new pattern choice alternatives resulting from each policy scenario. Initial analysis utilized only those individuals whose observed pattern's utility was greater than other patterns in that individual's choice set (i.e., that individual was "correctly classified" by the model.) Utilities and choice probabilities were computed for alternatives resulting from each policy scenario and compared with these base probabilities. If an individual's observed choice was eliminated from his choice set, then the individual would obviously be forced to change his/her desired pattern. If an additional pattern in an expanded choice set was defined by a higher utility, then that individual was also characterized as a potential convert to a different choice alternative.

The probable responses to the sample policy scenarios are presented in Figures 9 and 10. When the network speed was increased, additional patterns of greater utility were generated resulting in the classification of individuals as potentially choosing a different alternative. The greater the increase, the more substantial the change. Corresponding decreases in network speed, while eliminating many patterns form consideration, in most cases maintained the observed alternative in the final choice set, perhaps due to an inherent ranking of the observed patterns as optimal with respect to variables such as travel time in the base choice set. Accordingly, few individuals are listed as potentially changing their pattern choice under this scenario.

Temporal strategies such as flextime and an increased personal travel time budget produced similar probable response patterns (13% and 15% probable change, respectively). Each policy was associated with an expanded choice set, yet the increase in alternatives with greater utility is apparently not significant for the majority of the sample. Significant behavioral response is probable for the restrictive travel budget policy, with 71% of the sample associate with change in choice preference. This figure includes the half of the sample faced with no choice alternatives due to the imposed policy. These individuals would be forced to implement major changes in their activity program, such as alternate destination choice, or foregoing an activity, to produce a feasible activity pattern.

Probable responses patterns indicated a general trend in behavior shifts, yet are limited in their assumption that the alternative initially selected will no longer be if a new alternative characterized by a greater value of utility is generated. It is more appropriate to examine expected utilities, which are merely a weighing of each alternatives utility by the estimated choice probabilities. These values are aggregated for the sample and plotted relative to the base scenario for each policy under consideration (see Figure 11). The shift in overall utility toward greater expected values is evident for all expansive policies, yet the distribution is somewhat different in each case. Similar results are evident for restrictive policies, but toward reduced expected utilities. Furthermore, the distributions depicted for the latter policies each indicated a significant shift from the midrange of expected utilities toward the lower range of the scale. Full interpretation is beyond the scope of the present paper.

6. Summary

In addition to these transportation-related policies, the impacts of the introduction and utilization of new modes of travel (e.g., electric vehicles) could also be estimated, by specifying the vehicle design parameters such as speed, range (time that vehicle can be used before recharging), and recharge time (time before vehicle can be used again). As with the other policies discussed above, these design characteristics impose a new set of constraints on the individual which would be used to generate a new set of feasible activities patterns. Once these new opportunities are generated, the choice set is created and new choice probabilities can be estimated.

In addition to forecasting an individual's activity pattern changes in response to policy-induced alterations in the transportation and activity systems, the proposed methodology may also be used to provide information regarding the range of potential opportunities (and hence, possible choices) available to individuals. This information can then be used to identify segments of the population most impacted by policy alternatives.

These and a wide range of other policy issues may be analyzed using the STARCHILD model system. The present application serves to illustrate the potential of the activity-based approach for comprehensive policy evaluation involving behavioral responses.

REFERENCES

1. Burns, L.D. (1979) Transportation Temporal, and Spatial Components of Accessibility, Lexington Books, D.D. Heath.

2. Damm, D. (1983) "Theory and Empirical Results: A Comparison of Recent Activity-Based Research," in S. Carpenter and P. Jones (eds.), Recent Advances in Travel Demand Analysis, Gower, London.

3. Davis, C.F., W.H. Groff, and T.E. Steahr (1981) An Analysis of the Potential for Dynamic Ridesharing in a Low Density Area, Final Report to US DOT, Washington, D.C.

4. Fried, M., J. Havens, and M. Thall (1977) Travel Behavior - Synthesized Theory, Final Report for NCHRP, TRB.

5. Golob, J., and T.F. Golob (1983) A Classification of Approaches to Travel-Behavioral Analysis, Special Report 201, Transportation Research Board.

6. Heggie, I.G., and P.M. Jones (1978) "Defining Domains for Models of Travel Demand," Transportation, 7, pp. 119-125.

7. Hirsh, M., J.N. Prashker, and M. Ben-Akiva (1984) Dynamic Model of Weekly Activity Pattern, draft manuscript.

8. Jones, P.M. (1983) "The Practical Application of Activity-Based Approaches in Transport Planning: An Assessment, in s. Carpenter and P. Jones (eds.), New Approaches in Travel Demand Analysis, Gower, London.

9. Jones, P.M., M.C. Dix, M.I. Clarke, and I.G. Heggie (1983) Understanding Travel Behavior, Gower, London.

10. Landau, U., J.N. Prashker, B. Alpern, and M. Ben-Akiva (1980) The Application of Space Time Budgeting Concepts to Shopping Destination Choice, Research Report No. 80-24, Transportation Research Institute, Technion, Israel.

11. Lenntorp, B. (1976) Paths in Space-Time Environments, Lund Series in Geography, Series B., No. 44. University of Lund, Sweden.

12. Moore, E.F. (1963), "The Shortest Path Through a Maze," Proceedings International Symposium on the Theory of Switching, Harvard.

13. Recker. W.W., G.S.Root, M.G. McNally, M. Cirrimcione, and H.J. Schuler, (1983a) An Empirical Analysis of Household Activity Patterns, Final Report to U.S.DOT, Washington, D.C.

14. Recker, W.W., M.G. McNally, and G.S. Root (1983b) Chaining Behavior in Urban Trip Making, Interim Final Report for U.S. DOT, Washington, D.C.

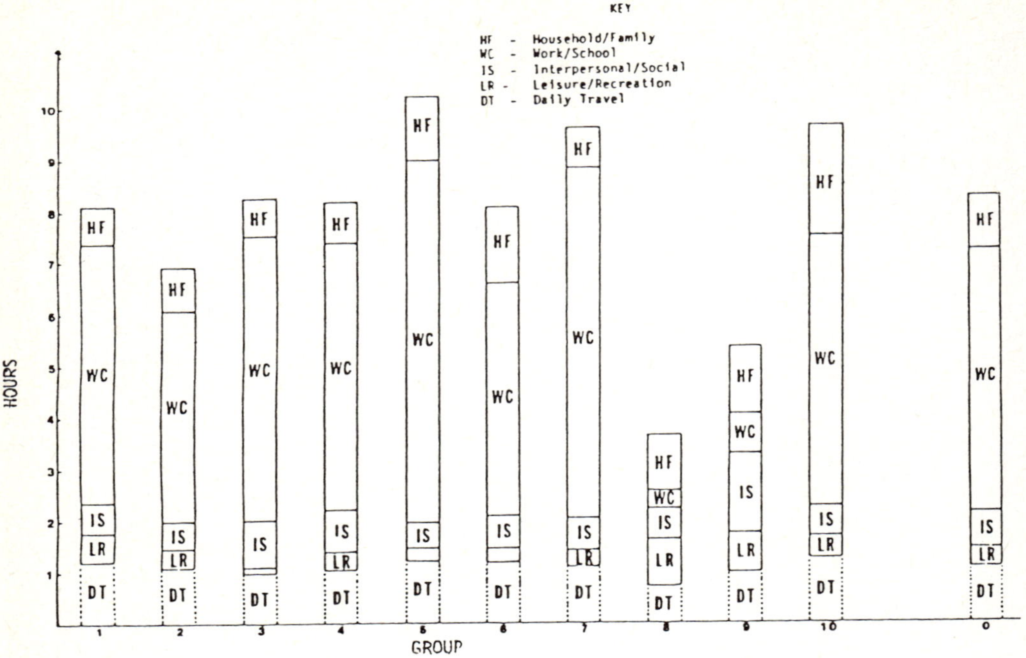

FIGURE 1.

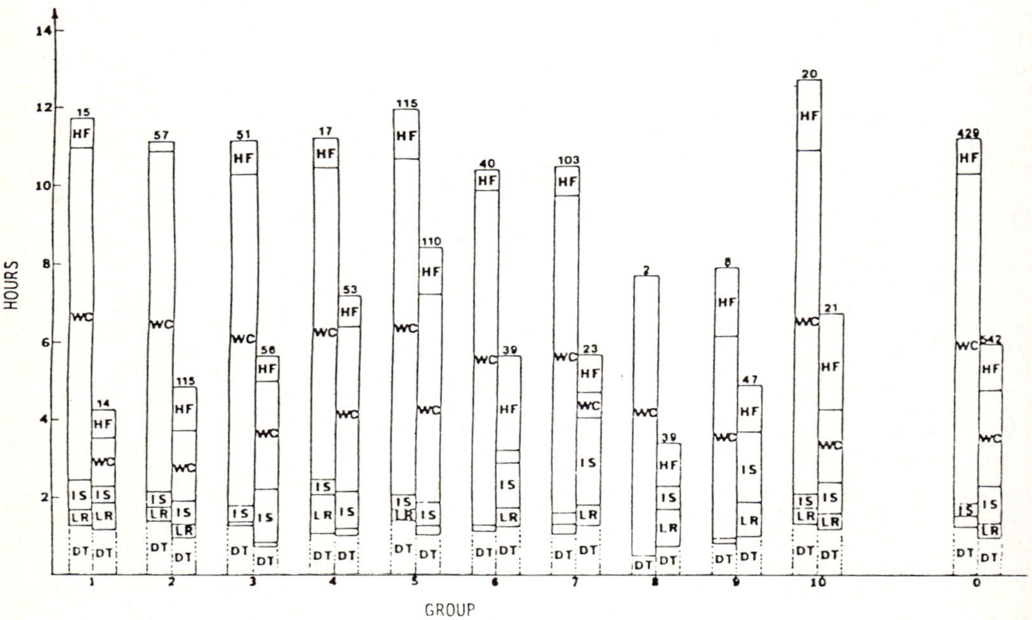

FIGURE 2.

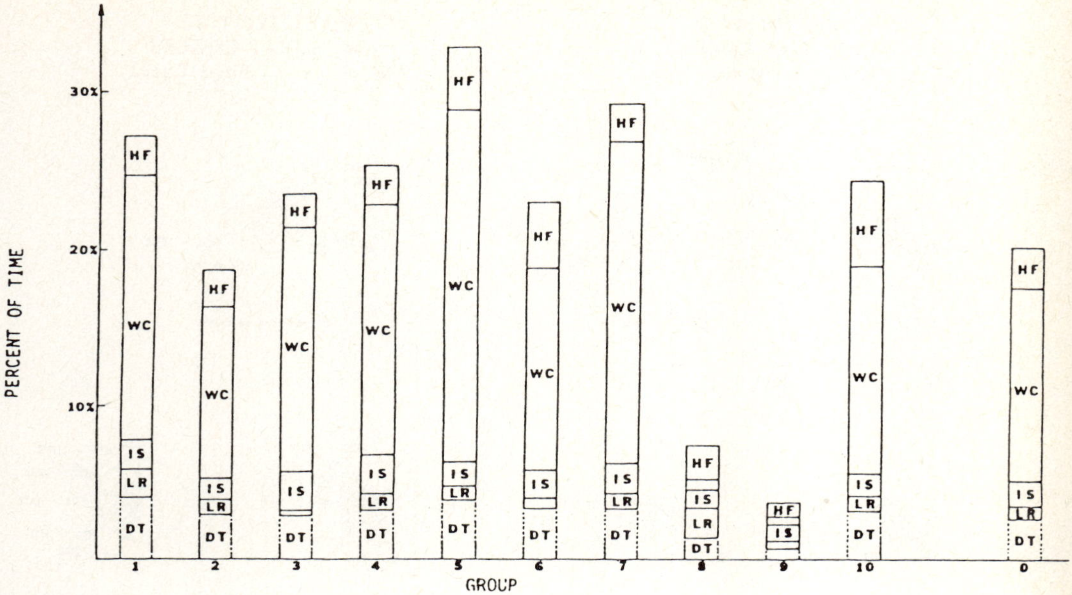

FIGURE 3.

TABLE 4.

	Group										
	1	2	3	4	5	6	7	8	9	10	0
number of HHs	36	67	38	26	70	56	65	48	153	25	600
number of people	36	264	156	94	291	115	171	83	329	65	1653
number of travelers	29	172	107	70	225	79	126	41	55	41	971
% travelers	80.6	65.2	68.6	74.5	77.3	68.7	73.7	49.4	16.7	61.2	58.7
total number of tours	34	217	129	91	309	98	152	45	63	55	1223
tours/trav	1.17	1.26	1.21	1.30	1.37	1.24	1.21	1.10	1.15	1.34	1.26
total number of trips	98	547	319	226	704	228	367	104	146	138	2939
trips/tour	2.88	2.52	2.47	2.48	2.28	2.33	2.41	2.31	2.32	2.51	2.40
trips/trav	3.38	3.18	2.98	3.23	3.13	2.89	2.91	2.54	2.65	3.37	3.03
avg tour duration (hrs)	6.07	4.93	6.30	5.51	6.62	5.44	7.38	2.33	2.86	5.94	5.75
avg. trip time (min)	21.29	20.24	19.54	18.90	22.87	24.18	22.14	16.44	21.72	21.73	20.84
avg daily travel (min)	72.0	64.4	58.2	61.0	71.6	69.9	64.4	41.8	57.6	73.2	63.1
number of complex tours	16	58	34	17	53	17	35	9	13	17	270
% tours which are complex	47.1	26.7	26.4	18.7	17.2	17.3	23.0	20.0	20.6	30.9	22.1

SENSITIVITY TO
TRAVEL TIME CHANGES
(relative to Base Case)

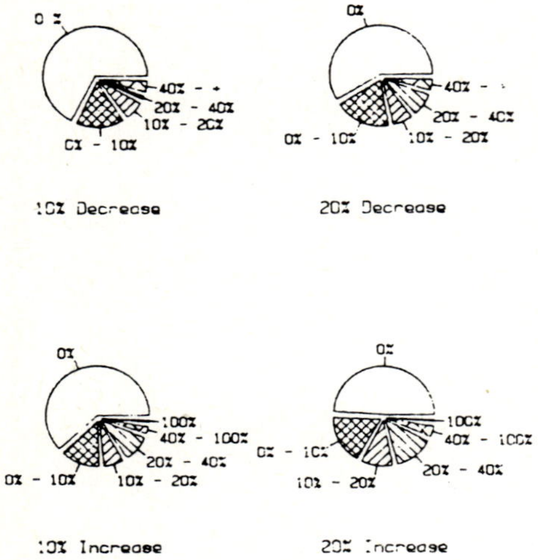

FIGURE 4.

SENSITIVITY TO
TIME BUDGET CHANGES
(Relative to Base Case)

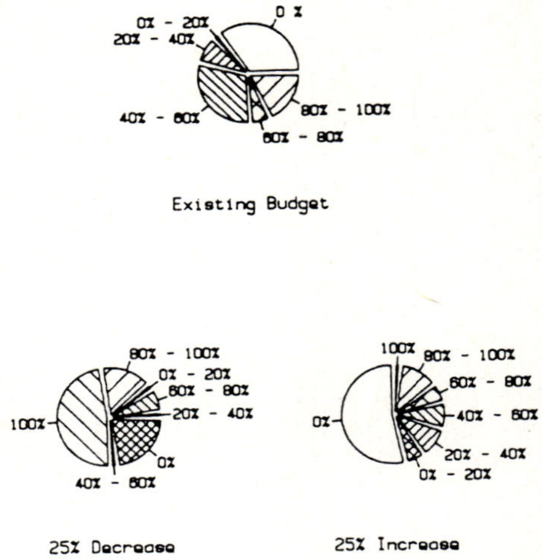

FIGURE 5.

SENSITIVITY TO
FLEX TIME POLICY
(Relative to Base Case)

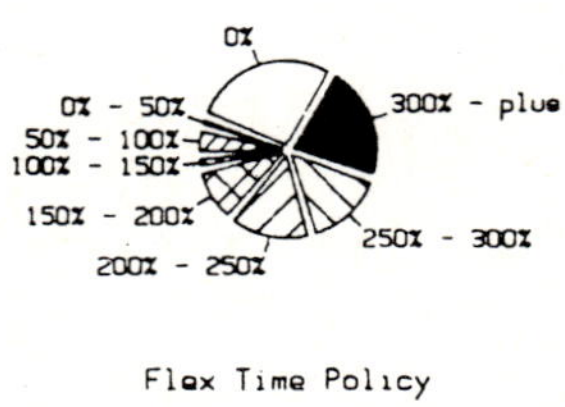

FIGURE 6.

FIGURE 7.

Changes in Choice Space Dimensionality Resulting from Changes in Network Speed

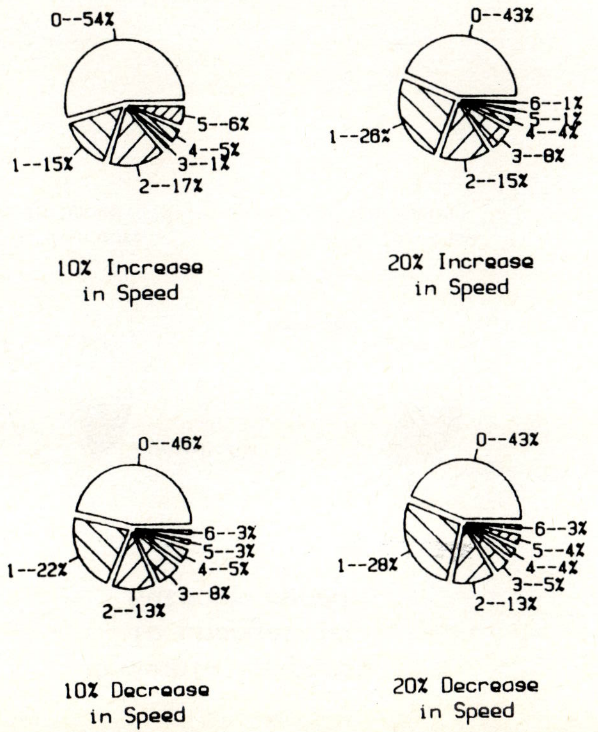

FIGURE 8.

Changes in Choice Space Dimensionality Due To Temporal Strategies

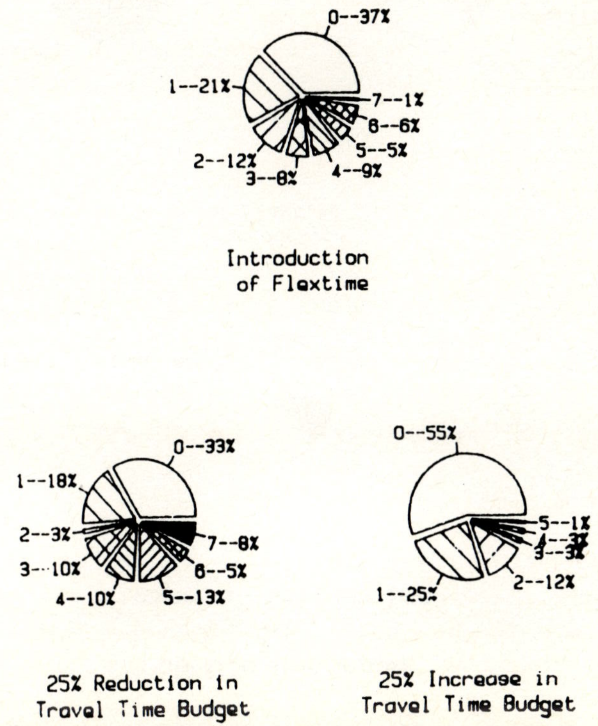

FIGURE 9.

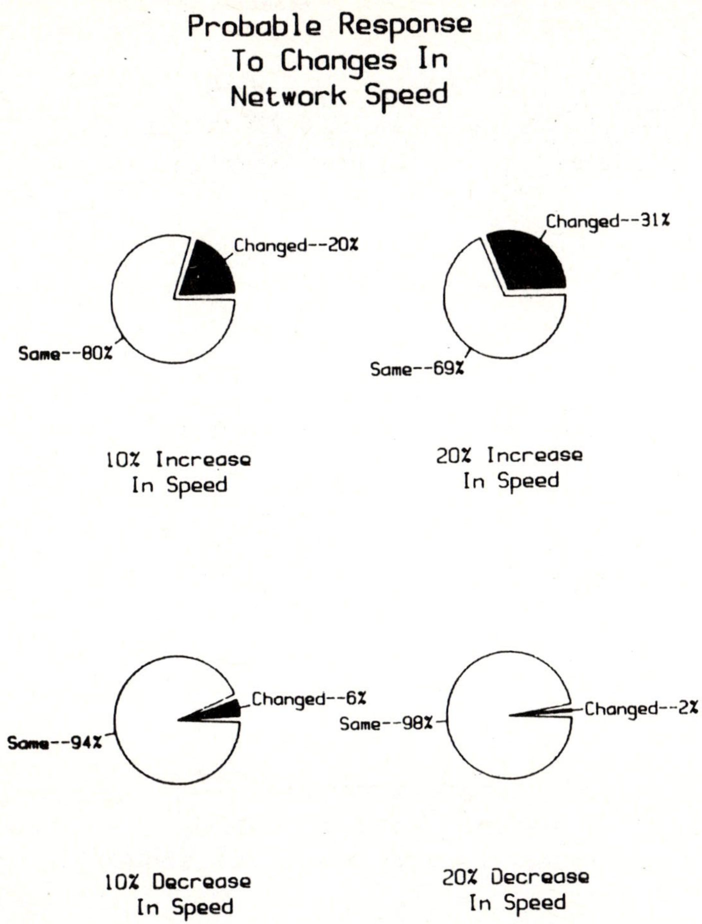

FIGURE 10.

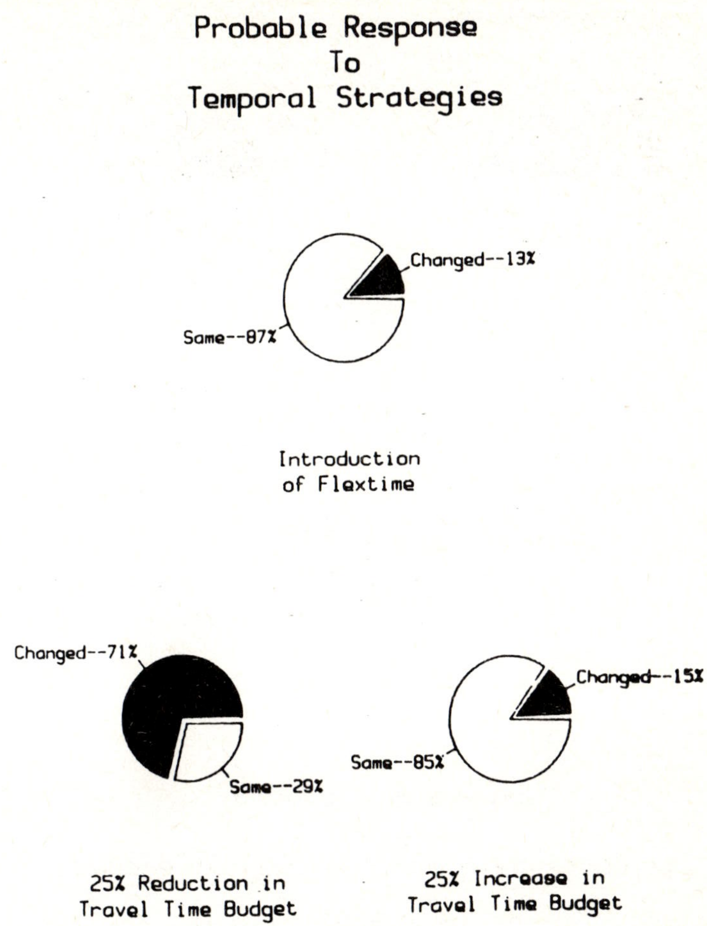

FIGURE 11.

(a) Distribution of Expected Utility with 10% Increase in Network Speed

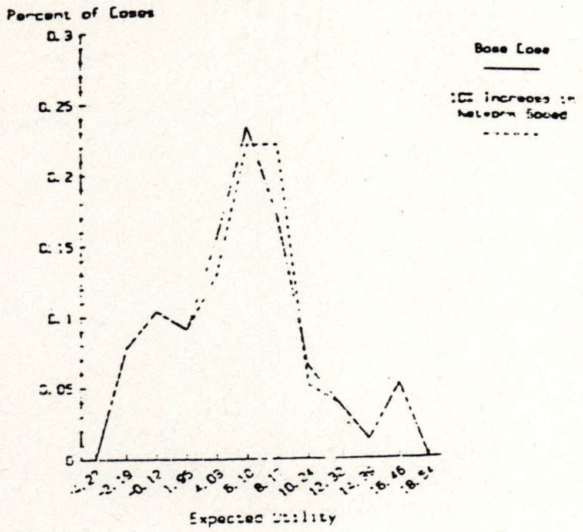

(b) Distribution of Expected Utility with 20% Increase in Network Speed

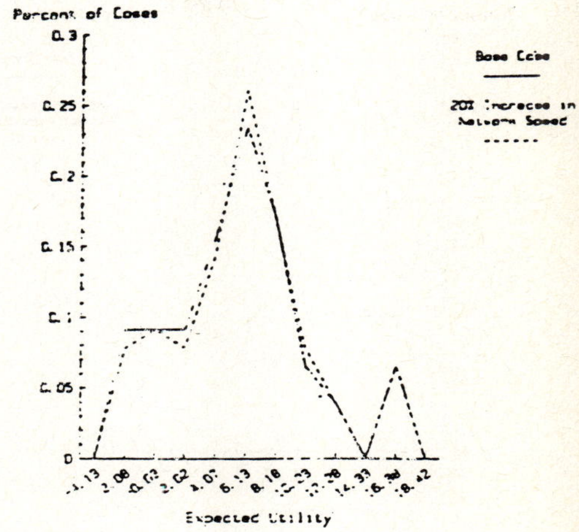

(c) Distribution of Expected Utility with 10% Decrease in Network Speed

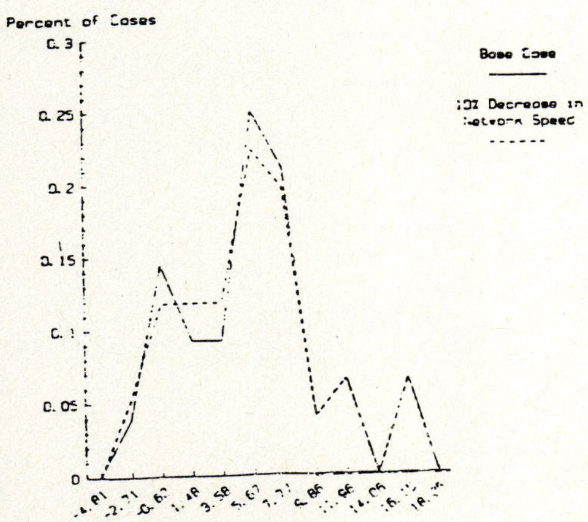

(d) Distribution of Expected Utility with 20% Decrease in Network Speed

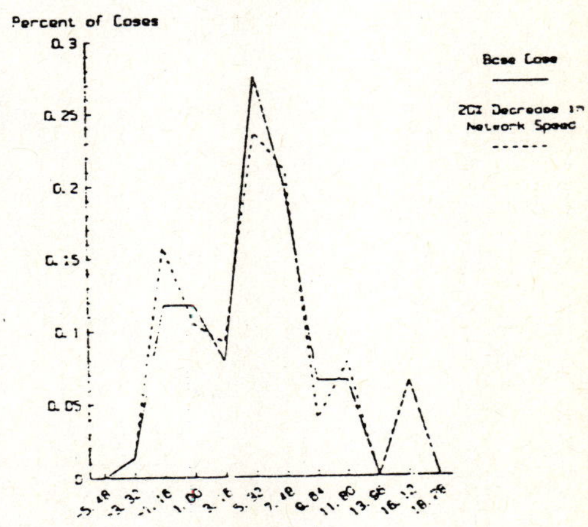

FIGURE 11. (cont'd.)

(e) Distribution of Expected Utility with 25% Increase in Travel Time Budget

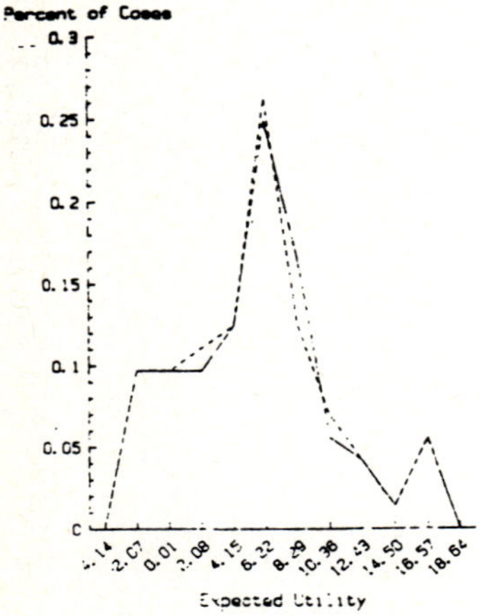

(f) Distribution of Expected Utility with 25% Decrease in Travel Time Budget

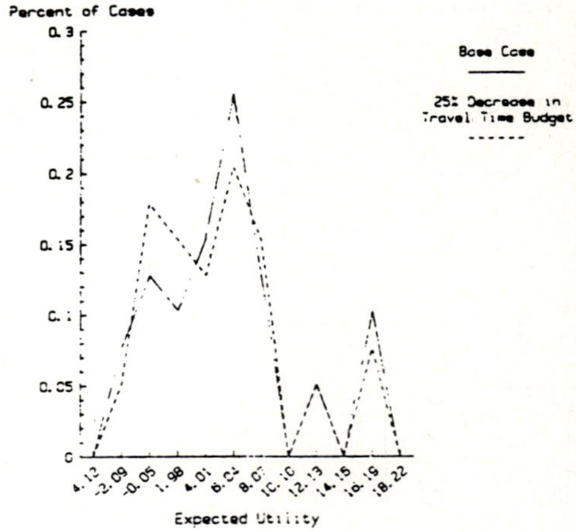

(g) Distribution Of Expected Utility With Flextime Policy

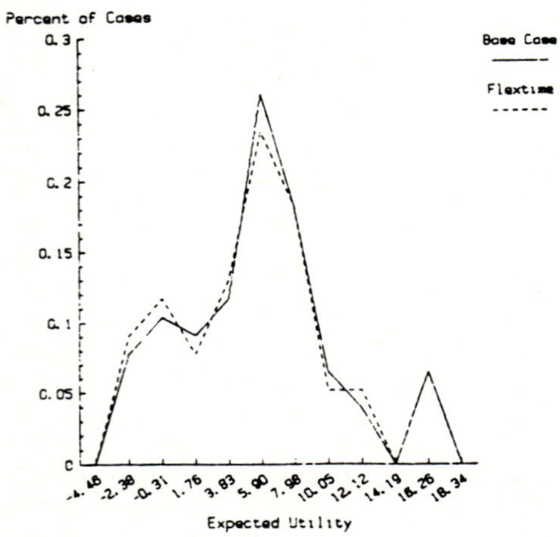

2
Longitudinal surveys and analysis methods

Behavioural Research for Transport Policy
pp. 55–79.

Behaviour Changes in Changing Populations: problems and proposals for policy-directed longitudinal studies.

M.C.Dix and A.D.Layzell

Transport Studies Unit, Oxford University

1. Introduction

Longitudinal designs typically need to be appropriate to particular policy contexts. This means that the designer will often be faced with difficult prior judgements as to the nature of policy developments and their consequences; together with their likely timescale, and possibly, their geographical diffusion.

The current use of longitudinal studies in transport has focussed largely upon addressing relatively short-term questions, for example monitoring the responses of individuals to interim policy changes. In the medium to long term, however, if such approaches are to continue to increase our understanding of travel behaviour, other and more fundamental influences are likely to become increasingly important. These relate at one level to movements of individuals or households between sub-populations (whether defined in terms of geographical areas, age groups or household characteristics), and at a higher level to changes in basic demographic variables, such as birth rates, mortality, family formation, unemployment etc.

This paper attempts to illustrate and clarify the form which such problems can take, in practice, and makes reference to policy questions which have recently been explored, using longitudinal designs, at TSU.

Processes of Household Turnover and Recomposition

Most longitudinal survey designs concentrate (very properly) on monitoring changing behaviour of the same people through time, but equally important, over all but the shortest term, are processes through which the population itself changes. For example, within any residential population, we can expect changes:

(a) of households in occupancy, through decisions by original residents to move out, and newcomers to move in to dwellings (turnover)

(b) of individuals within households, through 'decisions' of family formation; growth; disintegration and re-formation. (recomposition)

Both processes have complex social and demographic causes and either may be sensitive to a variety of policy influences.

Net outcomes of (b) are apparent at the level of the national population (see Figure 1, from OCPS repeat-cross-section surveys). Consequences for travel demand patterns are inevitable, although not always easily forecast: one role of monitoring studies can be to detect such trends and to assess their possible transport implications, at an early stage.

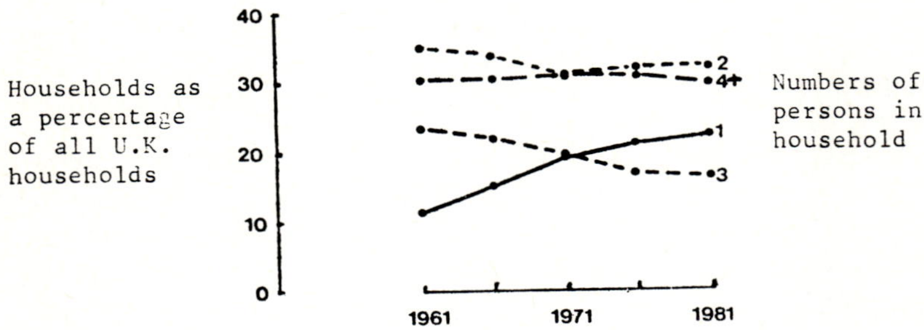

Figure 1 Changes of household size within the U.K. household population, 1961 to 1981. (Source: from Social Trends, 11, HMSO)

Within increasingly smaller (geographically defined) subpopulations, household relocation-replacement decisions ((a) above) generally become proportionately more important as a source of change. The smaller the area - mean relocation distance and other things being equal - the briefer is the °half life' of original household occupancy.

One data collection strategy which allows aspects of both processes above to be monitored is the dwelling-based panel. This (see Dix (1985)) has advantages over the more usual panel approaches (although even these are relatively new to transport research) and exploits the propositions that:

a) while the unit of analysis is the household and/or individual, the sampling unit need not be the same

b) choice of a (sampled cross-section) of dwellings as the sampling frame avoids the problem of having to trace people over time as they disperse geographically (which it might not be in the interest of, for example, a local transport authority so to do).

c) the fact that there will be 'natural' turnover of people within the dwellings sample means that an automatic sample-replacement mechanism is provided. This will tend to ensure ongoing representativeness of the local area population assuming that the sample of dwellings remains representative of housing stock in the area concerned. Such an assumption will be reasonable in the short term, although in the longer term, and where building renewal has been vigorous, the dwellings sample itself will need to be refreshed.

d) The unit of analysis is the household, and/or individual members of the household. Problems of deciding 'who to trace' over time (and as members of originally sampled households split up, remarry, form new households, etc) have led users of other longitudinal approaches to favour the individual as the unit of (both sampling and) analysis. These are clarified by the simple definition of households as °those resident in the selected dwellings at each re-survey interval'.

e) Analysis is possible of both changing behaviour of the same people and the patterns of turnover of people (including patterns of selective migration in and out of the area). It is likely to be in the interests of a transport/planning authority to monitor both such sources of changing travel behaviour locally. Joint monitoring of the two processes occurs 'naturally' within the dwellings-based panel, which - in effect - traces changes in the trip-generative and other properties of dwellings in the areas concerned, and allows their causes to be attributed to changing characteristics and decisions of the (recomposing) population in residence.

The next section of the paper aims to illustrate the dwellings-based panel survey design by reference to a repeat-interview exercise carried out on a small scale in Banbury last year. The exercise represents the simplest possible application of the sampling and measurement model. There were just two survey waves, with no adjustment for changes in housing stock. Details of the exercise follow.

The Banbury dwellings-based panel survey exercise

In the autumn of 1976 all members of some 200 households in Banbury completed activity-travel diaries which, together with in-depth interviews, provided the core data for analysis reported later in Jones et.al. (1983). Opportunities arose in 1983 to add a longitudinal dimension to this cross-sectional study, and a fifty per cent sample was drawn from the original list of respondents.

The pattern of household turnover during the seven year interval proved to be such that members and roughly half the original families had stayed in residence. Some but not all 'movers' could be traced elsewhere, and consequently, it was decided to collect new cross-section information on travel and characteristics of both stayers and incomers. Diary methods included four-day records of travel (self-completion activity diary for a Thursday and trip information for three additional weekdays).

The data thus collected did not represent an attempt at building a data set to support (for example) thoroughgoing testing of research or policy hypotheses. Primarily, the aim was to gather 'baseline' information as the first stage in an qualitative in-depth study involving 'backtrack' interviewing techniques. These aimed to understand the history of decisions taken by respondents to change their behaviour patterns, and are reported in Layzell, Goodwin and Dix (1985). Only the baseline information is used here. This indicates outcomes rather than decisions themselves, and is used simply to illustrate the basic sampling design; its potential and limitations; and the nature of change-processes which it is capable, at least in principle, of monitoring.

Figure 2 from the Banbury exercise gives a simple indication of the possibilities. The figure describes redistributions of people between household lifecycle groups (Table 1) and of households between dwellings, over time.

Within the dwellings sample there are tendencies (subprocess A) for 'stayer' households to change their characteristics, through continuous ageing of individuals and transitions by families into 'later' lifecycle stages. While deaths, births and home-leaving introduce some renewal of people within these households, the overall tendency is for 'ageing' in net and lifecycle terms. Partially counteracting this tendency, household turnover (subprocess B) introduces new and relatively younger families.The structure of population changes within a fixed set of Reading dwellings, which was sampled by chance in each of two cross-sectional transportation surveys, has been discussed, similarly, by Downes (1980).

Note that subprocess A is the only one to be captured by cohort or conventional panel sampling designs (except that these may trace movers-out into further dwellings and other areas, which ours did not). Subprocess B cannot be captured. Repeat cross-section surveys (which typically adopt independent, probability-based sampling designs) may accurately reveal the net outcome (path C) but cannot normally separate the constituent subprocesses.

Table 1 Household lifecycle-stage groupings

Group	Description
1	Adults of working age, no children
2	Families with children, youngest <5 years
3	Families with children, youngest ≥5 and <16 years
4	Families with children, youngest ≥16 years
5	Older (retired) adults >60/65 years

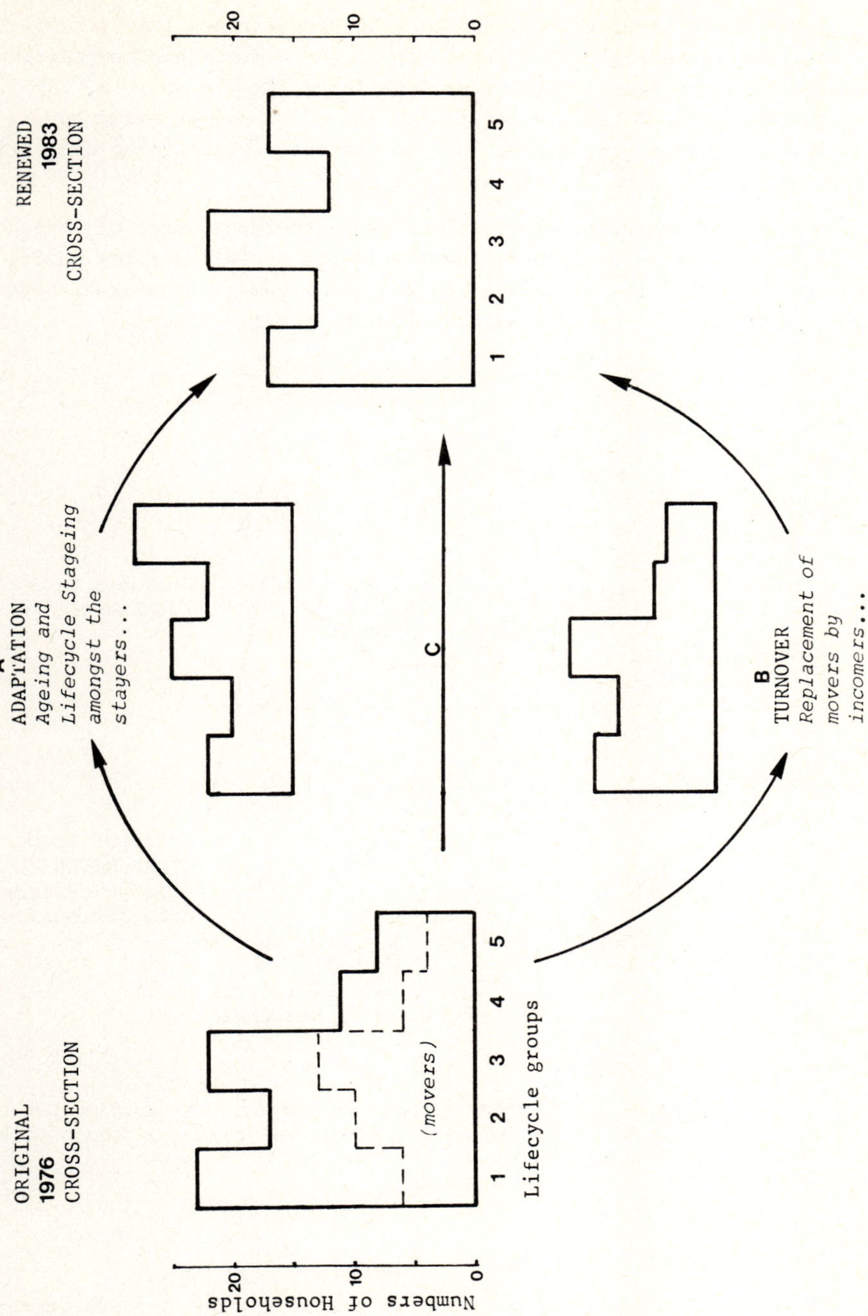

Figure 2 Changes in the frequency distribution of households by lifecycle group, within a fixed dwellings sample in Banbury, between 1976 and 1983.

The above illustration is lifecycle-based, since this variable can usefully help to explain aspects both of the demographic composition of households, and their travel demand patterns (Clarke and Dix, 1983). However, any socioeconomic descriptor may of course be examined similarly. Later, we consider car-ownership, and changes in licence-holding arising through turnover are identified in figure 3.

Plotted separately are outcomes from the replacement of households leaving the sample of Banbury town dwellings by (a) families moving from elsewhere within the sampling area and (b) incomers from areas beyond, which for the most part, meant surrounding rural villages.

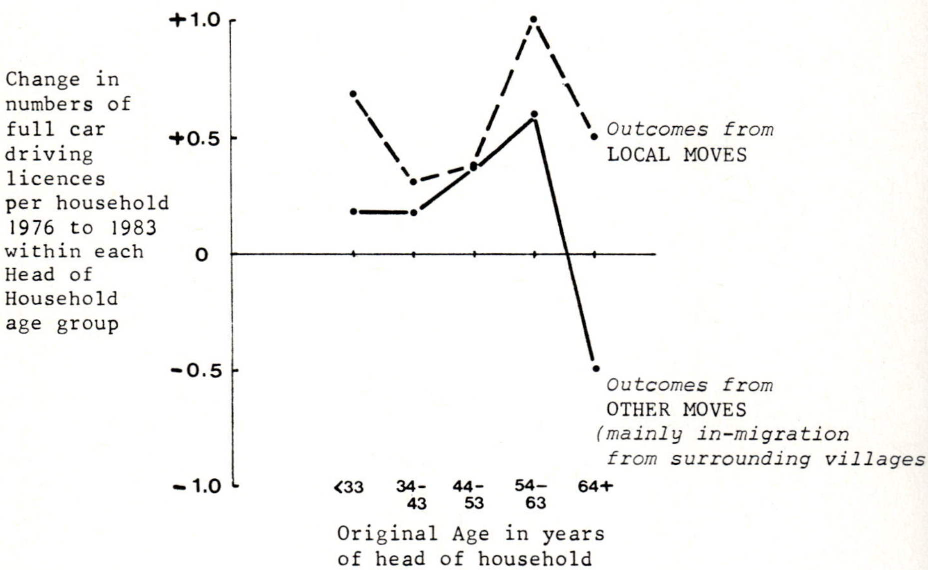

Figure 3 Turnover effects on car licence-holding per household, 1976 to 1981, within age groups of heads of household.

The interim policy context was of public transport fares increases affecting both town and village services, but additionally, with substantial network and frequency reductions in rural areas. While the small numbers involved in the re-survey exercise make policy conclusions extremely hazardous, the suggestions (further informed by accompanying in-depth interview accounts) were of:

- rural transport policy having distinct secondary impacts on the (not directly affected) urban area, which featured

-'forced' in-migration of previously bus-dependent village residents, in turn causing

- some suppression of the trend towards growth in licence acquisition, in the town's population, particularly within the youngest and oldest age groups.

There would follow from this a number of possible implications for policy; bus demand forecasting; and evaluation methodology.

No evidence was available as to the character of population turnover within the rural area itself. An extended sample, covering both urban and rural subpopulations, would be needed for this. With such an extended dwelling sample base, the panel design would clearly allow the concerned local transport authority (for example) to trace policy influences upon inter-relationships between country and town. We could also gauge (for example) the extent to which replacement of 'transport dispossessed' village residents was by people:

(a) from the town (the migratory 'exchange' model)

(b) from even more rural/bus-deprived areas (the migratory 'cascade' model)

Behavioural adaptation by people over time

Measures of behavioural changes which real people make over time are obtainable from all longitudinal designs, and only from these. Repeat cross-section sampling allows comparisons over time of the group behaviour of people occupying given stages at each measurement point (e.g. average trip rates within income or age bands). But the separate influences of (a) changing behaviour among people who remain in a given state, and (b) different behaviour introduced by turnover of people within each state, are confounded

There are converse patterns in tracking behaviour of the same people through time. The following influences may all be involved: ageing; passage of time; transitions between different states (e.g. income, lifecycle group); varying exposure to changes in policy and other exogenous variables. However, the strategy of tracking individuals over time lends itself to analysis at highly disaggregate levels, in which the aim is to separate each possible dynamic influence, by comparing behaviour of carefully selected groups/individuals, and exploiting the possibility of using individuals as 'their own control'. Layzell (1983) explains special opportunities for 'quasi-experimental' analysis, within longitudinal designs, and their advantages in terms of statistical precision of the estimators of change. One example of such analysis is presented in the following section.

Joint outcomes from adaptation and household turnover

One special feature of the dwellings-based approach is that it can detect effects of simultaneous processes, within the local area concerned, of adaptation and turnover. The following example from the Banbury exercise illustrates this with reference to inter-relationships between the two processes and car ownership.

The change in numbers of cars owned (i.e. the residential car stock) represented a 13% increase over the period. As part of an attempt to investigate causes of changing ownership patterns, the data was organised as follows.

First, the 118 households covered by the two survey waves were clustered into the groups shown in figure 4 (numbers refer to dwellings, not households). This clustering distinguishes, in the case of 'stayers', households which did and did not undergo lifecycle transitions. Where turnover had occurred, we distinguish cases in which the lifecycle category of replacement households matched, or did not match, that of originally resident households. Figure 5 gives more detail on the pattern of residential life-cycle transitions. It confirms that without turnover, these are predominantly of the 'stageing-up' kind. Where turnover had occurred, there is an apparently marked tendency either for the earlier stage to be 'perpetuated', or for 'stageing-down' to occur.

Factors explaining the tendency are partly rooted in theories of residential choice (see Mackett 1983, Carpenter 1983, and Layzell, Goodwin & Dix (1985)). The interest here is simply in considering any hints of systematic links between the patterns of turnover, stageing, passage of time, and car ownership levels.

Two simple indices of changing car ownership were constructed and applied within each cell of the basic table. The first indicates net change in car ownership thus

$$I_N = ((n_i - n_d) \times 100/N)$$

where:

n_i = number of dwellings in which increases in residential car ownership were recorded, 1976-83.

n_d = number in which decreases were recorded

N = number of dwellings in each cell.

Since this conceals offsetting changes (gains by some households cancelled by losses among others) we additionally examined the extent of change-car-

	Dwellings in which lifecycle group for 1976 = 1983	Dwellings in which lifecycle group for 1976 ≠ 1983
Dwellings in which there was NO TURNOVER *(stayers only)*	**21**	**21**
Dwellings in which there was TURNOVER *(movers replaced by incomers)*	**20**	**19**

<u>Figure 4</u> Clustering of the 118 households according to presence and absence of household turnover and lifecycle stageing within the 81 dwellings.

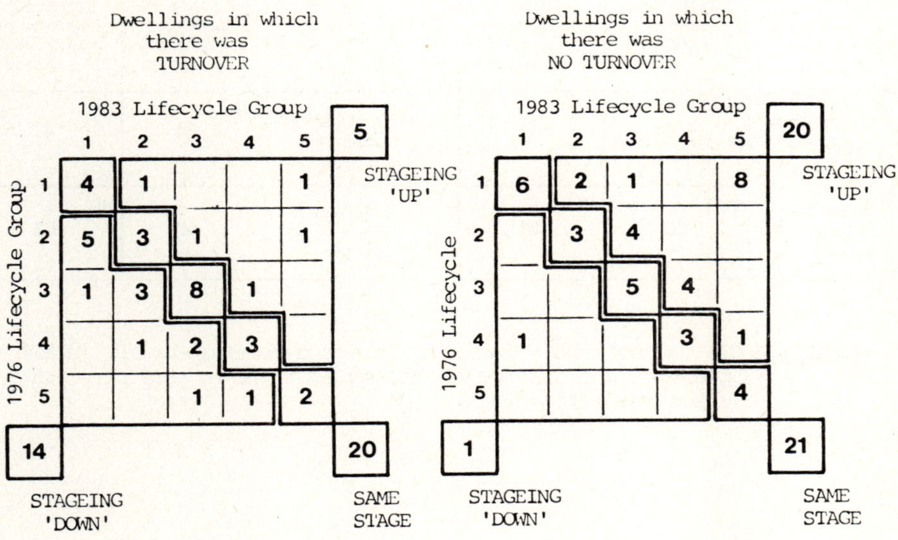

total dwellings = 39

total dwellings = 42

overall total dwellings = 81

numbers indicate numbers of dwellings in which each transition occurred

<u>Figure 5</u> Transitions between particular lifecycle groups in the presence and absence of household turnover within dwellings.

ownership-level decisions' (gains or losses) thus:

$$I_V = ((n_i + n_d) \times 100/N)$$

This second index always gives positive values and represents an index of 'volatility'. Note that in the absence of turnover, 'n' values refer to decisions among 'stayer' households to change from their 1976 car ownership state. Under conditions of turnover, these compare the 1976 state of original households with that of their replacements in 1983. Results are given in Figure 6 (to which the following, tentative interpretations refer).

I_N Net outcomes of changes between car ownership levels	NO STAGEING (*same lifecycle group*)	STAGEING (*change of lifecycle group*)
NO TURNOVER (*same people*)	(a) +24	(b) +38
TURNOVER (*change of people*)	(c) 0	(d) -31

I_V Extent of transitions made between car ownership levels	NO STAGEING (*same lifecycle group*)	STAGEING (*change of lifecycle group*)
NO TURNOVER (*same people*)	(a) 24	(b) 47
TURNOVER (*change of people*)	(c) 40	(d) 53

Figure 6 Net change and volatility in car ownership levels in the presence and absence of household turnover and lifecycle stage transitions.

cells (a) In the absence of experience of Lifecycle Transitions (ie., with ageing and passage of time alone) 'stayers' showed moderate volatility and all decisions to change car ownership level are accounted for by increases (since I_N = I_V)

cells (b) Among the remaining stayers who did make transitions between lifecycle groups (mainly of the 'stageing-up' kind) volatility of car-ownership decisions is greater. Decisions both to increase and decrease numbers of cars owned are evident (because $I_N > I_V$).

cells (c) The perpetuation of former residents' lifecycle stage, despite renewal of people in residence, is associated with moderate volatility of car ownership but there was apparently no net change between the two measurement periods ($I_N=0$) i.e., car stock remained stable.

cells (d) The greatest volatility is observed in the joint presence of turnover and stageing (which was shown mainly to comprise stageing-down). This subprocess is the only one leading to net decreases in ownership (negative I_N). These (of course) partially offset tendencies towards increase within cells (a) and (b).

A Framework for Analysing Replacement and Turnover Effects

It is useful when designing longitudinal studies for particular policy questions to have a general way of depicting the possible joint influences of renewal and turnover in modifying or channelling long term changes in travel behaviour. Some simple steps in this direction are taken below.

Movements between two sub-populations reflecting outcomes of a discrete choice'

1. We conceive of a general population (for example the U.K. or a region of the U.K.) and (initially) two exclusive sub-populations 'A' and 'B' (figure 7(a)). At one point in time (or, more accurately, within one time slice) everyone counted in the population is allocated - we assume unambiguously - to one of the two sub populations according to their state in respect of a particular characteristic or variable. For example, individuals are counted as car owners or non-car-owners.

2. At a later time, t2, we should be prepared to take account of three possible categories of change (figure 7(b)):

i) in the general population itself (size, composition, socio-economic characteristics)

ii) in the sub-population (size, composition etc...) and

iii) perhaps most important, in terms of a new set of relationships between the recomposed sub-populations.

3. We note, however, that when we are restricted (as in (2)) to two or more 'snapshot' views, it is not possible to specify the identity of people within the population (or therefore to distinguish buyers, keepers and sellers) or, more generally to analyse intermediate events and turnover processes. To move in this direction we need to consider successively more time periods, eventually conceiving of 'continuous time' information on all

Figure 7 A schematic framework

(a) Subpopulations in cross section

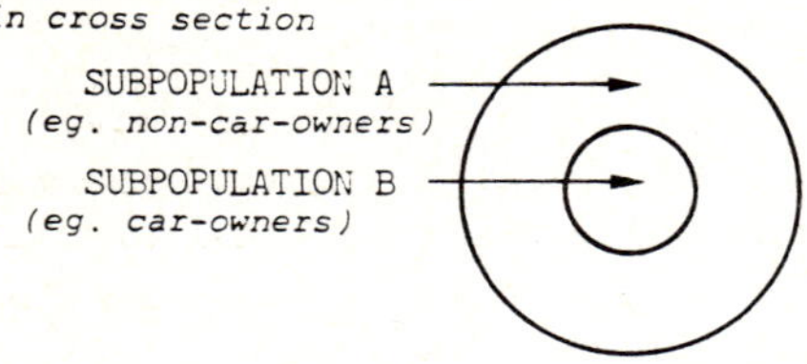

Comment *Static cross section demand models consider these relationships alone, for example*

$$p_B = f\ (X_{i_B}\ ,\ X_{j_B}\ \ldots.\ X_{n_B})$$

(probability of being in car-owning subpopulation)	*(values of socioeconomic characteristics of car-owning, versus non-car-owning, subpopulation)*

(b) Recomposition of subpopulations with the passage of time

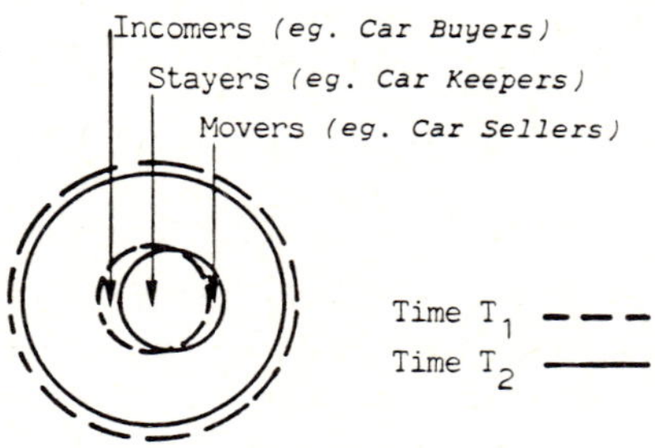

Comment *Static time series (repeat cross section) demand models consider relationships of the general form*

$$p_{B_T} = f\ (X^{T}_{i_B}\ ,\ X^{T-1}_{i_B}\ ;\ \ldots\ X^{T}_{n_B}\ ,\ X^{T-1}_{n_B}\ )$$

(c) Extension into continuous time framework

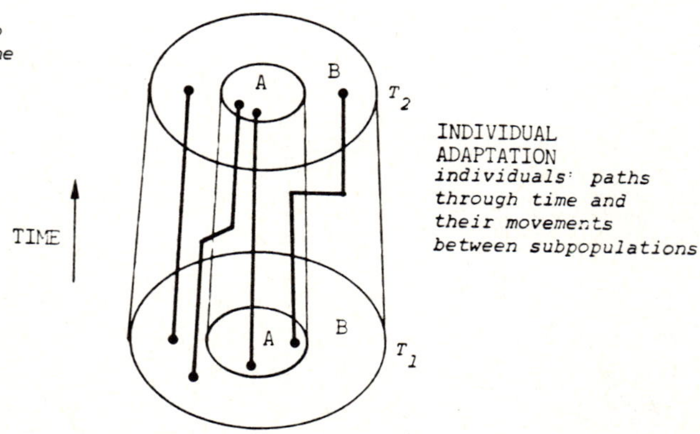

characteristics of the members of the population. In Figure 7(c) each trace represents the passage of an individual or household through time, and within and between the sub-populations (i.e. states of car-owning/ non-car-owning). Over all but the shortest time periods renewal of the general population needs to be accounted for - so that births, deaths and cross-migration are included, as in Figure 7(d).

4. We are now in a position to consider the special opportunities for analysis and modelling which are presented by the information contained in the extended framework.

5. Consider the boundary or interface between the two sub-populations (which we expect to have undergone a dynamic recomposition). Focus on transfers across this boundary (see figure 7(e)) enables the identification of three key dynamic indicators:

a) Differential rates of entry and exit from the sub-population, and indicators of duration within it.

b) Differences in the characteristics of entrants and leavers, and ratios of change in sub-population characteristics relative to those within the general population. Both these measures may provide a sounder basis for analysis of turnover and underlying market trends; and prospects for further market penetration. This focus, moreover, allows for the specification of:

c) Entry costs and exit costs. We can conceive of costs and perhaps conditions, being attached to transfer into car ownership, which would relate to the probability of the transfer being made. Costs would include purchase price, but in a broader sense, conditions of entry will apply - for example constraints on ownership below the minimum legal age. Similarly, costs and conditions for both staying and exit can be considered, and should be expected to differ from those applying to entry, both quantitatively and qualitatively (c.f. Goodwin and Mogridge, 1981). 'Staying' costs will reflect costs of running the vehicle. Exit costs should probably include searching for information about alternative modes of transport, and perhaps the value of disruption that doing without a car would cause. The existence of entry and exit costs will lead to a degree of irreversibility or inertia, which may be more powerful in one direction than another if there is any lack of symmetry between costs of entry and exit.

Extension of the Framework Towards a 'Continuous Choice' Context

So far a discrete choice (whether or not to be a car owner) has been considered. Many transport choices, however, relate to continuously variable quantities, such as choice between different levels of car ownership; or degrees of car use; or between using the bus for a proportion of trips ranging from none to all. We might therefore wish to consider moving from a discrete to a continuous framework.

Figure 7 (continued)

(d) Allowance for renewal and turnover (between this and a wider population)

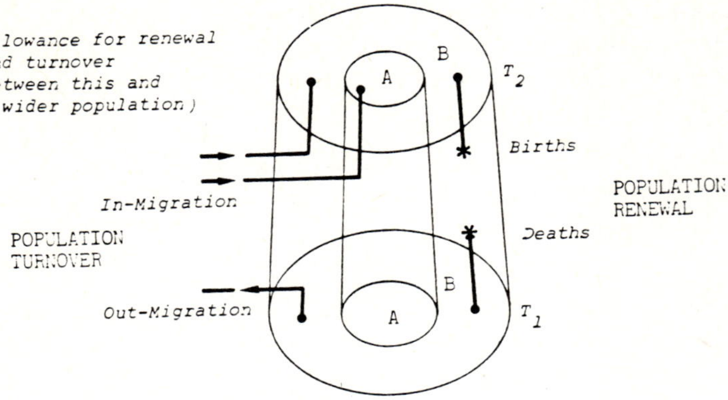

(e) Incorporating boundary costs and conditions

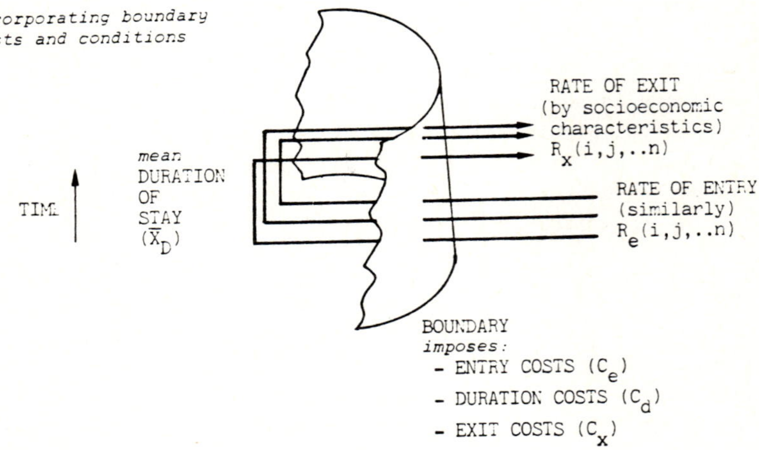

(f) Theoretical extension from two subpopulations... ...to an infinity

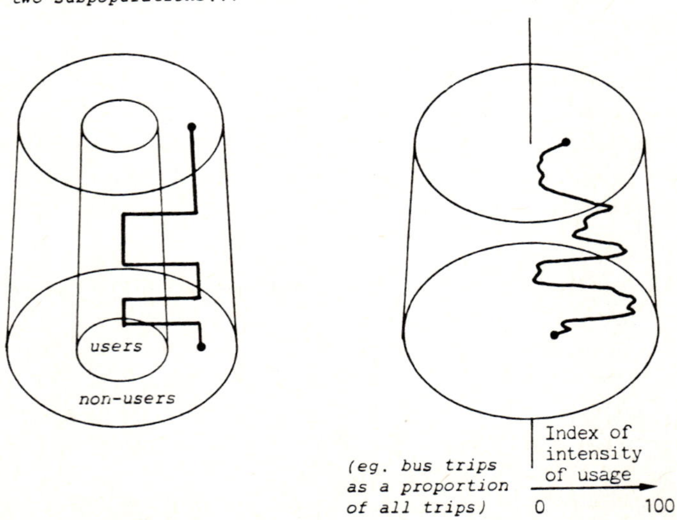

In figure 7(f), the left hand sketch shows the simplest solution: we simply impose discrete categories onto the continuous choice dimension. This of course suffers from problems of defining a cut-off point and allows for only the simplest measures of demand, expressed perhaps in terms of market allegiance or brand loyalty which would be of limited usefulness to the policymaker or operator.

A second approach (right hand sketch) is to conceive an infinity of choice states. Within this purely theoretical structure we would gain the maximum information on variations in bus use intensity, but lose the opportunities of tracing turnover of people as between identifiable 'states'.

A combination of the two offers advantages over each and is distinctly more practicable. This defines a limited set of sub-populations, defined in terms of (in this case) intensity levels e.g. one trip per week or less; 2 to 5 trips and so on, or bus kilometrage bands may be defined (figure 7(g)).

Figure 7 (continued)

(g) Definition of a limited set of subpopulations

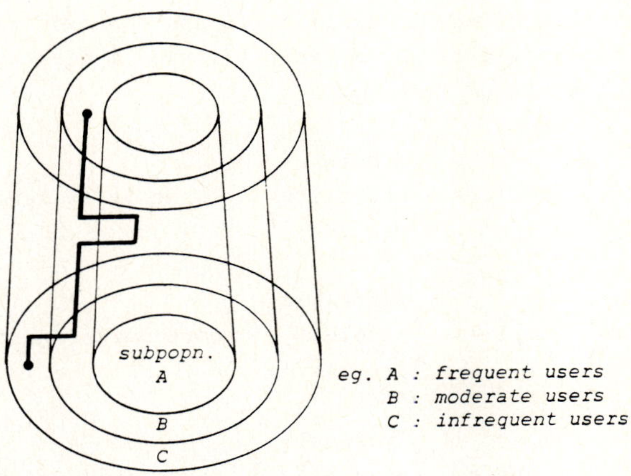

Turnover Ratios

Examination of bus trip rates per head, say, in successive cross section surveys, or operating records of ticket sales adjusted for population, would show whether bus use is increasing or decreasing. With some care, it would be possible also to use one-day trip data to show whether bus use was broadening or narrowing within the population. But neither indicator can say anything of use about the component processes which make up such trends. For this, we need to distinguish three elements, namely rate of entry, rate of exit, and duration as a member of the sub-population.

We define the rate of entry into the sub-population (Re), the rate of leaving or exit (Rl) and the average duration within the sub-population (AXd). To illustrate consider the ratio Re/Rl. Ceteris Paribus, and assuming a constant AXd:

$$\text{Re/Rl} \quad \left\{\begin{array}{c} <1 \\ \text{unity} \\ >1 \end{array}\right\} \quad \text{implies} \quad \left\{\begin{array}{l} \text{diminishing} \\ \text{stable} \\ \text{increasing} \end{array}\right\} \quad \text{market capture}$$

Constant monitoring of this ratio could, therefore, in the simplest case, alert operators or authorities to increasing or decreasing degrees of market capture. Similarly, changes in the <u>behaviour</u> of a sub-population (ΔKt) for example a change in public transport kilometres travelled per unit of time per traveller, may be examined in the context of turnover. In very general market planning terms, movements in key indicators can produce statements of the following kind:

$$\Delta\text{Kt} \quad \left\{\begin{array}{c} >1 \\ \text{unity} \\ <1 \end{array}\right\} \quad \text{implies} \quad \left\{\begin{array}{l} \text{growth} \\ \text{stability} \\ \text{decline} \end{array}\right\} \quad \text{in } \textbf{activity levels} \text{ within a ...}$$

$$\text{Xd} \quad \left\{\begin{array}{c} >1 \\ \text{unity} \\ <1 \end{array}\right\} \quad \text{implies} \quad \left\{\begin{array}{l} \text{increasing} \\ \text{stable} \\ \text{decreasing} \end{array}\right\} \quad \textbf{'core market'} \text{ under conditions of ...}$$

$$\text{Re/Rl} \quad \left\{\begin{array}{c} >1 \\ \text{unity} \\ <1 \end{array}\right\} \quad \text{implies} \quad \left\{\begin{array}{l} \text{increasing} \\ \text{stable} \\ \text{decreasing} \end{array}\right\} \quad \textbf{penetration} \text{ of the population}$$

Any trends within these key indicators must also be seen within the context of general population trends. The least favourable conditions that might be detected, would include, for example, a decreasing penetration of a diminishing population.

This type of analysis suggests that market analyses can benefit by further examination of the effects of turnover and renewal within the sub-population. For example, by relating the processes of adjustment and turnover, it would be possible to calculate empirically differential timescales of market response; either in retrospect (given a full monitoring exercise), or in prospect (given partial information). Critically these timescales of adjustment and turnover will vary with the context of the behaviour or policy being analysed. We can, for example, contrast the following situations:

a) Work by Hollings into British Rail's first class passenger market found that a majority of passengers were making their first such journey, and that for a high proportion of these people that journey would also be their last. Clearly overall response to policy changes, such as fare or service charges, is likely to be dominated by turnover.

b) However, with ordinary rail commuter traffic on the Southern Region where turnover is also evident, the majority of commuters (including a large proportion of season ticket holders) are captive to their choice of mode, over the short to medium term, for journeys linking home to work.

Thus response patterns are likely to be context-dependent, particularly with respect to the relative timescale over which they occur.

The types of insights - and applications - which may become evident using this sort of framework both help to explain and also go considerably deeper than the traditional elasticity measures. The final sections of this paper describe an initial attempt to use this framework to build an essentially descriptive simulation model of the processes of turnover and renewal, and of their impact upon the characteristics of sub-populations.

MIDCATS: A Model of the Impact of Demographic Change and Turnover on Sub-populations

Where, as with processes of turnover and renewal, the timescale of events being analysed is longer than can be normally (and cost-effectively) examined using collected data, some progress can be made using models which simulate changes in a population over time. MIDCATS is one such model which has been developed at TSU (based on earlier work by Clarke - see Clarke and Dix (1982)). It attempts to examine the implications for survey design and analysis of turnover and renewal. In essence, a sub-population of households is generated and then aged year by year, allowing for births, deaths, unemployment, changes in income and expenditure etc. The basic structure is set out below:

Structure of the model

1) GENERATE AN INDIVIDUAL

-generated randomly from the age distribution of an existing population and assigned head of household (hoh) status. Sex of hoh determined.

2)GENERATE A HOUSEHOLD AROUND THIS INDIVIDUAL

-marital status of hoh, age of partner (if appropriate), number and age of children, also desired ultimate number of children in relation to current number

-employment/schooling status of each individual

-type of employment (if employed) distinguished as between manual/non-manual, full-time/part-time

-given age of each individual and their type of employment, generate income for each individual

-calculate household income and adjust to take taxation and allowances into account

-anticipate household car ownership (based upon household income)

-calculate approximate household expenditure under headings of housing costs, food, travel, fuel etc. (taking into account size and age structure of the household, no. of cars, type of housing tenure, level of income itself etc.)

3)AGE THE HOUSEHOLD YEAR BY YEAR and look for:

-changes in household size and structure through births, deaths, marriages remarriages and children leaving home

-changes in employment/schooling status of each individual (taking into account higher education, duration of unemployment, retirement, early retirement etc.)

-changes in household income (dependent upon changes in employment and annual wage increases) and expenditure (dependent upon changes in income, age structure of household, changes in car ownership etc.)

-use the difference between household income and expenditure to define a resource level which can be accumulated from year to year and which has an impact upon the decision-making process in 4) below

4)DETERMINE MAJOR HOUSEHOLD DECISIONS. Those which have been investigated to date include:

a) House-moving decisions These are decisions which can move the individual household out of the current sub-population (and consequently out of the modelled population). The house-moving decision is dependent upon the current 'resource level' (to make an obvious allowance for an exit cost); age of the head of household; time in current dwelling, etc. The influence of different assumptions of the impact of these variables upon the house-moving process can then be examined.

b) Changes in car ownership These decisions may result in either increases or decreases in car ownership and again are dependent upon the current 'resource level', existing number of cars, and other conditions.

Output from the model

As each household in the sub-population is aged year by year, it can be interrogated at any point and data accumulated either for individuals or for households. Although it is still at an early stage of development, some of the results from MIDCATS will illustrate the potential of simulation to address longitudinal questions. Table 2, for example, attempts to forecast life-cycle group changes (of a particular, atypical sub-population) over a five-year period and allowing that over this time, some of the households will 'die out' (column A). As yet it has not been possible to test the effectiveness of such forecasts because of the absence of data from a real population.

Problems of longitudinal survey design have been well discussed elsewhere (see Layzell, Goodwin and Dix (1985) for a review). The effectiveness of different sampling strategies, however, can be examined using MIDCATS.

One of the essential paradoxes of panel surveys is that:

> "To prevent bias through attrition (i.e. as between respondents and non-respondents), and to guarantee a continued longitudinal survey, it is desirable to maintain as high a response rate as possible for each survey (presumably within certain resource restraints).
>
> However, a sample with no replacement of respondents will become progressively less representative of the sub-population from which it was drawn. Such a sample would simply age over time and, we may assume, the population will not. It may also become biassed in other ways if sample and sub-population do not follow the same growth paths or if respondents become conditioned by continued exposure to the survey."
>
> (Layzell, Goodwin & Dix (1985)

This can also be illustrated (figure 8). Year by year life-cycle distributions (for a 20-year period) are plotted and the increasing skew of age of head of households in the sample becomes apparent. Other sampling strategies which have been examined, such as a dwelling-based panel, show little (or at least a lot less) of this bias. Again see Layzell, Goodwin and Dix (1985) for further examples.

A final example considers the impact when both turnover and renewal are allowed to operate within MIDCATS. In this situation households may decide to 'move out' of the sample. To simulate a dwelling-based panel they are subsequently replaced by a new household according to (at present) one of three scenarios:

i) The new household may be randomly generated,

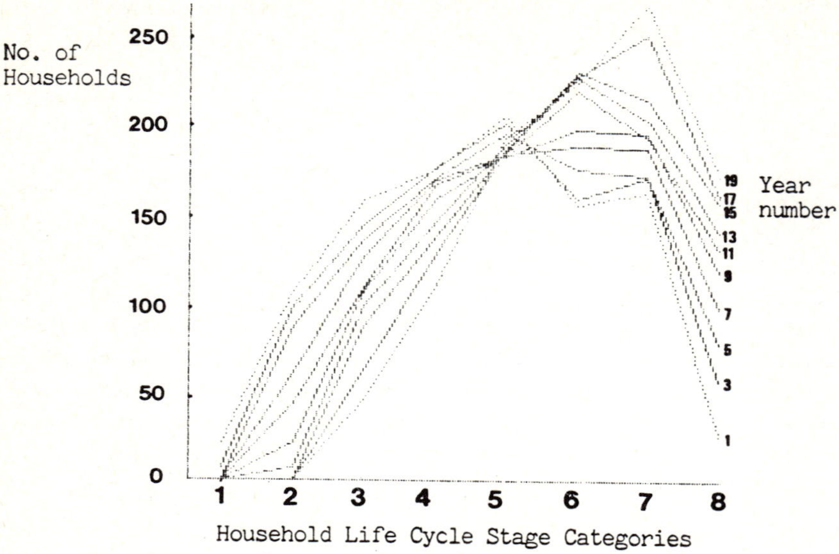

Figure 8

Year on Year Estimates of Life Cycle Structure over a Twenty Year Period.

(Probability of leaving subpopulation reduced to zero)

Table 2

MIDCATS simulation of life cycle transitions over 5 years

Life cycle group (year 6)

Life cycle group (year 1)		1	2	3	4	5	6	7	8	A
	1	46	12	0	3	0	0	12	0	0
	2	-	0	0	18	0	0	0	0	0
	3	-	-	0	24	0	0	0	0	0
	4	-	-	-	0	55	19	1	0	0
	5	-	-	-	-	0	41	17	14	0
	6	-	-	-	-	-	5	23	60	17
	7	-	-	-	-	-	-	68	36	5
	8	-	-	-	-	-	-	-	196	78

ii) The characteristics of the new household can be related to those of the outgoing household (for example ages of the two heads of household or size of household are associated),

iii) The characteristics of the new household can be related to those of the outgoing household at the time that they themselves entered the sample (the dwelling).

Strategies ii) and iii) are based upon the casual finding that housing stock in a given area tends to attract households of a similar size and composition.

Table 3 illustrates the combined effect of turnover and renewal upon the car ownership of a given stock of houses (perhaps a housing estate). Once households are tracked over more than two or three years the transition tables (as in table 2) become more difficult to analyse and resort has to be made to aggregation or the year by year transition tables which are shown here. Each matrix plots car ownership in successive years (years 2 to 6) against car ownership in year 1. Allowance is made that within this period some households will leave the sample (through turnover or renewal) - column A -, and that others will enter the sample - row B. Over time some of the households which enter the sample (row B) will subsequently (within the timescale of the survey) also leave the sample, resulting in the cell at the location AB.

The stabilising influence of turnover and renewal can now be simply illustrated (table 4) by contrasting the car ownership levels of households who remain in the sample throughout (car ownership tends to increase) with the levels for the whole sample (which remain fairly constant).

From these examples we can begin to see how the use of longitudinally structured models like MIDCATS, in combination with longitudinal surveys themselves, can offer a powerful complementary approach towards the assessment of partiular policy proposals. Survey information is necessarily retrospective, and a large investment lead time is required for building up an adequately timescaled data base. This will nevertheless be real - rather than simulated - and will itself yield continually more refined information, over time, for input to our dynamic models. The models themselves allow for policy assessment over longer, forward periods, and need not entail the same lead time. Using models we may also test, experimentally, the sensitivity of policy outcomes towards possible future changes in fundamental variables such as birth rates, preferred family structures, and trends in migration. Such modelling exercises could thus, in turn, usefully inform continuing decisions as to design and maintainence of our survey monitoring work, and particularly, the choices between alternative sampling and replacement strategies, as were discussed earlier.

By undertaking modelling and monitoring work concurrently, and interactively, the policy maker will be more comprehensively informed than if either approach was followed alone, or both followed in isolation.

Table 3

Changes in car ownership over 5 years (as predicted by MIDCATS) allowing for households to enter and leave the sub-population of dwellings.

where: row B = incoming households (since year 1)
column A = outgoing households (since year 1)

year 1

year 1	0	1	2+	TOTAL	A
0	119	0	0	119	0
1	0	232	0	232	0
2+	0	0	143	143	0
B	0	0	0	0	0
TOTAL	119	232	143	494	0

year 2

year 1	0	1	2+	TOTAL	A
0	104	7	0	111	7
1	0	210	11	221	13
2+	0	1	135	136	4
B	11	10	3	24	0
TOTAL	115	228	149	492	0

year 3

year 1	0	1	2+	TOTAL	A
0	95	10	0	105	13
1	0	178	20	198	31
2+	0	4	128	132	8
B	26	22	4	52	0
TOTAL	121	214	152	487	52

year 4

year 1	0	1	2+	TOTAL	A
0	81	17	0	98	20
1	2	153	33	188	41
2+	0	10	120	130	11
B	34	34	7	75	3
TOTAL	117	214	160	491	75

year 5

year 1	0	1	2+	TOTAL	A
0	68	21	0	89	28
1	3	138	33	174	56
2+	0	10	114	124	15
B	45	47	7	99	9
TOTAL	116	216	154	486	108

year 6

year 1	0	1	2+	TOTAL	A
0	59	24	1	84	34
1	6	127	33	166	60
2+	0	16	106	122	18
B	52	52	10	114	13
TOTAL	117	219	150	486	125

Table 4

The stabilising effect of turnover and replacement upon car ownership for a sub-population of dwellings (MIDCATS simulation).

A) Households which have remained in the sample for whole 5 years

Percentage of households with:

	0	1	2+ cars
year 2	22	47	31
3	22	44	34
4	20	43	37
5	18	44	38
6	17	45	38

B) The complete sample of households (stayers + incomers)

Percentage of households with:

	0	1	2+ cars
year 2	23	46	30
3	25	44	31
4	24	44	33
5	24	44	32
6	24	45	31

CONCLUSIONS

Recent interest in the use of longitudinal studies has concentrated upon examining relatively short-term processes and problems. Over time, and if the full potential of such studies is to be realised, other and more fundamental factors will need to be taken into consideration. These relate to changes in the basic structure of households and of populations. This paper has attempted to examine some of the issues involved. Three main points emerge:

1) Over time populations change, both in terms of their membership and of the characteristics of those members. For longitudinal studies which extend beyond anything but the short-term, processes of turnover and renewal within the relevant population will become important considerations.

2) The use of a framework which specifically recognizes turnover and renewal can provide non-conventional insights into market structures which are of potential importance to policy-makers. Development of this framework into a demographically-driven model can allow further scrutiny of the characteristics of sub-populations without (or before) recourse to relatively high cost, long duration surveys.

3) Where longitudinal studies are mounted, and in the presence of changing identity and characteristics of members of a sub-population (for example bus users, car owners etc.), panel surveys which are dwelling-based have some important advantages in terms of stability and in terms of their ability to describe the travel-generative potential of a given sub-population.

REFERENCES

Bourgin, C. and X. Godard (1981) Structural and Threshold Effects in the Use of Transportation Modes in Stopher, P.R., A.H. Meyburg and W. Brog (Eds) **New Horizons in Travel-Behaviour Research** Lexington.

Carpenter, S.M. (1983) House Movers in Oxfordshire **Transport Studies Unit** Special Report, Oxford University.

Clarke,M.I. and Dix,M.C. (1982) Simulating Demographic and Behavioural Change in a Population with the Progression of Time **UTSG Conference,** Leeds TSU ref(142)

Dix,M.C. (1985) Medium Term Dynamics of Change in Travel Behaviour and the Role for a Dwellings-based Panel Survey Design **UTSG Conference,** Birmingham

Downes, J.D. (1980) Life Cycle Changes in Household Structure and Travel Characteristics **TRRL Laboratory Report 930**

Goodwin, P.B. and A.D.Layzell (1983) The Difference between Age and Ageing for Public Transport Policy **International Transportation Planning** Research Colloquium The Netherlands.

Goodwin,P.B. and Mogridge,M.J.H. (1981) Hypotheses for a Fully Dynamic Model of Car Ownership **International Journal of Transport Economics** viii, pp.323-326 TSU ref(105)

Jones,P.M., Clarke,M.I., Dix,M.C. and Heggie,I.G. (1983) Understanding Travel Behaviour **Gower**

Layzell, A.D. (1983) Panel and Continuous-Time Data Sets; their scope and some initial results **UTSG Conference, London.**

Layzell, A.D., P.B. Goodwin and M.C. Dix (1984) Transport Policy and the Dynamics of Travel Demand **Gower (forthcoming).**

Mackett, R.L. (1983) The Inter-Relationship between Travel and Locational Decisions **in Carpenter, S.M. and P.M. Jones (Eds)** Recent Advances in Travel Demand Analysis Gower.

Behavioural Research for Transport Policy
pp. 81–95. © 1986 VNU Science Press.

THE DESIGN AND POLICY APPLICATIONS OF A PANEL FOR STUDYING CHANGES IN MOBILITY OVER TIME

by

J.M. Golob, The Project Bureau for Integrated Transport Studies

L.J.M. Schreurs, The Institute for Longitudinal Policy Research

J.G. Smit, Bureau Goudappel Coffeng

1. INTRODUCTION

This paper begins with a discussion of the policy relevance of panels as a research tool; explains their potential advantages over other survey methods and outlines the initial policy applications for which the Dutch panel is intended to be used. There then follows a summary description of the panel design, sample selection, survey instruments and an evaluation of the response to the first two survey waves. Results of the analysis are available for only the first wave and uses of these preliminary results are discussed. Finally general consideration is given to the anticipated use of the panel for policy purposes.

2. REASONS FOR A PANEL

A panel is a sample of individuals, households or institutions which are the subjects of repeated measurement. For some years the Project Bureau for Integrated Transport Studies in The Hague has been interested in the possibility of establishing a panel of Dutch households in order to study mobility patterns, how they change over time and why (see Baanders and Slootman, 1982). The need for this new longitudinal data source has slowly become apparent with the increasing number of questions which require both descriptions and explanations of changes in travel behaviour. For policy makers it is often essential to know how travellers have responded to policies in order to judge their efficacy. However it is often far from easy to isolate the effects of a particular policy from other changes that are taking place and which may have an overwhelming influence.

The Netherlands has a national travel survey, the Onderzoek Verplaatsingsgedrag (OVG) which has taken place annually since 1978 and this survey supplies useful descriptive mobility data for particular points in time (Moning, 1983). However it's sample size and cross-sectional character make the statistical variances too great to allow the measurement of change other than at a highly aggregate level. This means that for example changing patterns of car or public transport use cannot be explained in terms of a response to changes in household incomes, fares policy, petrol prices or any other salient factors. A means of overcoming this limitation is to survey the same individuals more than once in order to observe the changes that they make in their travel behaviour while simultaneously registering changes in the household environment and external factors such as petrol prices and fares.

By registering a wide variety of mobility and individual characteristics data will then be available which can be analysed for a variety of purposes. Also if required the panel members can be easily contacted with additional questions. To summarize, the policy justifications for establishing a panel with which to study changes in mobility over time were that:

- It would generate policy relevant insights unobtainable with cross-sectional surveys.
- It would give potential rapid access for answering short term policy questions.
- It could supply information for many types of policy questions.

In order to relate changes to causes it is necessary to see beyond the measurement of net changes. In panels change can be measured at an individual level and this allows insights which are obscured in cross-sectional analysis. The following simple example demonstrates "turnovers" in mode use which can take place and which become visible with panel data.

Figure 1. Cross-sectional statistics showing the use of public and private modes for the journey to work in Tyne and Wear. Source: Smart (1984).

	1982	1983
Public transport users Private transport users	288 630	281 637
	31.4% public transport users	30.6% public transport users

Figure 1 shows the results of a simple cross-sectional analysis of mode use implying a 0.8% net reduction in the use of public transport for the journey to work between 1982 and 1983. However, with the aid of figure 2, Smart (1984) shows that a much greater shifting in mode use at the individual level is visible through the use of panel data. As can be seen changes have taken place in both directions.

Figure 2. Panel-data statistics showing use and changes in use of public and private modes for the journey to work in Tyne and Wear. Source: Smart (1984).

		1983		
		public	private	
1982	public private	233 48	55 582	288 630
		281	637	918

What has happened is that 80.9% of the 1982 users of public transport remained public transport users in 1983, but 19.1% of these previous public transport users switched to private transport. However, 7.6% of the original users of private transport in 1982 switched in the opposite direction to become public transport users in 1983; while 92.4% of the 1982 private transport users remained private transport users. Smart (1984) points out that more than 12% of the panel sample has changed mode between 1982 and 1983. Even higher turnover rates have been found in other studies: Attanucci et al (1982) registered a 40% turnover in public transport users in a nine-month period using a telephone panel survey.

The analysis of turnovers is essential for relating changes to their causes. To take the example of public transport such data would allow us to follow those people who become users, those people who are and who remain users and those who cease to use public transport in the course of time. Changes can be related to causal factors such as changes in fares and services but also to changes in car ownership, job or home re-location, life cycle developments etc. Data about the changing make-up of the public transport market is clearly important for marketing purposes but it is also likely to offer insight into both long and short term fare elasticities and the role of external circumstances in changing public transport use.

Having made the case for the policy relevance of panels it is necessary to remark that they are not without problems. The analysis possibilities are highly dependent on the extent to which panel respondents can be persuaded to cooperate by taking part in repeated surveys over time. Securing long term funding is also a problem because the policy relevance of panel studies cannot be adequately tested and demonstrated until the survey has been underway for some time. Finally the analysis requirements and methods used for such data are different from those normally associated with cross-sectional data sets and therefore require the development and application of techniques currently little known in the transportation field.

3. POLICY APPLICATIONS

Feasibility studies were undertaken to investigate the methods, problems and costs of setting up a household based panel for the purpose of studying the manner in which Dutch mobility pattern change over time. These studies were in a final phase when the question arose of the possibility of using the panel within the context of an evaluation of the effects of a new national fares policy. In April 1984 fares on buses, trams and metros were due to be raised by 10% on average in real terms and train fares by 3%. Further sharp increases were foreseen annually until 1987 in order to reduce the growing burden of subsidies.

A major data collection exercise was already planned in which twice a year passenger counts and short on- and off-vehicle interviews would take place at selected points throughout The Netherlands. This cross-section would be used to establish general trends in public transport use but would be unsuitable for analysing the causes of changes in passenger demand. Consideration was given to taking a sub-sample of these passengers to interview in greater depth perhaps repeating the interviews more than once. However this sample would have been biased towards the frequent public transport user and would miss sampling those who had ceased to use public transport. The advantages in using the panel were as follows:

- By selecting a sample of households in which both users and non-users were randomly represented it would be possible to follow the full process in which individuals initiate and cease use of public transport.

- Because the mobility characteristics of all households would be known, changes in public transport use could be set against changing mobility patterns in the use of all modes.

- Causal analyses would be possible and the specific effects of the fare increases could be traced both for public transport users as a whole and for particularly sensitive groups within the population such as low income groups and retired people.

- Because the panel would be used for general purposes as well as for one specific policy purpose, joint funding from two different sources would be available.

It was thus decided that the panel should be adopted for use in the fares policy evaluation exercise which would be taking place throughout the period 1984-1987. The panel as established has two explicitly stated purposes:

- to describe and analyse general changes in mobility.

- to establish the effects of fare changes on mobility generally, and in particular for policy-sensitive groups of travellers.

4. THE PANEL DESIGN

During the preparatory phase of the study, much attention was given to quantifying the concept of mobility and the identification of segmentation variables for the panel sample which would help in identifying and discriminating between mobility homogeneous groups. This research was conducted using data from the 1980 nationwide OVG survey, and is reported by Van Wissen et al (1985). Analyses were undertaken at household level because of the assumed strength of household interactions with regard to car use, family based season tickets etc. First, sixteen measures of household mobility were calculated from trip diary information. Included were measures such as daily travel time per person by various modes, trips per household per day for various purposes and percentages of household members travelling for various purposes. These measures were then summarized in a set of statistically independent dimensions using principal component analysis. The lack of longitudinal data on mobility meant that the only guidance available was to use this secondary analysis of an existing cross-sectional data set. It is not yet known whether the explanatory factors of mobility obtained from the static data will prove to be the same as those to be obtained from the dynamic panel data.

The final decision on the panel segmentation reflects the joint objectives of the study. The three segmentation variables involved six household lifecycle categories, four household income categories, and five different types of geographical area which reflect different types of public transport operating areas and degrees of urbanisation. A sub-category of one parent families was included within two life cycle groups reflecting special political interest in this often under-privileged social group.

The total size of the panel was limited to circa 2,000 households by the extent of the available budget. This meant that much consideration had to be given to the numbers of households required in each cell of the lifecycle/income group of the segmentation matrix, in order to insure that sufficient observations would be available for analysis purposes. In particular it was necessary to forecast the incidence of public transport users who could be expected to occur randomly in order to have sufficient numbers to address the fares policy questions. Data from the nationwide cross-sectional OVG survey was used to calculate the likelihood of public transport using households occuring in each cell of the panel sample. One of the frequently cited advantages of panels is that it is possible to work with small samples because intra-personal variations have been reduced by repeatedly questioning the same individuals. The numbers required however will very largely be dependent upon the extent to which real changes take place in the mobility patterns of the panel members. Such information can however only become available after the panel has been established. There appears to be a paradox involving changes over time in a panel sample. For reasons of analysing cause and effect, the more changes in the panel sample the better. However, for reasons of describing changes in the overall levels of mobility for the sample between two points in time, the changes will be reflected in high variances in the standard errors of estimate of the mean level of change.

For the purpose of selecting the sample of households, The Netherlands was divided into five groups of local authorities. These groups reflect the different types of public transport operating areas and the extent of urbanisation. They include the big cities divided into two groups, a set of smaller cities, commuting areas lying outside the immediate environment of the cities and an "other" category which is essentially more rural in nature. The local authorities are spread throughout The Netherlands and care was taken in the selection to ballance the sample of communities such that the public transport policy objectives were not allowed to compromise the use of the panel for other purposes which would require a more nationally representative sample. In total 20 local authorities were selected which are considered to be relatively representative of The Netherlands. The Nationally representative OVG survey is used for establishing the weighting factors.

5. SAMPLE SELECTION FOR THE PANEL

As was seen from the previous section, the preliminary work on mobility homogeneous groups identified seven groups within the Dutch population. These groups can be described by their household income and lifecycle characteristics. In order to establish the correct household income/lifecycle phase of the potential panel members a limited screening survey was undertaken. This survey was also used to find a group of households willing to take part in the panel. For each of the 20 selected local authority areas a random sample of addresses was obtained from the Post Office and Telecommunications Service. For those addresses with a telephone number the screening survey was administered by phone. Those without telephones were interviewed at home. Based on a review of literature (Schreurs, 1983) and personal experience, it was assumed that the response rate for the screening survey together with agreement to take part in the panel (the sign up rate) would be about 50%. Using this working assumption 6,128 households were contacted yielding 2,886 households (47%) who were willing to be panel households. The 50% guideline was apparently a reasonable estimate.

From the 2,886 households who were prepared to take part in the panel a final selection was made. It was necessary for analysis purposes to insure that adequate numbers of public transport using households would be found randomly in each of the different public transport operating areas and in each income/lifecycle cell. Further particular policy relevant groups were over-represented. These groups were households with low incomes, households with school age children, pensioners and one parent families.

For 8.2% of the households in the screening survey the income question was unanswered. In order to avoid biasing the sample it was decided that households with unknown incomes would be selected in the same proportion as their occurence in the screening population. Taking into account the above and including an estimate of 10% non-response, 2,185 households were invited to take part in the first panel wave in the hope of obtaining a response from 1,957 households.

6. MEASURING TRAVEL BEHAVIOUR

On the basis of preliminary studies, see Baanders and Slootman (1982) and Schreurs (1983), a number of assumptions were made with regard to measuring changes in travel behaviour:

- panel households would be asked to register their travel behaviour twice a year in spring and autumn.

- all individuals over the age of twelve in each household would be asked to participate.

- all individuals who left the household and also households who moved house would be followed to their new address.

The latter assumption was considered to be particularly important because changes in home location and household structure are likely to be highly correlated with changes in mobility patterns.

In order to insure that the panel surveys were well understood and would achieve good response rates it was decided that the first wave in March 1984 would be administered by interviewers. Survey material would be taken to each home and discussed with the panel members. At the end of the week registration period the interviewer would return and check with the repondents for mistakes and difficulties. It was hoped that such intensive contact with interviewers in the first wave would secure the panel and allow the second wave to be administered more cheaply by post.

The survey instruments are composed of three elements: a household questionnaire, a personal questionnaire and a seven day travel/ activity diary. The household questionnaire contains questions about the household structure, income and vehicles owned. The personal questionnaires contain more detailed questions relating to income, employment, car availability, public transport ticket type etc. This questionnaire also includes retrospective questions about train travel in the preceeding seven day period. This extension of the registration of train trips is used to try to increase the amount of information available about train travel which takes place on a less than daily basis. In the seven day travel/activity diary respondents are asked to register all movements outside the home throughout the seven day period. The first day of travel registration by respondents is spread over the seven days of the week so that analysis of biases can be undertaken and if necessary corrected for.

As indicated the survey instruments are extensive in terms of the information collected about each household and it's individual members. The multiple objectives for the use of the panel made it impossible to adopt a simpler approach at this stage even though it might have been preferable from the research point of view. To help encourage the response to this extensive survey each respondent receives a gift on completion of the questionnaires and the travel diary.

7. THE RESPONSE TO THE FIRST WAVE

The estimated non-response between the screening and the first measurement proved in practice to be nearly twice what was expected: Of the total number of 2,185 households who agreed to take part in the panel when approached in the screening and who were subsequently invited to take part in the first wave, about 19% refused when approached with the actual survey. Consequently, the first wave sample size of 1,764 was about 10% less than expected.

Analysis of the non-response has shown that people who live on their own, one parent families, pensioners and the lowest income group have a lower than average response rate. There are also response differences between the various local authority groups. The response rates are lowest for the large cities. Looking at the reasons for non-response it is clear that certainly those living on their own and to a lesser extent one parent families frequently on more than three occasions were not at home. However the extent of the survey its self is likely to largely explain the higher than expected non-response.

8. THE SECOND WAVE

In preparation for the second wave new households were screened to be used to replace the anticipated dropouts and also to try to increase the panel size. Also all households from the first wave were contacted and asked to send back information about household changes. This resulted in information about individual members of the household and whole households which had changed their address, as well as whether or not new members had joined the household or had left it. There were 125 such reported changes whose net effect was to reduce the panel size by 9 households to 1,755. For the replacement sample, 552 new households were visited in their homes in the same manner as was used for the first wave. The 1,755 households from the first wave were sent their survey material by post. Reminders were sent first by post and later households which had not yet responded were telephoned or visited by an interviewer if they had no telephone.

9. THE RESPONSE TO THE SECOND WAVE

In September/Oktober 1984 the second panel wave took place and the main indications of the response are now known (the final response rates will not be known until the data have been completely processed). An immediate concern it to know which households have refused to take part in the second wave, the so-called dropouts, so that this information can be used in deciding the replacement strategy for the third wave. The sample compositions for the first and second waves are compared in Table 1. The cells marked with an asterisk (*) in Table 1 are those for which the relative difference between the numbers of respondents who remained in the panel and the number who dropped out is statistically different (in the direction of higher dropout rate) from the remainder of the sample at the p=.05 confidence level, as determined by F-tests. Likewise the cells marked by a minus sign (-) had a significantly lower dropout rate than the total sample.

Table 1. Comparison of panel sample composition after first wave (upper number) and after second wave (lower number); cells with significantly greater than expected dropout rates are indicated by (*), and cells with significantly less than expected rates are indicated by (-).

	Annual net household income x 1,000 guilders					
Lifecycle	< 17	17-24	24-38	> 38	unknown	total
Households < 35 years no children at home	23 15	47 37	86 62	52 39	2 1	210 154
One parent households with all children < 12 years	22 11	6 4	2 1	1 1	0 0	31 17
Households with all children < 12 years	22 13	117 83	181 143-	59 44	6 3	385 286-
One parent households with at least one child ≥ 12 years	16 12	16 11	4 1	4 4	1 0	41 28
Households with at least one child ≥ 12 years	4 4	54 37	139 103	116 86	10 7	323 237-
Households 35-64 years no child at home	22 11	73 44	116 82	101 68	9 7	321 212
Households ≥ 65 years	69 22*	83 42*	47 33	40 26	6 3	245 126*
Singles < 65 years	93 58	52 36	47 32	12 9	4 2	208 137
Total	271 146*	448 294	622 457-	385 277	38 23	1,764 1,197

The average dropout rate is 32% which is some 12% higher than forecast. In terms of lifecycle groups, it is primarily pensioners (49%) who have left the panel. These results are of course in keeping with the response to the first wave. The pensioner dropouts are strongly biased towards those who live on their own (63%). In general female heads of household have a tendancy to leave the panel more than male heads of household. Of the local authority groups the dropout rate is highest for the big cities. The lowest income group (f 0 - f 17,000) has the highest dropout rate and households whose income was unknown in the first wave have a higher than average dropout rate. Household size is also significantly correlated with a tendancy to dropout, one person households being much more likely to do so than five person households. The same tendancy is true for car ownership, non-car ownership being correlated with dropouts. (Since the statistics of Table 1 were tabulated, twenty-six more households responded late, bringing the total in waves 1 and 2 to 1,223.)

These response results are particularly disappointing given that low income groups, pensioners and one parent families were all identified as being particularly policy sensitive and were thus over represented in the sample. It has yet to be seen whether they remain sufficiently well represented to allow successful analysis of their changing mobility patterns. Given the difficulties encountered with the response rates for these important groups serious consideration will have to be given to means by which they can be helped and encouraged to take part in future panel waves. Response rates might be improved by reverting to the use of interviewers instead of administering the survey by post. Lightening the burden of the survey by administering the travel diary at only yearly intervals with less demanding surveys inbetween might also yield a positive result.

As a result of the selective nature of the dropouts it is anticipated that biases in the trip making characteristics of the more permanent panel members will be present. To gain insight into this aspect, analysis of variance techniques were used to compare the travel characteristics of the dropouts with the rest of the sample. In all instances the analyses refer to the numbers of trips per person made during the seven day week of the first panel wave. (The trip making characteristics of the panel members in wave two are not yet available.) For the modes bus, tram and metro (BTM) only shopping trips appear to show a significant difference. The trip rate for the whole population is 0.16, and for the dropouts 0.22 and for the rest of the panel members 0.13. This suggests that mobile BTM using shoppers are more likely to leave the panel and thus BTM shopping trips will be under represented. There were no significant differences found for train use. For the other modes there is a significant relationship for all journey purposes. The general tendency is for the dropouts to have a much lower trip making frequency (18.0/week) than for the rest of the panel members (22.0/week). Taking all modes together the dropouts have a trip making frequency of 19.1 trips per week and the rest of the panel 24.0 trips per week. It is clear from these results that the bias introduced by the selective nature of the dropouts will cause the panel to become more mobile the only exception being in the registration of shopping trips made by BTM. Experience elsewhere suggests that this bias in the Dutch panel is by no means unique and would tend to indicate that mobility surveys in general tend to over represent the more mobile members of the population.

10. SOME FIRST-WAVE ANALYSIS RESULTS

A descriptive cross-sectional analysis of the first panel wave has been undertaken of BTM and train users separately. These data have been used to give a detailed profile of the public transport users for one point in time. Excluding non-response the total sample was 3,763 individuals over the age of twelve. For the weighted sample 84,949 trips were registered some 22.6 trips per person for the fourteen day registration period. Taking only the seven day diary period 83,949 trips and 3.19 trips per person per day were recorded. 793 individuals some 21.1% of the sample made use of BTM in the seven day period and 390 individuals some 10.6% of the sample made use of the train in the fourteen day registration period. The additional week registration of train use is judged to have been a successful exercise not only because it increased the number of train trips but also because it increased the number of non-work, non-school trips which occur with a less than weekly frequency.

Analysis-of-variance and loglinear modelling techniques were used to investigate the following questions:

- What are the household and individual characteristics of public transport users and what are the characteristics of the trips they make and the tickets that they use?

- Do public transport users differ from non-public transport users and which categories of the population make much and which categories little use of public transport?

To summarize very briefly, the population groups that make relatively more than average use compared with all BTM users are those under 26 and those over 60/65 (women/men), the lowest income group and the group non-car owning households. The relatively highest train user groups compared with all train travellers are those between 19 and 26 years old, the lowest income group and non-car owning households.

Looking at the frequency of BTM use those between 12 and 18 years old make the most trips per week. Men make more trips than women and these mainly for work and education purposes. The high income group make relatively frequent use of BTM for work journeys and relatively infrequent use for visiting purposes in comparison with the average BTM user. This is reversed for the lowest income group who make relatively few work trips and relatively more trips for visits with BTM. The frequency of use of the train is lowest for the age group 60/65 plus. Men are more frequent users than women and then mainly for work and educational purposes. The highest income group make clearly more train trips than the average train traveller.

There are clear differences between BTM users and non-BTM users. Comparing users with non-users one finds that young people below 25 years old and people over 60/65 are over represented in the BTM user group as are also women, the two lowest income groups and households without cars. In comparing train and non-train users one finds that 19-25 years olds, the lowest and highest income group and households without cars are over represented in the user group.

These results of the first panel wave have been used as background material in a first round evaluation of the fares policy. The real value of the results will however only begin to become apparent once the second and third wave become available.

An entirely different analysis topic has been the investigation of biases in the temporal reporting of trips in the seven day travel diary. This work is fully reported in Golob and Meurs (1985). Biases of a significant nature were found in the temporal reporting of trips. The principal causes of these purely temporal biases was found to be the increasing tendancy over time for respondents to report no travel at all on a given day. In the present dataset there was an approximately 4.8% per day increase in the number of respondents inferred to be immobile. However certain biases cannot be explained by increases over time in the number of respondents inferred to be immobile. There is for example a day to day increase in the under reporting of short walk trips. Other biases of a more complex nature were also found. There were no significant biases found between population groups and their reporting of trips. It is currently proposed that correction factors be applied to the seven day diary data in order to adjust data from later days to the level of the first day. This analysis work demonstrates the value of the seven day diary in comparison with shorter registration periods. Such biases are also likely to exist in diaries of two and three days duration but cannot be investigated and corrected for. This analysis work will be repeated for later waves.

An entirely different form of analysis using the first wave panel data has been concerned with trip chains. A trip chain data set has been constructed from the seven day activity diaries. The intention is to study changing patterns in linked activities in relation to the personal and household characteristics of the travellers. These data will be most useful for studying the following types of topic:

- the effects of changing working hours and the use made of increasing free time for out of home activities.

- the increase in working women over time and the impact on their activity patterns of the need to shop or collect children outside their working hours.

- changes in job and home locations and the impact on journey linkages.

The application of the results of such studies would help to improve forecasts of the likely effects of shifts in population characteristics and the evaluation of whether or not such changes are likely to have consequences of relevance for policy makers.

11. PLANNED MULTI-WAVE ANALYSES

It is planned that the first two waves of the panel will be analysed using a three-fold approach: First, a generalised turnover analysis will be conducted. Turnovers will be defined in terms of several modes - bus/tram/metro, walking, bicycling, train, car driver and car passenger - and usage in terms of levels of travel over the one-week diary period. This analysis is at the level of the person, rather than the household. This is a generalisation of the type of analyses presented by Smart (1984) to multiple modes, multiple travel purposes, and multi-day travel data. But more importantly, the descriptive analysis will be extended in an attempt to identify certain explanations of change. Descriptions will be sought of significant differences among the population segments which do and do not switch use of modes, on the basis of the socio-demographic characteristics of the segments compared at the two points in time, and the characteristics of the modes (ticket prices, etc.) compared at two points in time. This can be accomplished by performing a type of discriminant analysis in which the dependent population segments are defined, in terms of switchers or non-switchers of mode use, and the potential explanatory variables include lifecycle, income, car ownership, etc. and changes in these socio-demograph variables, as well as changes in fare structure, service levels, etc. It is necessary to use a non-metric (or non-linear) version of discriminant analysis because many of the socio-demographic variables of interest (such as lifecycle), are nominally scaled, or categorical (such a technique is described in De Leeuw, 1973 and Gifi, 1984).

The second phase of the analysis focusses on overall levels of mobility. The objective is to quantify the general concept of mobility by factor-analysing a comprehensive set of travel variables, grouping households on the basis of similarities on the resultant mobility factors measured at the two points in time, then investigating changes in the group characteristics from time one to time two. This methodology is described in Golob and Perdok (1983), and was tested for one point in time using the cross-sectional nationwide OVG travel survey (Van Wissen et al, 1985). It is hoped that this phase of the analysis will provide information about changes in general mobility levels for various population segments. It is a primarily exploratory approach.

The third phase of the analysis represents a search for a causal structure of change. Estimations of simultaneous equations will be attempted, with the results from the first two analysis phases as guides. It is intended that the LISREL (Linear Structural RELationships) technique as described in Jöreskog (1973) and Jöreskog and Sörbom (1984) be used to identify the best representation of causality and to measure the goodness-of-fit for this model. Specifically, the LISREL technique, which can be viewed as an extension of path analysis, has been applied to panel data in fields other than travel demand analysis (Jöreskog and Sörbom, 1977). The approach represents the empirical associations among the mobility (dependent) variables and changing socio-demographic and situational (explanatory) variables in terms of specific cause and effect linkages. That is, the models can be used to test the degree to which the observed associations among the variables can be accounted for by a pre-determined theoretical cause-and-effect structure. An application of LISREL modelling to the explanation of mobility levels at one point in time using cross-sectional data is provided by Golob and Zondag (1983). With panel data it is possible to use variables which measure changes, and it is intended that such variables include changes in fare structure as well as changes in socio-economic characteristics.

12. FUTURE POLICY APPLICATIONS FOR THE PANEL

As has been said earlier in this paper a first policy objective is to use the data from the panel for the purposes of tracing the effects of the national fares policy within the context of changing mobility patterns generally. There are many aspects of individual mobility which are of interest for wider policy planning purposes. In particular changing patterns of car ownership and use, driving licence holding and multiple household car ownership will certainly be important research topics whose results will be used for general policy planning purposes and car ownership forecasting improvements. More specific topics are likely to be the study of changing income levels and their relationship to changing travel patterns. Also the effects of changing working hours on the structure of activities, mode use and peak and off peak travel. Both of the latter topics are the subject of current conjecture in The Netherlands. In the case of working hours and changes in both male and female participation in the work force, it is anticipated that sufficient data will be available from a limited number of waves. The quality of data on income levels has yet to be assessed however.

Finally it is important to remember that early waves of the panel can be analysed in the future for as yet unknown reasons. The information content will be building steadily over time and the extent of the data collected should give considerable scope for analyses.

13. CONCLUSIONS

Assessment of the full potential of the Dutch panel as a valuable policy relevant research tool is premature. The indications from the study to date suggest that exciting analysis possibilities will be available which will allow the study of changes in mobility patterns over time, including the investigation of the causes of those changes. Undoubtedly new insights will be obtained in the dynamic aspects of mobility. On the negative side, non-response bias appears to be a greater problem in the panel than in cross-sectional surveys. The initial non-response which is common to both panels and cross-sectional surveys is compounded in panels by a bias in dropouts for later waves of the survey. Resolving the difficulties which have been encountered with the response rates of certain groups is likely to remain a challenge which can only be partially be resolved. The continued under-representation of certain groups will mean that compromises will be required with regard to the level of analyses which can be reliably performed and this will exert certain limitations on the policy sensitivity of the available data. Account will also have to be taken of the biases introduced by the selective nature of the dropouts causing the panel to over-represent the more mobile members of the population.

The authors wish to acknowledge the thoughtful comment and contributions received from T.F. Golob during the writing of this paper.

REFERENCES

Attanucci, John, David Vozzolo, Imogene Burns, 1982, Evaluation of the July, 1980 S.C.R.T.D. (Los Angeles) Fare Increase, U.S. Department of Transportation, Urban Mass Transportation Administration, Report No. DOT - TSC - UMTA - 82-30, Washington.

Baanders, A. and K.C.P. Slootman, 1982, A panel for longitudinal research into travel behaviour, in New Approaches to Analysing Travel behaviour, S. Carpenter and P.M. Jones, eds, Gower, London.

Gifi, A., 1984, Non-linear Multivariate Analysis, DSWO Press, Leiden.

Golob, Thomas F. and Henk Meurs, 1985, Biases in response over time in a seven-day travel diary, submitted for publication to Transportation.

Golob, Thomas F. and J. Perdok, 1983, A methodology for analysing changes in travel mobility using panel data, in Social Research and Prediction, pp 61-80, Amsterdam: European Society for Opinion and Marketing Research (ESOMAR).

Golob, Thomas F. and E.R. Zondag, 1983, A causal model of mobility, paper presented at the Tenth Transportation Planning Research Colloquium, Zandvoort, The Netherlands, 14-16 December.

Jöreskog, K.G., 1973, A general method for estimating a linear structural equation system, in A.S. Goldberger and O.D. Duncan, eds, Structural equation models in the social sciences, pp 85-112, New York: Seminar Press.

Jöreskog, K.G. and D. Sörbom, 1984, LISREL VI user's guide, Chicago: Chicago National Educational Resources.

Jöreskog, K.G. and D. Sörbom, 1977, Statistical models and methods for analysis of longitudinal data in D.J. Aigner and A.S. Goldberger, eds, Latent variables in socio-economic models, pp 285-325, Amsterdam: North Holland.

De Leeuw, J., 1973. Canonical analysis of categorical data, Ph.D. dissertation, Leiden University, Leiden.

Moning, H., 1983, The National Travel Survey in The Netherlands, Central Bureau of Statistics, Report H. 4308-83-E4, Heerlen, The Netherlands.

The Project Bureau for Integrated Transport Studies, 1983, Een panel voor longitudinaal verplaatsingsonderzoek: beschrijving en analyse van de mobiliteitsontwikkeling, The Hague.

Schreurs, L.J.M., 1983, Panelbeheer, non-response en migratie, The Project Bureau for Integrated Transport Studies, The Hague.

Smart, H.E., 1984, The Dynamics of Change - applications of the panel technique to transportation surveys in Tyne and Wear, in Traffic Engineering and Control, December 1984, pp 595-598.

Van Wissen, Leo, Hans Smit, Thomas F. Golob, 1985, Determination of Difference among household mobility patterns in G.R.M. Jansen, C.J. Ruygrok, P. Nijkamp, eds, Transportation and mobility in an era of transition, Amsterdam: North Holland.

Project Bureau publications are available on request from: Netherlands Ministry of Transport and Public Works, The Project Bureau for Integrated Transport Studies, Koninginnegracht 16-21, 2514 AB The Hague, The Netherlands.

Behavioural Research for Transport Policy
pp. 97–116. © 1986 VNU Science Press.

STATISTICAL MODELLING OF DISCRETE CHOICES IN DISCRETE TIME WITH PANEL DATA

David A. Hensher, Macquarie University (Australia)
Neil Wrigley, University of Bristol (United Kingdom)

1. INTRODUCTION

Despite the major advances during the last ten years in procedures for understanding and predicting *individual choice behaviour*, with rare exception the emphasis has been strictly cross-sectional. Available non-cross sectional studies are typically characterised by either highly aggregate data series (e.g. quarterly nationwide economic and demographic time series, annual automobile registration series) or repeated *single* cross-section data on the same closed population but with no planned comparison at the time of the initial survey (with the consequent lack of control over sample specification; question range, wording and order; and conduct).

Although these studies will continue to have an important role, especially at the level of univariate descriptive representation (e.g. general changes in trip frequency rates) and in long-range forecasting (e.g. stock of motor vehicles in 2010), such data is not sufficiently rich intertemporally to be able to distinguish the many sources of influence on individual (and hence market) choice behaviour. Improvements in predictive capability now require us to identify the causes (or at least sequences) of change. For example, to be able to separate the effects of persistent interindividual differences from real intertemporal relationships, a panel can provide the desired statistical leverage to sort out the causal patterns.

The major focus of this paper is on identification of sources of heterogeneity, nonstationarity and intertemporal dependence, and ways of incorporating these sources of behavioural variation into a computationally tractable model specification. Although this approach will in general require us to distinguish the nature of time (continuous, discrete), the nature of choice (continuous, discrete) and the nature of intertemporal linkage (pooling, structural) this paper is specialised to discrete choice in discrete time. The other specifications are detailed in a longer paper (available on request).

2. BEYOND MYOPIC MODELS OF DISCRETE CHOICE BEHAVIOUR: THE PANEL APPROACH

A dynamic representation of individual choice behaviour *ideally* requires a complete history of the sequence and timing of relevant

---

Acknowledgement: This paper is one in a series prepared for a study on the dimensions of automobile demand, supported financially by the Australian National Energy Research Development and Demonstration Program, the Australian Research Grant Scheme, and a Macquarie University Research Grant. This paper was written during Neil Wrigley's visit to Macquarie University (financially supported by The British Council and Macquarie University) and serves as a context paper for the methods being used in the automobile project. Discussions with James Heckman, Richard Dunn, Richard Davies, Rod McDonald and George Cooney are appreciated.

events. Completeness requires that the analyst identifies the starting date of the process and matches the sampling period accordingly. The process is then continuously monitored. In practice such data are rare. To observe the initial conditions of the process typically requires a long-term panel; such panels become the victim of attrition (which may not be a source of bias in itself) which necessitates rotation in order to maintain sample size. Given the specific (rather than general) orientation of recent panels and the cost of data collection it is likely that the growth of longitudinal surveys will be in the area of purpose-specific (once-off) data, at least until we have enough experience in statistical modelling to identify what types of data are of sufficient importance that a more general-purpose panel may be able to meet the diverse needs of the social sciences. Hence we must recognise the short time period of most panels and tailor our statistical methods to suit these conditions. We also assume that there exist observed and unobserved explanatory variables, and make only passing reference to methods which predict occurrence levels without such variables (e.g. Wrigley and Dunn 1984a,b; Goodhardt et al. 1984).

In developing an argument in support of a panel data-based approach to modelling the dynamics of choice behaviour we are ipso facto exposing the inherent weaknesses of existing static cross-section efforts. In saying this we add the caveat that the inherent weaknesses need only be of empirical concern where the static approach has been used as a basis for studying inherently dynamic processes *and* where any permissable 'change of variables' is unable to generate an intertemporal specification that is formally equivalent to the myopic specification. These caveats are non-trivial; what they suggest is the possibility of formal equivalence between static and dynamic models; the equivalence however requires demonstration and not assumption (see Phlips 1983, Chapter 11.9 for one example).

The reasons advocated for panel analysis methods are well documented (see Hensher 1984b, Heckman 1981, Markus 1979, Coleman 1981). The essential argument centres on an interest in change, especially economic and social change, and the belief that intra-individual effects (both observable and unobservable) are of importance in understanding 'response'. The use of inter-individual variation as a proxy, to justify less-rich data designs (A, B, C in Table 1) is thus inappropriate. Identifying the influences on change at different points in time requires a clear definition of the relationship between current, past and future behaviour (i.e. true state dependence). When some influences on change are unobserved in the sense of being omitted and/or are unmeasurable (e.g. motivations), with the further possibility that they are temporally invariant and hence correlated with the subset of time invariant observable variables, the only way to identify and account for variation between individuals due to the influence of the exogenous observable and unobserved variables is to use a panel. This will also avoid confoundment with true state dependence, which otherwise leads to spurious state dependence.

Since most panels will commence subsequent to the natural beginning of the choice process, some influences will be unobservable. Since duration status is inextricably linked with intertemporal dependence (and habit formation or cumulative inertia), the failure to measure duration (e.g. elapsed time since a vehicle was purchased) and to account for unobservables specific to pre- and post-sample period commencement will result in biased estimation of the other.

TABLE 1 LONGITUDINAL SURVEYS - A TAXONOMY IN ASCENDING ORDER OF RICHNESS

[See Hensher 1984b for an extended discussion of this taxonomy]

[A] SINGLE CROSS SECTION

A single observation of many individuals in a single cross-section sample survey

[A] → [t]

[B] (i) [STATIC] REPEAT CROSS-SECTIONS

Two or more observations of cohorts at fixed time intervals. Different samples from same closed population

[B] → $[t_0, t_1, \ldots, t_T]$

[B] (ii) [STATIC] PANEL

Two or more observations of many individuals at <u>fixed</u> time intervals, involving the same individuals

[C] (i) QUASI-DYNAMIC REPEAT CROSS SECTIONS

Two or more observations on cohorts with information about preceding changes in a fixed time period

[C] → $[t_0, t_1, t_{01}]$

[C] (ii) QUASI-DYNAMIC RETROSPECTIVE CROSS SECTION

Two or more observations on individuals sampled at a single point in time but providing information on changes during fixed prior periods (and the current period)

[C] (iii) QUASI-DYNAMIC PANEL

One or more observations on a sample of the same individuals with information about preceding changes in a fixed time period

[D] SEQUENCE-DYNAMIC EVENT HISTORY PANEL

A history of the sequence of states of the individual over a period of time. <u>Known</u> elements are

(i) Type of each event (moving from state i to state j)

(ii) Position in sequence

[D] → $[S_1, S_2, \ldots S_J]$ <u>Unknown</u> element is the time at which an event occurred

[A]- [D] are a PARTIAL RECORD of the history of events ranging from
[A] a record of the current state $[t_0]$
TO
[D] the history of sequence of states $S_1, S_2, \ldots S_J$

[E] COMPLETE HISTORY OF EVENTS [DYNAMIC PANEL]

. Sequences and time at which events occurred

$$[tS_{01}, tS_{02}, \ldots, tS_{J-1,J}]$$

Separating out heterogeneity from intertemporal dependence, feasible only with a panel, enables the analyst to study the influences on change, and thus the dynamics of choice behaviour. Simply adding a series of lagged exogenous variables does not guarantee accounting for true intertemporal dependence.

2.1 Structural vs Single Equation Panel Modelling

The majority of research on discrete choices using panel data in discrete time is single equation. A logical extension is to structural equation approaches. Developments have occurred in simultaneous equation approaches using panel data where the choice variables are continuous (e.g. vehicle usage - Hensher 1984, 1984a) - allowance for heterogeneity, nonstationarity, state dependence and initial conditions is relatively less complicated (but still complex). The difficulty with extending the logit specification to structural equations is that the structure of the covariance matrix in the usual form of a logit model precludes estimation of the cross-equation dependence. Only if the disturbance covariance is unnecessary to achieve identification (i.e. where a reduced form is not required) may the logit form be used as a simultaneous equation system. In discrete time this has appeal in preference to the current approach of pooling the data. Whereas a 'time series of cross sections' approach views each wave of observations as a set of new measurements on the same observations, a simultaneous equation system (SES) (one equation per wave) treats each wave of observations as a more extended set of data on the same observation. When the number of waves is small and the number of observations is large, pooling is much less attractive than SES since the *flexibility* gained by having T = 4 waves say is much more valuable than the *precision* gained by increasing the sample size to 4N. SES modelling of unordered discrete choices appears currently only feasible with choice sets up to 3 alternatives, where multinomial probit is feasible.[1] A promising approach however is via the use of selectivity correction procedures where the error terms are correlated across discrete (or mixed continuous-discrete) choices *i.e.* the influences on one choice are endogenous influences on the other choice. This procedure facilitates separate estimation of each equation in a staged procedure, correcting for selectivity. To date such methods have been restricted to static model systems with rare exception (e.g. Mannering and Winston 1983). In pursuing this approach one has to satisfy recursivity conditions to obtain consistency (where all dependent variables are truncated) - see Schmidt (1981, 427).[2] The problem appears to be due to the inclusion of truncated observable variables as right hand side variables. Pooling discrete choices over time avoids this problem but at a cost.

2.2 Discrete-Choices in Discrete-Time

The discrete-time model for recurrent (discrete) choices[3] is likely to take its starting point as the cross-sectional multinomial logit model. Although multinomial probit is a more general discrete-choice model, with an unconstrained error covariance matrix, it is currently computationally intractable for more than 4 alternatives.

---

1 More progress has been made for *ordered* discrete choices (see Winship and Mare 1983).

2 Schmidt speculates 'It is by no means clear that the best way to model truncated and/or qualitative variables is to embed them (essentially by analogy) into the usual simultaneous equations model' (Schmidt 1981, 474).

3 Hereon, when we refer to discrete choice, we imply unordered discrete choices; i.e. polychotomous variables which require more than one dependent variable to characterise the response. There are J alternatives in a choice set with J-1 response variables, taking the value zero, the remaining response for the chosen alternative, taking the value 1.0. Ordered discrete choices require only one response variable, with the values taken by the discrete dependent variable corresponding to intervals within which an unobservable continuous random variable falls (e.g. 1 = low income, 2 = medium income, 3 = high income).

Existing software for handling probit discrete choices over time in a panel are limited to binary choices at each time period (Avery et al. 1983). The prospects for nested decision structures using panel data via the probit route are not encouraging.

There are (at least) three ways of studying unordered discrete choices in discrete time. *Approach 1* involves estimating wave-specific models separately for each wave and using the estimated choice probabilities to identify a choice sequence probability. Although the approach can include exogenous variables to represent previous period outcomes or propensities to occupy states in previous periods, strong assumptions such as zero serial correlation are invoked. This approach has been used in Hensher and Le Plastrier (1983) and Mannering and Winston (1983). In *Approach 2* the data is pooled, with inter-period linkages built in through a lagged index, one for each exogenous variable. The use of a lagged index, defined as a weighted sum of each period's level of the variable[4] is a way of handling state dependence without the serial correlation attributable to right-hand side lagged endogenous variables. The index serves as a mechanism for adjusting the current levels of 'consumption determinants' to allow for 'stock' effects (Hensher 1984a). *Approach 3* views the data as an explicit sequence wherein the likelihood function associated with the choice sequence probability has two major components, one which accounts for the time invariant influences (including the initial conditions) and the other which incorporates the time varying influences. The association between states (in a sequence) may be due to an intrinsic characteristic of the state that changes the likelihood of that state, but it may also occur because some unmeasured and unchanging set of characteristics of an individual causes observed states to persist. The latter phenomenon is heterogeneity (see Section 3). The separation of the time-varying and time non-varying components provides the formal mechanism for explicitly handling heterogeneity.

The method increases in complexity as we move from Approach one through to Approach three. The essential features are summarised below with more detail given in Hensher (1984) for approach two and Davies (1984) and Hensher, Smith and Wrigley (1984) for approach three. The critical issues in a dynamic (intertemporal) specification are associated with proper allowance for true state dependence (also called feedback or cumulative inertia), nonstationarity (varying choice probabilities over time) and the presence of heterogeneity. Approaches two and three provide differing ways of addressing these issues. The approaches summarised below are implemented at each level of any (nested) set of decisions, with an appropriate inclusive value carried forward to higher-level choices.

The approach two model is summarised in equations (1) to (5).

$$y^*_{qt} = \beta b^*_{qt}(\theta) + \varepsilon_{qt}, \quad q = 1, \ldots, Q; \quad t = 1, \ldots, T, \qquad (1)$$

---

4 With the weight given as the stationary parameter (to a power of the time periods back from the current, with the current power = 0) associated with a lagged endogenous choice dummy variable.

with $y_{qt} = 1$ if $y^*_{qt} > 0$

$= 0$ if $y^*_{qt} \leq 0$.

$$b^*_{qt}(\theta) = \theta^0 b_{qt} + \theta b_{q,t-1} + \theta^2 b_{q,t-2} + \theta^3 b_{q,t-3} + \ldots. \quad (2)$$

$$= \sum_{z=0}^{t-1} \theta^z b_{q,t-z} .$$

$$\varepsilon_{qt} = \theta^t \alpha_{q0} + w_{qt} \quad (3)$$

$$w_{qt} = \frac{\xi_{qt}}{1-\theta L} \quad (4)$$

$$\alpha_{q0} = \beta \sum_{z=0}^{\infty} \theta^q b_{q,-z} = E_q(y^*_{q0}). \quad (5)$$

where y^*_{qt} is the choice probability of an alternative associated with individual q in period t,

$b^*_{qt}(\theta)$ is a lagged index operator conditioned on the state dependence parameter θ associated with a lagged endogenous $y^*_{q,t-1}$, (i.e. $y^*_{qt} = \theta y^*_{q,t-1} \forall t, t-1 \in T$), and

w_{qt} is the traditional error term with a suitable lag operator.

The random component, ε_{qt}, includes an initial conditions element, $\theta^t\alpha_{q0}$ to recognise the presence of left censoring. Subscripting by q, means that prior to the sample origin individuals have differing 'experiences' (which may be grouped into a manageable set of homogenous classes), and that failure to accommodate these individual-specific differences is a failure to account for one element of heterogeneity. Two strategies may be adopted to handle the initial conditions problem; we can include a set of group-specific constants to partially account for the heterogeneity of the initial conditions, and can adjust for heteroscedasticity (a special case of heterogeneity) due to the variation in θ^t. Two-stage estimation is required to identify the optimal value of θ, which is required in the calculation of the correction for heteroscedasticity so that the random term has equal variance for each observation.

Approach three assumes that the random component should represent errors in the specification or approximation of the representative indirect utility function and not in the random individual variation. Thus $\theta^t\alpha_{q0}$ and any other sources of heterogeneity should be included *in the* numerator and denominator of the logit model (see equation 6) and not in the error component associated with the derivation of the logit form. In fact this is the interpretation placed on $\theta^t\alpha_{q0}$ (and any α_q) in approach two. Whereas approach three specifies a parametric distribution for α_q and integrates this individual-specific effect out (producing the beta or Dirichlet distribution - see below) approach two which can be used to establish that the α_q (incidental parameters) do not drop out of the conditional likelihood function of the intertemporal logit model (as Chamberlain's evidence suggests it should (Hensher 1984)) is another way of accommodating the prior presence of heterogeneity in the random component.

The difference between approaches two and three is essentially the way in which the individual-specific effects are handled for unobserved influences inside and outside of the sample period. We can now turn to a more detailed discussion on sources of behavioural variation, emphasising Approach three as a means of illustrating various concepts.

3. MODELLING THE SOURCES OF BEHAVIOURAL VARIATION

3.1 Heterogeneity and Initial Conditions

3.1.1 Definitions

Heterogeneity (see Figure 1) was initially described in this paper as the variation which exists between individuals in the population/sample due to both observed and unobserved exogenous influences.

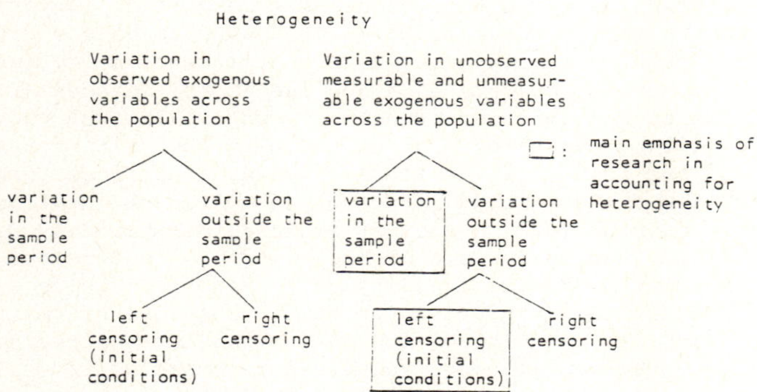

Fig. 1 Sources of Heterogeneity

However, in statistical modelling terms it is the unobserved (omitted) variables, both those which are potentially measurable and those which are normally unmeasurable (e.g. motivations and tastes), which are crucial to an understanding of the heterogeneity 'problem'. An intuitive understanding of why this is so can be obtained from Fig. 2. In Fig. 2(b) the probability of choosing a particular alternative (type of automobile, travel mode, shopping destination, etc.) is shown for all individuals in a subgroup of the population, where each member has *exactly* the same values of the observed exogenous variables. Although the expected value of the probability for all members of the subgroup, conditional upon the observed exogenous variables, is constant and equal to 0.5, the individuals will differ in terms of the levels of the omitted variables. As such, the true probabilities for members of the subgroup will have a distribution around the expected value, and it is this distribution which is the unobserved heterogeneity resulting from the omitted variables. As Figure 2 shows this unobserved heterogeneity can take many forms, with very different behavioural implications.

Omitted variables can be associated with both the sample period and the periods prior to and subsequent to the sample period (see Figure 1), but whenever they occur their effect is likely to be a common one: the introduction of a spurious time dependence effect

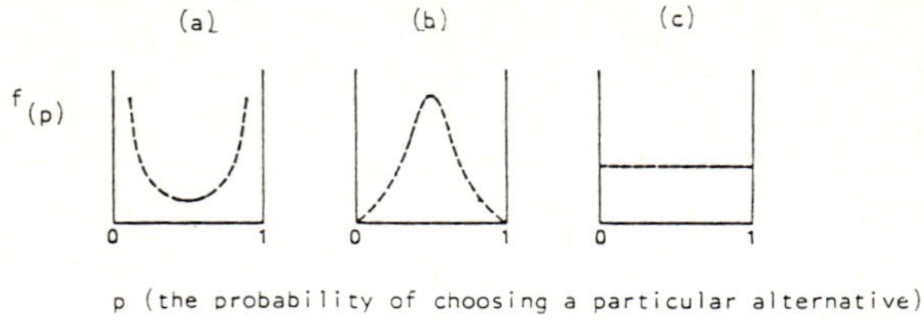

Fig. 2 Three probability distributions with the same expected value of 0.5. Shows potential range of heterogeneity caused by omitted variables

and bias into the parameter estimates of the observed exogenous variables. A classical example of this is described by Lancaster and Nickell (1980) in the context of unemployment duration. If in this case there is an omitted variable termed 'motivation', which is independent of the observed exogenous variables but is positively correlated with the chances of obtaining work, then successive cohorts of longer and longer duration in the unemployed population are bound to have a higher and higher proportion of individuals who are deficient in 'motivation'. This will show up in the data as a decline in the chances of obtaining work with increasing duration of unemployment. Time already spent unemployed will be negatively correlated with the omitted variable 'motivation' and this will inevitably produce a spurious time dependence effect and bias the estimated impact of unemployment duration on the chances of obtaining work. Another example, described by Davies and Crouchley (1985) and Davies and Pickles (1984), concerns the extent to which declining residential mobility rates with increasing duration of residence are due to 'cumulative inertia'. Because of omitted variables which are correlated with the probability of residential mobility, the observed movement rates can be expected to decline across successive cohorts characterised by increasing duration of stay, as these cohorts are bound to have a greater and greater proportion of individuals with a low probability of movement. As a result, this will produce a spurious time dependence or 'cumulative inertia' effect which Davies and Pickles (1984) believe is almost certainly the result of omitted variables and which does *not* indicate any process of growing locational attachment with increasing duration occurring at the level of the individual household. It is clear, therefore, that because of omitted variables we may draw incorrect inferences concerning the effect of the observed exogenous variables in choice probabilities even though, in some cases, the omitted and observed variables may be uncorrelated.

Omitted variables which are associated with time periods prior to the sample period are part of a specialised facet of the problem of uncontrolled heterogeneity; a facet which is usually referred to as the problem of 'initial conditions'. Essentially this problem occurs because most panels will commence at a point beyond the natural starting point of the behavioural process under study. As a result, the history of the sequence and timing of events will be *left-censored* and bias may occur in the parameter estimates of the observed exogenous variables unless explicit consideration is given to the unobservable influences on choice behaviour which result from the censoring of the data base.

The problem of heterogeneity in the statistical modelling of longitudinal choice data can be summarised, therefore, as the problem of separating out and identifying the important determinants of choice behaviour in the presence of unobserved influences on behaviour which result from the censoring of the longitudinal data base and/or the omission of important time varying or time invariant influences during the sample period. If statistical modelling of such data is to succeed it is essential therefore, that appropriate methods for controlling heterogeneity are adopted.

3.1.2 Some methods of controlling heterogeneity in discrete-time models

Following Nickell (1979) and others we can write, in the dichotomous (two alternative) case, the probability, P_{qt}, that individual q chooses the first alternative at time t in logistic model form

$$P_{qt} = \frac{\exp[\underline{x}'_{qt}\underline{\beta} + \varepsilon_q^*]}{1 + \exp[\underline{x}'_{qt}\underline{\beta} + \varepsilon_q^*]} \tag{6}$$

where $\underline{x}_{qt}$ is a vector of explanatory variables describing socioeconomic characteristics of individual q and characteristics of the choice alternatives,

$\underline{\beta}$ is a vector of parameters,

$\underline{\varepsilon}^*_q$ is an individual specific error term representing the omitted variables.[5]

The probability of an observed binary choice sequence $\underline{f}_q = (f_{q1}, f_{q2}, \ldots, f_{qT})$ for individual q where $f_{qt} = 1$ if individual q chooses the first alternative in time period t, 0 otherwise, is then

$$\text{Prob}\,[\underline{f}_q | \underline{\beta}, \varepsilon_q^* : \underline{x}_q] = \prod_{t=1}^{T} [P_{qt}^{f_{qt}} (1 - P_{qt})^{f_{qt}}]$$
$$= \prod_{t=1}^{T} \frac{[\exp(\underline{x}'_{qt}\underline{\beta} + \varepsilon_q^*)]^{f_{qt}}}{1 + \exp(\underline{x}'_{qt}\underline{\beta} + \varepsilon_q^*)} \tag{7}$$

For parameter estimation purposes the individual specific error terms must normally be integrated out giving rise to the general expression (Davies and Crouchley 1985)

$$\text{Prob}[\underline{f}_q | \underline{\beta}, \underline{\beta}^* : \underline{x}_q] = \int_{\underline{\varepsilon}_q} \prod_{t=1}^{T} [P_{qt}^{f_{qt}} (1 - P_{qt})^{f_{qt}}] f(\underline{\varepsilon}_q) d\underline{\varepsilon}_q \tag{8}$$

where $f(\varepsilon_q)$, which is usually referred to as the 'mixing distribution', is the joint density function of the individual specific error terms, with parameter vector $\underline{\beta}^*$. In practice, alternative approaches to the handling of the heterogeneity in discrete time models can then be characterised by the various ways in which they treat the mixing distribution in (8).

The first (and most widely adopted) approach is to assume a parametric form for the mixing distribution. This is equivalent to the approach normally adopted in continuous time models, the only difference being that most attention in discrete time models has focussed on the beta rather than the gamma distribution, and, in practice, rather than specifying a parametric form for a mixing distribution across the error terms, it is assumed that (as a result of omitted

---

[5] It should be noted that this error term is *additional* to the random/error component of utility found within the random utility model and which is assumed to have a type I extreme value distribution to derive the standard logistic/logit form (see Amemiya and Nold, 1975; Parks, 1980; Wrigley 1985, section 8.5 for wider discussion of this issue).

variables) the outcome probabilities, P, will be beta distributed at each particular level g of the exogenous variables. In other words, a mixing distribution of probabilities is assumed and this usually has the flexible two-parameter beta form

$$f(P_g | a_g, b_g) = \frac{\Gamma(a_g + b_g)}{\Gamma(a_g)\,\Gamma(b_g)} P_g^{a_g - 1} (1-P_g)^{b_g - 1} \tag{9}$$

with parameters $a_g = \exp(\underline{x}_g' \underline{\beta}_1)$ and $b_g = \exp(\underline{x}_g' \underline{\beta}_2)$ (10)

and with mean and variance

$$E(P_g) = \frac{a_g}{a_g + b_g} \qquad Var(P_g) = \frac{a_g b_g}{(a_g + b_g + 1)(a_g + b_g)^2} \tag{11}$$

If the mixing distribution in (8) is assumed to be the beta distribution (9), then (8) takes the form

$$\begin{aligned} Prob\,[\underline{f}_q | \underline{\beta}, \underline{\beta}^*; \underline{x}_q] &= \int_0^1 P_g^{f_{q+}} (1-P_g)^{(T-f_{q+})} \frac{\Gamma(a_g+b_g)}{\Gamma(a_g)\Gamma(b_g)} P_g^{a_g - 1} (1-P_g)^{b_g - 1} dP_g \\ &= \frac{\Gamma(a_g + f_{q+})\Gamma(b_g + T - f_{q+})}{\Gamma(a_g + b_g + T)} \cdot \frac{\Gamma(a_g+b_g)}{\Gamma(a_g)\Gamma(b_g)} \end{aligned} \tag{12}$$

where $f_{q+} = \sum_{t=1}^{T} f_{qt}$. (For further details see Davies and Crouchley, 1985; Davies and Pickles, 1984; Dunn and Wrigley, 1985; Hensher et al., 1984). By substituting (10) into (12) and (11), and by estimating the $\underline{\beta}$ parameters using a likelihood function based on (12), the mean probabilities can be shown to be

$$E(P_g) = \frac{a_g}{a_g + b_g} = \frac{\exp[\underline{x}_g'(\underline{\beta}_1 - \underline{\beta}_2)]}{1 + \exp[\underline{x}_g'(\underline{\beta}_1 - \underline{\beta}_2)]} \tag{13}$$

This can readily be seen to be a logistic model, hence the expression (12) is referred to as the *beta-logistic* model (Heckman and Willis, 1977). Equation (10) may also be substituted into equation (9) to fully describe the form of the heterogeneity resulting from the omitted variables, or into equation (11) to provide a value for the variance around the mean.

In the polytomous (multiple alternative) case, a multivariate parametric mixing distribution is required to handle the heterogeneity. In this context, the beta mixing distribution can be replaced by its multivariate beta or Dirichlet distribution, and this provides a direct polytomous extension of the dichotomous beta-logistic model. Full details of this extension and an empirical application of the model to urban shopping travel using consumer panel data from Cardiff U.K. is provided by Dunn and Wrigley (1985).

Although the assumption of a parametric form for the mixing distribution is the most commonly adopted approach in discrete time models, Heckman and Singer (1982) have provided some rather disturbing evidence, (in the continuous time context), of the sensitivity of parameter estimates to the assumed form of the distribution. As a result, attempts have been made to relax the need to specify the full form of the mixing distribution and to proxy the distribution with a small number of what are termed *mass points*. In other words, in this approach the mixing distribution is replaced by a limited number of points which it is believed are adequate to fully control for the heterogeneity resulting from the omitted variables. The work of Laird (1978) provides the theoretical underpinnings of this approach, and the limited evidence currently available in both the continuous and discrete time contexts (Heckman and Singer, 1984; Davies and Crouchley, 1984, 1985; Davies and Pickles,

1984) suggests that the number of mass points required may often be quite low. Davies and Crouchley (1985) suggest that if this can be confirmed for more complex models, the mass point approach may provide a very attractive method for handling multivariate mixing distributions, and the same view has also been expressed by Heckman (in personal discussions).

3.2 State Dependence and Nonstationarity

3.2.1 State dependence

State dependence is the term used in longitudinal data analysis to summarise the intertemporal relationships exhibited in choice behaviour; that is to say, the dependence of current behaviour on past behaviour, and of future behaviour on current behaviour. As Figure 3 shows, there are several possible sources of state dependence (or 'feedback' as it is sometimes termed). In general, we can say that choice outcomes and the timing of choices are unlikely to be independent of previous choices (first-order or higher-order Markov effects), of time since last choice (duration dependence effects), of previous interchoice times (lagged duration dependence effects), and/or of the number of times the various choice outcomes have been selected within a specified time period back from the present (occurrence dependence effects). However, for practical modelling purposes it may prove valuable to demonstrate, or assume that one or more of these sources is unimportant for the choice process under consideration.

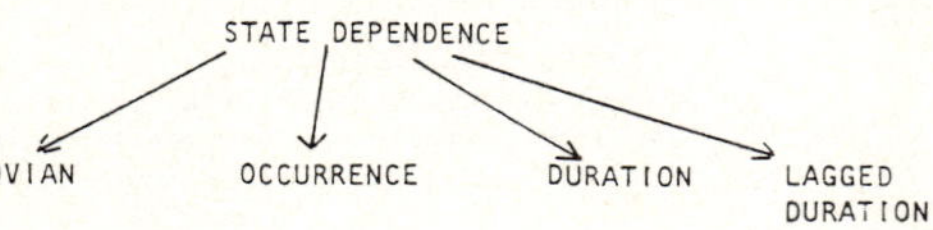

Definitions:

a) Markovian — Probability of a change in state is dependent on the current occupied state

b) Occurrence — Probability of a change in state is dependent on the number of times different states have been occupied

c) Duration — Probability of a change in state is dependent on the length of time the current state has been occupied

d) Lagged duration — Probability of a change in state is dependent on the lengths of time previous states have been occupied. Two variations:

i) All previous types of state considered
ii) Only those lengths of time spent in the same state as the current state considered

Fig. 3 Different forms of state dependence

A key issue in any longitudinal data analysis is the need to distinguish true state dependence effects from spurious state dependence effects. We have already noted in Section 3.1.1 how omitted variables are likely to introduce a spurious time dependence effect and bias into the parameter estimates of the observed exogenous variables, and we considered both the so-called 'cumulative inertia' effect identified in residential mobility studies, and the apparent strength of the unemployment duration effect on the chances of obtaining work, as classical examples of spurious time dependence effects resulting from omitted variables. It is clear, therefore, that identification of true state dependence effects must be conditional upon satisfactory handling of any heterogeneity which results from either omitted variables or censoring of the longitudinal data base.[6] Furthermore, it is clear that any overestimation or underestimation of the influence of heterogeneity because of inappropriate modelling techniques will lead to possible under- or overestimation of the role of other factors including state dependence effects. Finally, it should be noted that nonstationarity (see the discussion below) may also generate spurious state dependence effects.

3.2.2 Nonstationarity

Nonstationarity can be defined as the variation in individual choice probabilities over time which results from temporal changes or shifts in the values and significance of observed and unobserved exogenous variables. For example, in the context of shopping travel, Figure 4 shows that variation over time in the probability that an individual chooses particular shopping destinations may result from changes in *observed* exogenous variables affecting either: (a) the individual choice maker (e.g. his/her total working hours or the flexibility of those hours), or (b) the choice alternatives themselves (e.g. changes in the opening hours of stores because of retail deregulation of the type recently experienced in New South Wales, or of the type proposed for Britain by the Auld Committee). In addition, the probabilities of selecting particular shopping destinations may change over time because of variation in *unobserved* exogenous variables, both those which are potentially measurable (e.g. product availability at the destinations) or those which are normally unmeasurable (e.g. tastes, motivations, habits).

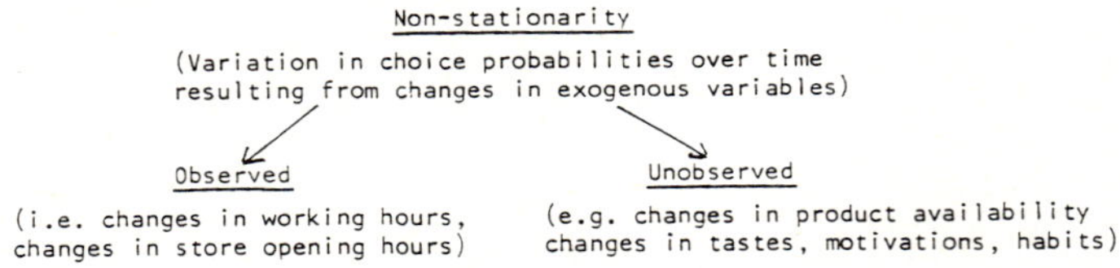

Fig. 4 Sources of nonstationarity

An assumption of stationarity used for analytical convenience in the modelling of longitudinal data implies, therefore, the absence of time-varying individual, social, or environmental factors, both observed and unobserved. In addition it implies that the effects of any special conditions associated with the actual start of the behavioural process (i.e. initial conditions), or of any intervening shock, have died out before observation of the process begins. In most longitudinal studies it follows, therefore, that an assumption of stationarity will not be justified (though there are some famous panel-data models of consumer purchasing behaviour, notably the NBD and Dirichlet models - see Section 4.2 - which exploit a stationarity assumption and which are specifically calibrated at a level of temporal aggregation at which relative stationarity can normally be shown to be a reasonable assumption). As a result, some attempt must normally be made to handle nonstationarity, in the simplest case by the inclusion of dummy variables or possibly, polynomial/Fourier trend terms, or in the more advanced case, by the additional use of the scaling procedure to be discussed in the next subsection.

---

[6] With state dependence (or feedback) effects in the model, information will normally be required on choice outcomes and the timing of choices *prior* to the sample period. However, such information will often be unavailable because of left-censoring of the longitudinal data base. Explicit consideration must therefore be given to the possibility of correlation between previous behaviour and omitted variables (i.e. to a wider definition of the 'initial conditions' problem).

3.2.3 Accounting for state dependence, nonstationarity, initial conditions and heterogeneity in a single model

The beta-logistic model can be thought of as providing a panel-data version of the conventional dichotomous logistic model; a version which is capable of handling the heterogeneity which results from omitted variables. However, in the form described in Section 3.1.2, the beta-logistic model is stationary and zero-order. Clearly, it would be valuable to generalise the model so that it can be applied to more complex processes, can incorporate time varying exogenous variables and state dependence effects, and can make proper allowance for initial conditions.

Recently, Davies (1984) and Davies and Pickles (1984) have provided such a generalisation and have developed an extended beta-logistic model in which a logistic scaling procedure is employed to take account of changes over time in both the exogenous and feedback variables. This extended beta-logistic model is composed of two major elements. The first element is a beta distribution which incorporates the heterogeneity which results from time-invariant observable and unobservable influences, and which is also the mechanism for handling the initial conditions.[7] The second element is a logistic scaling function which incorporates the time *varying* influences on the choice probabilities. In this function the probability that individual q chooses the first alternative at time t is expressed as

$$P_{qt} = \frac{\gamma_{qt} P_{qr}}{(1 - P_{qr} + \gamma_{qt} P_{qr})} \tag{14}$$

That is to say, P_{qt} is a logistically scaled function of a 'reference' probability $(P_{qr})^{qt}$ at the base (or reference) period r, and $\gamma_{qt} = \exp [(\underline{x}_{qr} - \underline{x}_{qt})'\underline{\beta}]$ is a scaling factor representing changes in the observed exogenous or feedback variables relative to the level of the variables in the reference period. Unobserved time varying influences are controlled for in the logistic scaling function by the use of time-specific dummy variables.

This approach to modelling the sources of behavioural variation is that described in Section 2.3 as 'Approach Three' and further details of it and an empirical application of the model to the problem of single versus multiple vehicle holdings by households in the Sydney Metropolitan Area are given in Hensher, Smith and Wrigley (1984).

4. ISSUES OF POLICY RELEVANCE: TWO EXAMPLES

To illustrate the relevance of the ideas for policymakers and the ways the results may be presented, we draw on the recent experiences of the authors in the development and analysis of panels in Australia and Britain. The Australian panel of 1436 Sydney households comprises four-waves of data on automobile possession (holdings and transactions) and usage during the period 1981-1985 (Hensher 1984a, 1984b). The British data relate to the daily purchasing records of 454 Cardiff residents collected via a continuous 24 week consumer panel in 1982 (Guy et al. 1983, Wrigley et al. 1985) with the emphasis on shopping patterns (travel, destinations, purchases).

4.1 Automobile Energy Consumption in the Household Sector

The relevance of the Australian panel study is best demonstrated in the context of predicting fuel consumption levels. The two identities

---

7 This is achieved by adding a set of group specific constant terms into the specification of the beta-logistic model described in Section 3.1.2.

used to calculate the level of fuel consumed are the fundamental energy formula (15) and the behavioural fuel efficiency formula (16) for period t.

$$\begin{array}{l} \text{annual} \\ \text{litres} \\ \text{consumed} \\ [L]_t \end{array} = \begin{array}{l} \text{efficiency} \\ \text{of} \\ \text{technology} \\ \text{(litres per} \\ \text{100 km)} \\ [e]_t \end{array} * \begin{array}{l} \text{utilisation} \\ \text{rate per} \\ \text{period} \\ \text{(kilometres)} \\ [k]_t \end{array} \quad (15)$$

$$\begin{array}{l} \text{efficiency} \\ \text{of} \\ \text{technology} \\ [e]_t \end{array} = \begin{array}{l} \text{fuel} \\ \text{usage} \\ \text{cost} \\ \text{(cents per km)} \\ [c]_t \end{array} \Bigg/ \begin{array}{l} \text{unit retail} \\ \text{price of} \\ \text{fuel} \\ \text{(cents per litre)} \\ [p]_t \end{array} \quad (16)$$

Prediction of L requires predictions of the distribution of e and k in the projection period, while predictions of e requires predictions of the distribution of c and the level of p. Since the levels of energy consumed are a consequence of the household's choice behaviour it is necessary to place the assessment in the broader context of the household's reasons for choosing particular types and quantities of vehicles and levels of utilisation. Furthermore, actions in the current period cannot be isolated from actions and experiences in the past and from expected future actions. The intertemporal perspective is essential where decisions on consumer durable acquisition (and disposal) constrain the desired level of consumption (i.e. usage) and indeed where the latter is used in the selection of an appropriate quality in a durable. Since automobiles are lumpy commodities with relatively long and varying life horizons within a household, one can expect the outcomes of durable purchase decisions to influence levels of consumption for some time into the future. This requires us to introduce habits, lags and leads, which act as macrodynamic modifiers (Clarke et al. 1982) on current choice behaviour of past and future behaviour.

To predict the level of vehicle utilisation and its fuel efficiency during a discrete period of time requires identification of the structure of the time path, which includes knowledge of the precise dates of adjustment (i.e. response) to the vehicle stock as well as the dates the household commenced the adjustment process leading up to the response. Since vehicle fuel efficiency is embodied in the household's vehicle choice which is determined by a broader set of influences (see Hensher 1984 for a review),[8] including experiences in vehicle usage and ownership, it is necessary to determine the level of vehicle efficiency via a discrete-choice model of choice of vehicle type (as conditioned on fleet size - see Hensher (1984) for details). The choice of vehicle type is assumed to be influenced by the stock of consumption habits as measured by previous levels of vehicle usage. The level of utilisation can be explained in a continuous choice model of intertemporal usage in which the level of use during each discrete time period is specified as a separate equation and simultaneously solved, accounting for the role of previous and subsequent levels of use either explicitly or via initial and final conditions. The incorporation of initial conditions is a way of ensuring that resistance to change and delayed reaction are properly handled[9] where the sample period is too short to pick up the full effect. By identifying the household's experiences with vehicle purchases from the very first time each individual was associated with a vehicle purchase (even before the formation of the household) we are able to impose explicit empirical restrictions on the initial

---

[8] We exclude any discussion of issues that are not directly relevant to the topic of panel methods. Other aspects of the automobile study are discussed in Hensher 1984, 1984a, 1984b, 1984c.

[9] subject to the caveats in Section 3 on the lack of robustness of parameters to different assumptions on the distribution of the initial conditions.

conditions. We have also obtained expectations of the period of vehicle retention to assist in specification of the right censoring conditions. The intertemporal vehicle usage model is linked to the vehicle choice model using the method of selectivity (Hensher 1984) which is a way of recognising the endogeneity of vehicle attributes (e.g. fuel efficiency) on the level of utilisation. It is linked to an intra-household vehicle usage model which in multi-vehicle households account for the influence that one vehicle's usage level has on the usage of other vehicles. The overall framework is summarised in Fig. 5.

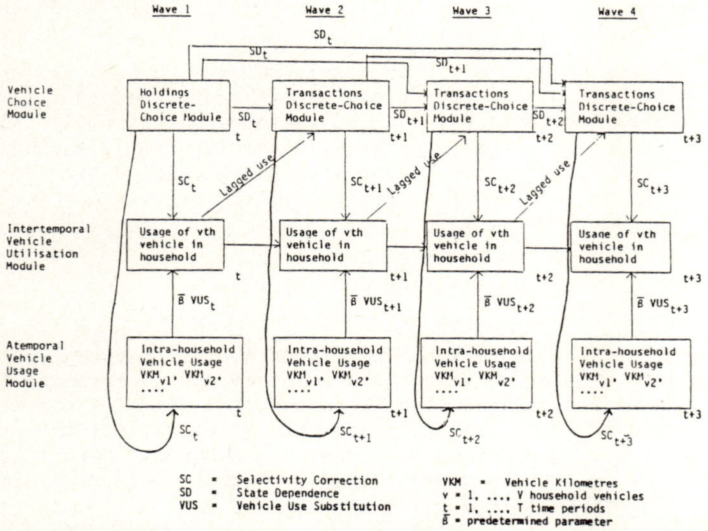

Fig. 5 The Overall Framework of the Automobile Model System

The dynamic vehicle choice and utilisation model is currently being estimated; the only results available relate to the contemporaneous specifications, which are reported in Hensher (1984c) and Hensher and Stonecash (1984). A specialised 3-wave discrete choice sequence model has been estimated for the binary choice of single vs multiple-vehicles, as a means of testing the method, and is reported in Hensher et al.(1984). The application of the model system reported in Figure 5 will, given projections on the distribution of the exogenous variables, produce predictions of vehicle usage, and the probability of selecting vehicles associated with particular fuel efficiencies. Total annual litres of fuel consumed by the household sector's automobiles is given as

$$L_t = \sum_{q=1}^{Q} \sum_{v=1}^{V} \text{Prob}_{vqt} \, [e_{vqt} * k_{vqt}], \quad q=1,\ldots,Q \text{ households}, \; v=1,\ldots,V \text{ vehicles in vehicle choice set} \qquad (17)$$

Forecasts of fuel consumed can be obtained for each type-class of vehicle, together with forecasts of vehicle holdings and usage by type class. Given projections of fuel prices, vehicle characteristics, socioeconomic structure of the population and general economy-wide effects, changes in the levels of these input variables will result in different forecasts. The model will be used for policy analysis for a variety of 'what if' situations. The application of the model will occur during 1985-87.

4.2 Household Shopping Travel

The Cardiff Consumer Panel Survey is believed to have been the most comprehensive panel survey of urban shopping behaviour ever carried out in the U.K. It provides a major resource for understanding the nature of shopping travel in Britain in the 1980s and for understanding the scale and type of change in shopping behaviour which has accompanied the significant shifts in the structure of the U.K. grocery retailing industry over the past twenty years (see Wrigley et al., 1985 for a wider discussion).

The relevance of this British panel-data research can best be illustrated by exploiting a formal linkage which exists between the 'extended Dirichlet model' and the beta-logistic models discussed in Section 4.1.2. The Dirichlet provides the multivariate parametric mixing distribution which is required to handle the heterogeneity which results from omitted variables.

In the dichotomous beta-logistic model, the two parameters (a_g and b_g) of the beta distribution are written as functions of a set of exogenous variables (see eq. 10). In the extended beta (or Dirichlet)-logistic model there are as many Dirichlet parameters (α_{rg}) as choice alternatives.[10] For example, in a four-alternative choice problem such as that considered by Dunn and Wrigley (1985), the expressions

$$\begin{aligned} \alpha_{1g} &= \exp(\underline{x}'_g\, \underline{\beta}_1) \\ \alpha_{2g} &= \exp(\underline{x}'_g\, \underline{\beta}_2) \\ \alpha_{3g} &= \exp(\underline{x}'_g\, \underline{\beta}_3) \\ \alpha_{4g} &= \exp(\underline{x}'_g\, \underline{\beta}_4) \end{aligned} \qquad (18)$$

replace those in (10). One task of the researcher is then to specify the composition of the set of exogenous variables and to estimate the associated β parameters using a likelihood function which is the multivariate equivalent of (12) - see Dunn and Wrigley (1985, eq. 20 for details). Having obtained estimates of the β parameters in (18) the mean probabilities of selecting each choice alternative, r=1,...R, can then be estimated from the expression

$$E(P_{rg}) = \frac{\alpha_{rg}}{S_g} \qquad (19)$$

where S_g is the sum of the Dirichlet parameters α_{rg}, r=1,...R, and the variances around the mean probabilities can be estimated from the expression

$$\mathrm{Var}(P_{rg}) = \frac{\alpha_{rg}(S_g - \alpha_{rg})}{(S_g + 1)(S_g)^2} \qquad (20)$$

In addition, a graphical display of the form of the heterogeneity for each choice alternative r within each subgroup g can be obtained.

In the Cardiff study (Dunn and Wrigley, 1985), choices of four shopping centres by a small group of 84 panellists from one sub-area of the city were related to four potential explanatory variables: whether or not the household owned a car; whether or not the housewife was in full or part-time employment; income (measured on an ordinal scale) and the number of children aged 15 or under in the household. The β and Dirichlet parameters (α_{1g}, ..., α_{4g}) in (18) were estimated to take the values

$$\begin{aligned} \hat{\alpha}_{1g} &= \exp(0.098 - 0.036\,INC_g + 0.592\,CAR_g - 0.233\,WWF_g + 0.086\,CHILD_g) \\ \hat{\alpha}_{2g} &= \exp(-0.621 - 0.143\,INC_g + 0.619\,CAR_g + 0.100\,WWF_g + 0.151\,CHILD_g) \\ \hat{\alpha}_{3g} &= \exp(-0.964 + 0.043\,INC_g + 0.526\,CAR_g + 0.623\,WWF_g - 0.311\,CHILD_g) \\ \hat{\alpha}_{4g} &= \exp(-3.522 + 0.098\,INC_g + 1.715\,CAR_g + 0.076\,WWF_g + 0.195\,CHILD_g) \end{aligned}$$

and all explanatory variables except income were found to be significant at conventional levels.

---

[10] These parameters are identified, see Heckman and Willis (1977) and Dunn and Wrigley (1985).

As an illustration of the type of information which can be obtained from the parameter estimates of beta (or Dirichlet)-logistic models, Fig. 6 shows the predicted probability distributions of choosing to shop at choice alternative 1 (Countisbury Avenue - a local shopping centre in Cardiff) for three distinct household types.[11] In the case of type I households (car owning households with the housewife in employment and no children) it can be seen that the expected probability of choosing to shop at Countisbury Avenue is 0.37 and the distribution of probabilities around the mean is unimodal with a mode of 0.22. In contrast, for type II households (non-car owning with the housewife not in employment and with no children) the expected probability of choosing to shop at Countisbury Avenue is 0.54, but the probability distribution around the mean is bimodal with modes at 0 and 1. Finally, for type III households (non-car owning with the housewife not in employment and four children) the expected probability of choosing to shop at Countisbury Avenue is 0.63 and the probability distribution is unimodal with the mode at 1. From this figure, it can clearly be seen that the capacity which beta (or Dirichlet)-logistic models possess to estimate the form of the heterogeneity in observationally equivalent subgroups is an important asset. The differences between the choice probability distributions of the household types in Figure 6 have important behavioural implications for any study of urban shopping travel.

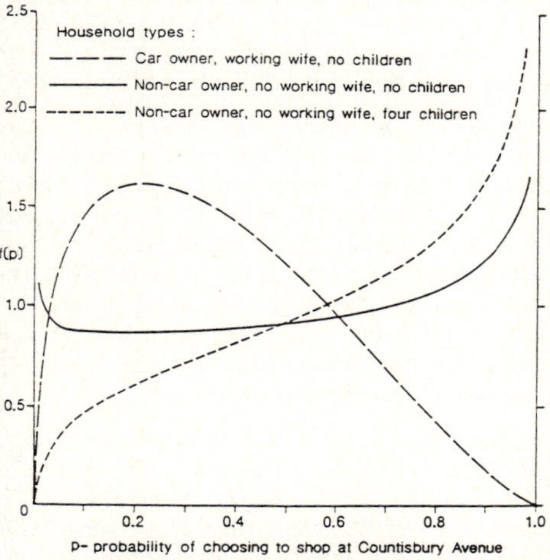

Fig. 6. Mixing Distributions for Different Household Types

It can be seen, therefore, that beta (or Dirichlet)-logistic models provide a valuable method of extending conventional cross-sectional MNL models to deal with choice patterns in panel data which incorporate heterogeneity. Moreover, they provide a wealth of information which is additional to that provided by cross-sectional discrete choice models. As such, these models represent an extremely valuable tool for the analysis of behaviour which involves recurrent choice and movement. Despite these advantages, however, two potential problems remain.

The first problem relates to the parametric assumptions adopted to handle heterogeneity. As noted in Section 3.1.2 it could be the case that the parameter estimates derived are unduly sensitive to the particular

---

[11] It should be noted that the income of each household type is assumed to be the sample average, since the influence of that variable was not found to be a statistically significant influence on the choice process. In addition, the three household types should be regarded as merely illustrative of the wide range of types for which similar distributions could have been obtained.

distributional forms (beta or Dirichlet) assumed and, in addition, only limited theoretical justification exists for the behavioural notions implied by such distributions. As a result, the models of shopping centre choice described above are currently being re-estimated (see Wrigley and Reader, 1985) using the non-parametric mass points method in which a non-parametric characterisation of the mixing distribution is obtained from a small number of empirically identifiable mass points.

The second problem relates to the lack of state dependence effects in the models and to the fact that the models have so far been calibrated over a time horizon which enables them to assume relative stationarity in the exogenous explanatory variables influencing shopping behaviour. Generalised panel-data models of shopping travel would, ideally, take account of both these sources of behavioural variation, and attempts are currently being made by Reader and Wrigley, and Davies and Pickles to develop and calibrate such generalised shopping models using the Cardiff panel data.

6. CONCLUSIONS

The focus of this paper is on the nature and scope of panel data methods as a means of incorporating important influences on individual discrete choice behaviour in discrete time which cannot be identified using a single cross-section approach. The relevance of the ideas for behavioural research in policy decision making is apparent in the discussion of the two examples.

The success of a panel in meeting the claims outlined above is very dependent on maintenance of the representative sample and the necessary financial support which inevitably transcends political budgets and often the life of a government. If dynamic models of behaviour are deemed important (and a significant improvement over static models in the context of relevant issues)then panel studies have a healthy future. Not only do they provide an empirical basis for the study of changes in the macro-environment (often long-term) they also provide an *immediate* empirical base for studying short-term responses due to their continuing presence. This last point is all too often overlooked.

Finally, the mere presence of panel data or the real possibility of collecting such data has generated a serious interest in more dynamic (and realistic) models of individual choice behaviour. Ask any researcher in the area of dynamic choice modelling why they are pursuing this approach and the response will be 'we now have the opportunity to access (or collect) panel data'.

REFERENCES

Allison, P.D. (1982) 'Discrete-time methods for the analysis of event histories', in Leinhardt, S. (ed.), *Sociological Methodology*, Jossey-Bass, San Francisco, 61-98.

Amemiya, T. and Nold, F. (1975) 'A modified logit model', *Review of Economics and Statistics*, 57, May, 255-257.

Avery, R.N., Hansen, L.P. and Hotz, V.J. (1983) 'Multiperiod probit models and orthogonality condition estimation', *International Economic Review*, 24(1), 21-35.

Broom, D. and Wrigley, N. (1983) 'Incorporating explanatory variables into stochastic panel-data models of urban shopping behaviour', *Urban Geography*, 4, 244-257.

Chamberlain, G. (1984) 'Panel data' in Griliches, Z. (ed.) *Handbook in Econometrics*, North-Holland, Amsterdam.

Chatfield, C., Ehrenberg, A.S.C. and Goodhardt, G.J. (1966) 'Progress on a simplified model of stationary purchasing behaviour', *Journal of the Royal Statistical Society, Series A*, 129,317-67.

Chesher, A. (1984) 'Testing for neglected heterogeneity', *Econometrica*, 52(4), July, 865-872.

Clarke, M., Dix, M.C. and Goodwin, P. (1982) 'Some issues of dynamics in forecasting travel behaviour - a discussion paper', *Transportation*, 11(2), 153-172.

Coleman, J.S. (1981) *Longitudinal Data Analysis*, Basic Books Inc., New York.

Crouchley, R., Davies, R.B. and Pickles, A.R. (1982) 'Identification of some recurrent choice processes', *Journal of Mathematical Sociology*, 9, 63-73.

Davies, R.B. and Pickles, A.R. (1983) 'A moments approach for omitted variables in residential histories and other panel data', *Journal of Mathematical Sociology*, 9, 227-241.

Davies, R.B. (1984) 'A generalised beta-logistic model for longitudinal data with an application to residential mobility', *Environment and Planning A*, 16, 1375-86.

Davies, R.B. and Pickles, A.R. (1984) 'Accounting for omitted variables in a discrete time panel data model of residential mobility', UWIST Papers in Planning, Working Paper 77, Cardiff, Wales (mimeo).

Davies, R.B. and Crouchley, R. (1984) 'Calibrating longitudinal models of residential mobility and migration', *Regional Science and Urban Economics*, 14, 231-247.

Davies, R.B. and Crouchley, R. (1985) 'Controlling for omitted variables in the analysis of panel and other longitudinal data', *Geographical Analysis*, 17, 1-15.

Dunn, R. and Wrigley, N. (1985) 'Beta-logistic models of urban shopping centre choice', *Geographical Analysis* (forthcoming).

Goodhardt, G.J., Ehrenberg, A.S.C. and Chatfield, C. (1984) 'The Dirichlet: a comprehensive model of buying behaviour', *Journal of the Royal Statistical Society, Series A*, 147, 621-44.

Griliches, Z., Hall, B.H. and Hausman, J.A. (1978) 'Missing data and self-selection in large panels', *Annales de L'Insee*, (30-31), 137-176.

Guy, C.M., Wrigley, N., O'Brien, L. and Hiscocks, G. (1983) 'The Cardiff consumer panel: a report on the methodology', *Papers in Planning Research 68*, University of Wales Institute of Science and Technology, Cardiff.

Heckman, J.J. and Willis, R. (1977) 'A beta-logistic model for the analysis of sequential labour force participation by married women', *Journal of Political Economy*, 85, 27-58.

Heckman, J.J. (1981) 'Statistical models for discrete panel data', in Manski, C.F. and McFadden, D. (eds.) *Structural Analysis of Discrete Data: with Econometric Applications*, M.I.T. Press, Cambridge, 114-178.

Heckman, J.J. and Robb, R. (1984) 'Alternative methods for estimating the impact of training on earnings', paper presented at Australasian Econometric Society Conference, University of Sydney, August (mimeo).

Heckman, J.J. and Singer, B. (1984) 'Econometric duration analysis', *Journal of Econometrics*, 24, 63-132.

Hensher, D.A. and Le Plastrier, V. (1984) 'Towards a dynamic discrete-choice model of household automobile fleet size and composition', *Transportation Research* (forthcoming).

Hensher, D.A., Smith, N.C. and Wrigley, N. (1984) 'Accounting for heterogeneity, state dependence and nonstationarity in a study of discrete-choice sequences in discrete time', *Dimensions of Automobile Demand Working Paper No. 18*, School of Economic and Financial Studies, Macquarie University, Sydney, Australia.

Hensher, D.A. (1984) 'Model specification for a quasi-dynamic discrete-continuous choice automobile demand system in discrete time using panel data: theory and method', *Dimensions of Automobile Demand Project Working Paper No. 11*, School of Economic and Financial Studies, Macquarie University, Sydney, Australia.

Hensher, D.A. (1984a) 'An overview of the theoretical, methodological, empirical and policy bases of the dimensions of automobile demand project', *Dimensions of Automobile Demand Project Working Paper No. 12*, School of Economic and Financial Studies, Macquarie University, Sydney, Australia.

Hensher, D.A. (1984b) 'Longitudinal surveys in transport: an assessment', in Ampt, L., Richardson, A.J. and Brog, W. (eds.) *New Survey Methods in Transport,* Gordon and Breach, New York (in press).

Hensher, D.A. (1984c) 'An econometric model of vehicle use in the household sector', *Transportation Research* (forthcoming).

Hensher, D.A. and Stonecash, R.E. (1984) 'Vehicle usage in the household sector with selectivity', *Dimensions of Automobile Demand Working Paper No. 15,* School of Economic and Financial Studies, Macquarie University, Sydney, Australia.

Hocherman, I. (1984) 'A dynamic approach to modelling of the demand for automobiles', Transportation Research Institute, Technion, Israel (mimeo).

Laird, N. (1978) 'Nonparametric maximum likelihood estimation of a mixing distribution', *Journal of the American Statistical Association,* 73, 805-811.

Lancaster, A. (1979) 'Econometric methods for the duration of unemployment', *Econometrica,* 47, 939-956.

Lancaster, A. and Nickell, S. (1980) 'The analysis of re-employment probabilities for the unemployed', *Journal of Royal Statistical Society A,* 143, Part 2, 141-165.

Mannering, F.L. and Winston, C. (1983) 'Dynamic models of household vehicle ownership and utilisation', Centre for Transportation Studies, M.I.T. (mimeo).

Markus, G.B. (1979) *Analyzing Panel Data,* Sage University Paper 18, Sage Publications, Beverly Hills.

Nickells, S. (1979) 'Estimating the probability of leaving unemployment', *Econometrica,* 47(5), 1249-1266.

Parks, R.W. (1980) 'On the estimation of multinomial logit models from relative frequency data', *Journal of Econometrics,* 13, 293-303.

Pickles, A.R. (1983) 'The analysis of residence histories and other longitudinal panel data: a semi-Markov model incorporating time varying exogenous variables', *Regional Science and Urban Economics,* 13, 271-285.

Schmidt, P. (1981) 'Constraints on the parameters in simultaneous tobit and probit models', in Manski, C.F. and McFadden, D. (eds.) *Structural Analysis of Discrete Data with Econometric Applications,* M.I.T. Press, Cambridge, 422-434.

Spinnewyn, F. (1981) 'Rational habit formation', *European Economic Review,* 15, 91-109.

Vaupel, J.W., Manton, K.G. and Stallard, C. (1979) 'The impact of heterogeneity in individual fraility on the dynamics of mortality', *Demography,* 16, 439-454.

Winship, C. and Mare, R.D. (1983) 'Structural equations and path analysis for discrete data', *American Journal of Sociology,* 87, 54-110.

Wrigley, N. (1985) *Categorical Data Analysis for Geographers and Environmental Scientists,* Longman, London.

Wrigley, N. and Dunn, R. (1984a) 'Stochastic panel-data models of urban shopping behaviour: 1. Purchasing at individual stores in a single city', *Environment and Planning A,* 16, 629-50.

Wrigley, N. and Dunn, R. (1984b) 'Stochastic panel-data models of urban shopping behaviour: 2. Multistore purchasing patterns and the Dirichlet model', *Environment and Planning A,* 16,759-78.

Wrigley, N. and Dunn, R. (1984c) 'Stochastic panel-data models of urban shopping behaviour: 3. The interaction of store choice and brand choice', *Environment and Planning A,* 16,1221-36.

Wrigley, N. and Dunn, R. (1985) 'Stochastic panel-data models of urban shopping behaviour: 4. Incorporating independent variables into the NBD and Dirichlet models', *Environment and Planning A,* 17,(3) March.

Wrigley, N. and Hensher, D.A. (1984) 'Collection of panel data: issues and experiences', *Dimensions of Automobile Demand Project Working Paper No. 20,* School of Economic and Financial Studies, Macquarie University, Sydney, Australia.

Wrigley, N. and Reader, S. (1985) 'Estimating panel-data models of urban shopping behaviour: a comparison of parametric and non-parametric approaches to the problem of heterogeneity, Working Paper, Department of Geography, University of Bristol.

Wrigley, N., Guy, C.M., Dunn, R. and O'Brien, L.G. (1985) 'The Cardiff Consumer Panel: Methodological aspects of the conduct of a long-term panel survey', *Transactions of Institute of British Geographers, New Series 10,* forthcoming.

3
Model structure and applicability

Behavioural Research for Transport Policy
pp. 119–134. © 1986 VNU Science Press.

Analysis of the Effects of Captivity on Travel Time and Cost Elasticities

by

Joffre D. Swait, Jr.
Dept. of Organization
Instituto Tecnologico
de Aeronautica
12.200 São Jose dos Campos, SP
BRASIL

Moshe Ben-Akiva
Dept. of Civil Engineering
Massachusetts Institute of
Technology
Cambridge, MA 02140
USA

1. Introduction

The prediction of users' response to changes in transport supply is a necessary input into the resource allocation procedures of any transport agency. The need for the predictions to be as accurate as possible results from the many competing alternative investments open to decision-makers. To meet this need, researchers in travel demand modelling have progressed from the aggregate level models of the 60's to the disaggregate models, based on the theory and methods of discrete choice analysis, of the 70's (see, for example, Domencich and McFadden, 1975; Ben-Akiva, 1973; Adler and Ben-Akiva, 1976; Westin and Gillen, 1978).

The theory behind disaggregate choice models requires that analysts, who usually have knowledge only of the revealed choice, be able to specify the set of alternatives actually compared during the choice process. Specification of this set of alternatives, known as the choice set, has usually been dealt with in one of two ways: (1) ignoring the issue and making all alternatives available to all decision-makers; (2) using logical rules to impute the set, e.g. in mode choice models, no auto drive alternative for individuals without a license.

The effect of ignoring or mis-specifying choice sets is especially apparent in modelling problems where it is expected that major changes in society, family structure, and economy will occur over time. For example, improvements in the transport infrastructure will not only result in a modification of the propensity to utilize one or another mode, it will also affect the availability of the modes to individuals. Hence, their choice sets and choices will both be potentially modified. It is to be expected that traditional discrete choice models, indeed demand models of any type that do not consider choice set generation, will display a lack of robustness (in terms of stability of parameters, hence forecasts) as the time horizon of the application increases.

In this paper we address the impact of ignoring or mis-specifying the effect of captivity (i.e. the state in which decision-makers' choice sets have but one alternative) on predicted users' response to transport supply changes, at both the disaggregate and aggregate levels. We first present a brief review of the literature dealing with the (mis-)specification of choice set generation, due to its fundamental relation to the specific problem we address herein. Swait(1984), Swait and Ben-Akiva(1984a,b), and Ben-Akiva and Swait(1984) review the literature on choice set generation, as well as motivating behaviorally, formulating, and estimating models of choice that incorporate the set formation stage. The interested reader is referred to these works for further information on this subject.

Second, we analyze theoretically a prototypical binary choice situation to gain insights into the effects of ignoring captivity, given its existence in the population in question, on users' response measures. It should also be noted that Kitamura and Lam(1984) provide a Monte Carlo study of the properties of a binary choice model which incorporates choice set generation, similar to the one analyzed in this paper.

Finally, we present empirical travel time and cost elasticities obtained from a model that explicitly allows for captivity in work mode choice for São Paulo, Brazil. As a standard of comparison, we also present analogous measures derived from a standard discrete choice model. The impacts of the differences in predictions on policy analysis are discussed.

2. The Importance of Choice Set Formation in Travel Demand Models

Transportation demand researchers have been aware of the ramifications of incorrect representation of choice sets. The work of Williams and Ortuzar(1982) and Stopher(1980), for example, give numerical and empirical verification of the problems that can arise when this stage of the full choice process is ignored or mis-specified.

Stopher(1980) studied empirically the impact of captives on the estimation of and forecasting with a binary mode choice model. It is clear that the choice of captive consumers need not be modelled, insofar as there is no need to model captivity itself, of course. However, if consumers who are deemed captive (according to a set of deterministic rules) to a given choice are assumed to have more than one alternative in their choice sets, and are included with non-captives in the data to estimate a choice model, Stopher states that the estimated coefficients for all attributes (both of alternatives and consumer-related) were smaller and less significant than in the "true" model (defined by him as the model estimated on data excluding the captives); and the alternative-specific constants were larger and more significant than in the "true" model. Theoretical treatment of a binary choice model leads to similar conclusions concerning the existence and direction of an estimation bias (see the following section).

Among other issues concerning discrete choice models, Williams and Ortuzar(1982) chose to investigate the impact of the analyst assuming that the choice set generation process is such that consumers have all alternatives in their choice sets, when in reality the process follows some known probability law. Using randomly generated data from a known model, they show that if choice set generation is actually a function of variables in the choice model itself, biased predictions of behavior due to policy changes will result, even though the fit to the base case data is quite good.

Meyer(1979) argues that demand model parameters may be as much influenced by variations in choice sets among individuals (not fully accounted for in standard discrete choice model structures) as by variations in preferences (these are accounted for, at least in theory). Other authors with this same concern are Louviere(1979) and Pirie(1976). Meyer is of the opinion that the indeterminacy of these effects may result in a lack of stability of destination choice model parameters over time and space; it seems reasonable to extend the coverage of this opinion to other choice contexts.

The valuation of travel time has long been a subject of interest to transportation researchers and practitioners alike (see Hensher, 1978, for a survey of that literature) due to the important role that travel time savings play in establishing the benefits of investing in transport infrastucture. A recent paper by Heggie(1983) deals with the empirical estimation of the value of non-work travel time, but he rightly points out the estimation sample must include only those individuals who are, in his terminology, the "choosers". In the terminology we employ here, it is necessary to isolate the captives and not permit them to bias the estimate of the value of time. Heggie utilizes information solicited from the user of the transport system to distinguish the choosers from the non-choosers.

Heggie's approach is inspired by the research of Brog and Erl(1979). The latter utilized a series of filters on a sample of 1200 urban households in West Germany to determine who is actually a chooser (i.e. explicitly compares alternatives) for the mode to work decision. Using objective criteria (availability of mode, type of employment,...), information availability (is the individual aware of objectively available alternatives?), and subjective availability (does the person consider an alternative a possibility?), Brog and Erl estimate that only 10% of the workers actually make rational comparisons to determine their modal choice. While the use of subjective data in such analyses makes rigorous interpretation more difficult, it is not easy to escape the conclusion that ignoring or mis-representing the choice set generation stage in a discrete choice model can lead to serious impacts, as we have described above.

3. An Analytical Investigation of Choice Set Specification Errors

This section will present a specification error analysis for a simple binary choice context. Besides yielding information about this specific problem, the analysis will also provide insight into the general process of mis-specifying choice sets and the consequent incorrect predictions of users' responses to changes in the transport system.

The choice situation we shall study is identical to that of Stopher(1980): namely, the subject population, faced with a choice between discrete alternatives 1 and 2, is composed of two subgroups, I_1 and I_2. The first of these, I_1, has no restrictions placed upon it and chooses between the two alternatives according to the true model, which has a binary logit form with two parameters and one explanatory variable:

$$P(1|B^*,X_n) = 1/(1 + \exp(-b_1^* - b_2^*X_n)) \quad , \quad \forall n \varepsilon I_1, \tag{1}$$

where $P(1|B^*,X_n)$ is the probability individual n chooses alternative 1; $B^*=(b_1^*,b_2^*)$ is the true vector of choice model parameters; X_n is the single explanatory variable of the choice between alternatives 1 and 2, for individual n.

Subgroup I_2, however, is captive to alternative 1, that is,

$$P(1|B^*,X_n) = 1 \quad , \forall n \varepsilon I_2. \tag{2}$$

Of course, $P(2|B^*,X_n) = 1 - P(1|B^*,X_n)$, $\forall n \varepsilon I$, where $I=I_1 U I_2$.

This model can also be expressed jointly as follows:

$$P(1|B^*,X_n) = d_n + (1-d_n)[1/(1+\exp(-b_1^*-b_2^*X_n))], \forall n \varepsilon I, \tag{3}$$

where d_n is equal to 1 if $n \varepsilon I_2$, and 0 otherwise. Suppose that the state of captivity or not is unknown to us when we randomly draw an individual from the subject population. Thus, d_n is a random variable from our viewpoint. Assume that

$$Pr(d_n = 1) = k^* \quad , \tag{4}$$

where k^* is a constant parameter, $0 \le k^* \le 1$. This parameter is equal to the fraction of the population in subgroup I_2. Thus, the true model for the probability that a random individual from the subject population chooses alternatives 1 and 2 becomes

$$Q(1|B^*,k^*,X_n) = k^* + (1-k^*)H(1|B^*,X_n),$$
$$Q(2|B^*,k^*,X_n) = 1 - Q(1|B^*,k^*,X_n), \quad \forall n\varepsilon I. \qquad (5)$$

where $H(1|B^*,X_n)=[1/(1+\exp(-b_1^*-b_2^*X_n))]$.

This specification is a binary choice version of the logit captivity model of McFadden(1976) and Ben-Akiva(1977), which is identical to the so-called "Dogit" model of Gaudry and Dagenais(1979).

Suppose now that, as analysts, we have drawn at random from this population N individuals. Suppose further that we know the form of the true choice model (5), and wish only to estimate the values of the unknown parameters, k^* and B^*. Expressed as a maximum likelihood estimation problem, we are faced with solving

$$\max_{B,k} \Gamma_N(B,k,X) = \sum_{n=1}^{N} \sum_{i=1,2} \delta_{in} \ln Q(i|B,k,X_n) \qquad (6)$$

where $\Gamma_N(B,k,X)$ is the log likelihood function; B and k correspond, respectively, to B^* and k^*; and

$$\delta_{in} = \begin{cases} 1 & \text{if person n observed choosing alternative i, i=1,2} \\ 0 & \text{otherwise} \end{cases}$$

Up to this point in the analysis no specification error has been committed. We assumed knowledge of the correct specification of the model, but not of the specific values of model parameters; for the purpose of estimating these values we have collected a sample of N individuals, and by solving the mathematical program (6) we will obtain the maximum likelihood estimates (MLEs) of the captivity and choice model parameters (<u>i.e.</u> k and B). With this level of knowledge, these MLEs will be asymptotically unbiased estimates of k^* and B^*.

Now we will commit a single but crucial error: namely, we will not use the knowledge that part of the population is captive to alternative 1. The estimation problem we are now faced with is a special case of (6) in which parameter k is forced to be zero, that is,

$$\max_{B} \Gamma_N(B,0,X) = \sum_{n=1}^{N} \sum_{i=1,2} \delta_{in} \ln Q(i|B,0,X_n) , \qquad (7)$$

Clearly, ignoring the captivity phenomenon in the population must yield inconsistent estimates of the choice model parameters; a question of greater interest lies in whether we can obtain further insight about the resulting bias in terms of its direction, even of its magnitude.

Maximization of (7) is equivalent to the maximization of

$$D_N(B,0,X) = \Gamma_N(B,0,X)/N \qquad (8)$$

since division by the positive constant N does not change the form of the function. Now define D(B,0) as

$$D(B,0) = \operatorname*{plim}_{N\to\infty} D_N(B,0,X) \quad . \qquad (9)$$

Since $Q(i|B,0,X_n)$ is continuous and everywhere positive, both $Q(i|B,0,X_n)$ and $\ln Q(i|B,0,X_n)$ are bounded. This guarantees that D(B,0) exists and is finite for all values of B. The function D(B,0) is the expectation of function (7) with respect to attribute X as the sample size grows without bound, i.e. over the population. Thus,

$$D(B,0) = \operatorname*{plim}_{N \to \infty} \left[(1/N) \sum_{n=1}^{N} \sum_{i=1,2} \delta_{in} \, \ell n Q(i|B,0,X_n) \right]$$

$$D(B,0) = \sum_{i=1,2} E_X[Q(i|B^*,k^*,X) \, \ell n Q(i|B,0,X)] \tag{10}$$

where $Q(1|B^*,k^*,X)$ and $Q(2|B^*,k^*,X)$ are given by (5) and $E_X[\,]$ represents the expectation operator with respect to attribute X. Thus, the estimation problem, originally stated as (7), has the equivalent asymptotic form below:

$$\max_{B} D(B,0) = \sum_{i=1,2} \int_X Q(i|B^*,k^*,z) \, \ell n Q(i|B,0,z) f(z) dz, \tag{11}$$

where f(X) is the probability density function for the attribute X. Substituting (5) into (11), we obtain

$$D(B,0) = (1-k^*) \, D_1(B,0) + k^* \, D_2(B,0) \, , \tag{12}$$

where

$$D_1(B,0) = \sum_{i=1,2} \int_X H(i|B^*,z) \, \ell n H(i|B,z) f(z) dz \, , \tag{13}$$

$$D_2(B,0) = \int_X \ell n H(1|B,z) f(z) dz \quad . \tag{14}$$

Consider now the following theorem:

Theorem: Let g(x) and h(x) be real-valued, concave functions on R^n, the n-dimensional argument space. Further, let x_g^* and x_h^* represent the points where g and h attain their global maxima, respectively. Then $f(x)=u_1 g(x)+u_2 h(x)$, where $u_i \geq 0$, i=1,2, $u_1+u_2=1$, will attain its global maximum at $x^*=u_1 x_g^* + u_2 x_h^*$.

Pf. See Swait(1984), Appendix 1.

Thus, if $D_1(B,0)$ and $D_2(B,0)$ are both concave with respect to B, then the optimal solution to (12) can be found by obtaining the global maximum for function (13) and that of function (14), and combining the two optima in the manner described in the theorem. As shown in Swait(1984), $D_1(B,0)$ and $D_2(B,0)$ are indeed both concave in B, so that the theorem is applicable. Thus, if we let $\hat{B}$ be the optimal solution to (12) we can write

$$\hat{B} = (1-k^*) \, \tilde{B}_1 + k^* \tilde{B}_2 \tag{15}$$

where $\tilde{B}_1$ is the solution to

$$\max_{B} \; D_1(B,0) \tag{16}$$

and $\tilde{B}_2$ is the solution to

$$\max_{B} \quad D_2(B,0) \quad . \tag{17}$$

To find the solution of (16), we make use of the following result:

Lemma: Let $g(s,x)$ be a real-valued function over a space $S \times X$ such that g is integrable with respect to a measure u over S and $g(s,x)>0$, all $s \in S$, $x \in X$. Let x^* be an element of X such that $g(s,x^*)>0$ for almost every $s \in S$ and $\int_S [g(s,x^*)-g(s,x)]du \geq 0$, all $x \in X$. Then,

$$f(x) = \int_S g(s,x^*)[\ell n g(s,x)]du$$

attains its maximum at $x = x^*$.

Pf. See Rao(1973), p.59.

The straightforward application of this lemma to function (14) leads us to conclude that the optimum of (16) is

$$\tilde{B}_1 = B^* \quad , \tag{18}$$

that is, the estimator of the choice model parameters in $D_1(B,0)$ is asymptotically equal to the true parameter values.

Now let us turn our attention to program (17), in which we must find the optimum of $D_2(B,0)$. By substituting the definition for $H(1|B,X)$ into (14), we obtain

$$D_2(B,0) = b_1 - \int_X \ell n[\exp(b_1) + \exp(-b_2 z)]f(z)dz \quad . \tag{19}$$

To find the stationary points of (19), we differentiate with respect to the two parameters, which results in the following first-order conditions:

$$\int_X H(1|B,z)f(z)dz = 1 \quad , \tag{20a}$$

$$\int_X z\ H(1|B,z)f(z)dz = \int_X z\ f(z)dz \quad . \tag{20b}$$

Expression (20a) states that the expectation of $H(1|B,X)$, with respect to attribute X, must be equal to 1 at any stationary point. Since $H(1|B,X)$ is a proper cumulative probability density function, it is bounded by 0 and 1. Thus, to satisfy (20a) it is necessary that the following hold at the point: $H(1|B,X)$ must be equal to 1 for all values of X. This, in turn, implies two conclusions: (1) for $H(1|B,X)$ to be equal to 1, it is necessary that b_1 grow arbitrarily large (i.e. $b_1 \to \infty$); (2) for this result to hold for all values of X, b_2 must be equal to zero. This solution also satisfies (20b), which indicates that a stationary point has been identified. Indeed, this degenerate solution is to be expected since $D_2(B,0)$ is monotonically increasing in b_1; since b_2 must be zero to satisfy conditions (20), we must simply increase b_1 towards infinity to maximize $D_2(B,0)$.

Since $D_2(B,0)$ is concave, we conclude that the maximum of (17) is attained at

$$\tilde{B}_2 = \begin{bmatrix} \infty \\ 0 \end{bmatrix} . \qquad (21)$$

This, then, is the result of using the improper estimator given in (7); by bringing the two solutions (18) and (21) together according to expression (15), we find that the erroneously estimated parameters for the choice model exhibit the following characteristics:

(1) $\hat{b}_1$, the estimate of b_1^* according to estimator (8), is a convex combination of b_1^* and an arbitrarily large number, whence we conclude that $\hat{b}_1 \to \infty$;
(2) $\hat{b}_2$, the estimate of b_2^* by (8), can be expressed as

$$\hat{b}_2 = (1-k^*)b_2^* .$$

Thus, the estimated alternative specific constant grows arbitrarily large, and the estimated coefficient of the attribute is smaller, in absolute terms, than the true value by a factor dependent upon the degree of captivity in the population (the greater the captivity, the smaller the estimated coefficient).

These conclusions are exactly those reached empirically by Stopher(1980), now supported by theoretical analysis. It is also helpful to note that the asymptotic results above are also identically valid for a binary probit, rather than logit, choice model, assuming also a linear-in-parameters specification. In fact, it is correct for any binary choice model of the form $Pr(1|B,X_n) = F[\Delta V(B,X_n)]$, where F[] is a proper cumulative density function, $\Delta V(B,X_n) = b_1 + b_2 X_n$, and F[] is monotonically increasing in $\Delta V()$.

The specific form of the bias deduced for this binary choice context lends credence to Meyer(1979)'s spatial and temporal parameter instability argument, mentioned in the previous section. The bias of attribute related parameters is a function of the degree of captivity in the population; the mechanisms that result in captivity will likely vary from place to place, across cultures and in different time periods, so it becomes even more critical that we model these mechanisms before attempting spatial and temporal transfer of parameters.

Besides quantifying the direction and the magnitude of the bias as we have done above, it is instructive to note that any function of the estimated parameters $\hat{B}$ will also, of course, be biased. Consider, for example, the elasticity with respect to attribute X of the choice probability for alternative 1 using the true model $Q(1|B^*,k^*,X)$, defined in (5):

$$\eta_X^* = \frac{\partial Q(1|B^*,k^*,X)}{\partial X} \cdot \frac{X}{Q(1|B^*,k^*,X)}$$

$$\eta_X^* = \frac{(1-k^*)H(1|B^*,X)H(2|B^*,X)b_2^* X}{k^* + (1-k^*)H(1|B^*,X)} \qquad (22)$$

where $H(i|B^*,X)$, i=1,2, is defined following expression (5). Analogously, one can calculate the same elasticity for the parameter vector $\hat{B}$ and $k^*=0$, yielding

$$\hat{\eta}_X = \frac{\partial Q(1|\hat{B},0,X)}{\partial X} \cdot \frac{X}{Q(1|\hat{B},0,X)}$$

$$\hat{\eta}_X = (1-k^*)H(2|\hat{B},X)b_2^* X \qquad (23)$$

Dividing (23) by (22), we find that

$$\frac{\hat{\eta}_x}{\eta^*_x} = \frac{H(2|\hat{B},X)}{H(2|B^*,X)} \left[(1-k^*) + \frac{k^*}{H(1|B^*,X)} \right] \tag{24}$$

To interpret (24), it is useful to first restrict out attention to the non-captive group of the population. For this group, we have shown above that the estimated parameters $\hat{B}$ make alternative 2 a less attractive choice than it actually is. Or in other words, $H(2|\hat{B},X) \leq H(2|B^*,X)$, $n\varepsilon I_1$. Therefore,

$$\frac{\hat{\eta}_x}{\eta^*_x} \leq 1 + k^* \left[\frac{H(2|B^*,X)}{H(1|B^*,X)} \right] \quad , n\varepsilon I_1. \tag{25}$$

For the non-captive population, then, the estimated elasticity of the choice probability of alternative 1 with respect to attribute X is less than than or equal to the true elasticity. Further, as one would expect, the distance between the two measures is an increasing function of the degree of captivity in the population. The term within brackets is the odds of choosing alternative 2 over 1, for a particular value of X, at the true parameters. This odds can be either less than, equal to, or greater than one, hence it can both dampen or amplify the effect of k^* depending upon the value of X.

A similar analysis can be performed for the captive group of the population, except that the $\leq$ in (25) is replaced by a $\geq$ since $H(2|\hat{B},X) \geq H(2|B^*,X)$, $n\varepsilon I_2$.

Such comparisons can also be performed for the corresponding aggregate elasticities. The expressions resulting from this exercise are more difficult to interpret than (25), and hence will not be presented here.

4. The Effects of Work Mode Choice Captivity in São Paulo

In this section we compare empirical elasticities obtained from a logit model of work mode choice in São Paulo, Brazil, with those obtained from a specification that explicitly allows for the existence of captivity in the population. This latter specification is given by

$$P(j|\beta,\delta,X_n) = \frac{1}{1 + \sum\limits_{i\varepsilon M_n} e^{\delta' X_{in}}} \left[e^{\delta' X_{jn}} + \frac{e^{\beta' X_{jn}}}{\sum\limits_{k\varepsilon M_n} e^{\beta' X_{kn}}} \right] \tag{26}$$

where β and δ are parameter vectors; X_{jn} is a vector of socio-economic characteristics and alternative-specific attributes; M_n is the set of alternatives deterministically allocated to individual n.

Ben-Akiva and Swait(1984) demonstrate the derivation of (26) from the assumption that each individual is either captive to one of the alternatives in M_n, or is otherwise free to choose (according to a logit model) from among all the alternatives of M_n. The first term on the right hand side of (26) expresses the probability of individual n being captive to $j\varepsilon M_n$, while the remainder of the expression is the probability of choice of j from the full choice set, given that the individual is not captive to any one alternative. This formulation is identical to the logit captivity model independently derived by McFadden(1976), Ben-Akiva(1977), and Gaudry and Dagenais(1979), except that in those references the odds of captivity to an alternative is modelled via a single parameter per alternative; in (26), a more flexible specification is achieved by defining a function, $\exp(\delta' X_{jn})$, to represent the odds of captivity to $j\varepsilon M_n$. Ben-Akiva(1977) suggested this extension to the logit captivity.

Swait(1984) (see also Swait and Ben-Akiva, 1984b), who presents the first known applications of the logit captivity model to disaggregate data, calibrated a work mode choice model of the form of (26), calling it the parametrized logit captivity model. The models we utilize here are those calibrated by Swait(1984). Before presenting the models and their corresponding elasticities, we briefly describe the data utilized for estimation.

4.1 The Estimation Data

As described in EMPLASA(1978b), the data utilized to calibrate these models were collected during the 1977 home interview origin-destination survey. The sampling frame of buildings spanned the geographical area of Greater São Paulo (26 municipalities are included). The sampling of units was random within strata defined by architectural standards (size, type of finish, state of maintenance). All households in a selected building were interviewed to determine socio-economic attributes and transportation-related behavior on the previous day. A total of 31,380 households were thus sampled, which provided a coverage of 108,000 persons who made a total of 182,000 trips (EMPLASA, 1978b).

Whereas the 1977 O/D survey data relates trip-making behavior to a detailed 633 traffic zone system, the impedance measures we have available are from a more coarse 125 zone network that was utilized in a posterior aggregate policy analysis (CET, 1979). Impedance measures were made available from separate walk, auto, bus, and rail networks. The latter is a combined network for the physical metro and suburban rail modes; since the two are non-overlapping, they are represented in one network.

The choice we shall examine in our empirical application is mode choice for work trips. More specifically, we shall use home-based, morning peak period (arrival time before 10:30 AM) trips. Of the 17,088 such one way trips contained in the 1977 São Paulo Origin/Destination Survey, 1,746 were randomly selected for inclusion in the estimation data set.

The O/D survey provides a rich source of information on modal choice, including access, egress, and principal modes. However, the currently available impedance measures (which represent the shortest path over all access, egress, and main mode combinations) have led us to perform an aggregation of the elemental choices into the following alternatives: (1) Bus, (2) Auto Drive, (3) Auto Passenger, (4) Train, (5) Metro, and (6) Walk. No two zones are served by the two rail modes since the physical networks are non-overlapping. For our study, these two modes have been combined into a single rail alternative.

The deterministic allocation of alternatives to individual trip makers followed certain rules: (1) the network connection for the observed and generated modes had to exist; (2) the maximum allowed one-way travel time was 3 hours for bus, auto, and walk, and 4 hours for the rail mode (clearly the latter restriction applies, in practice, to the suburban train mode, not the metro alternative). Swait(1984) was unable to restrict the auto drive alternative to individuals from auto-owning households since a significant number of observed drivers declared an automobile ownership level of zero.

The application of the above rules eliminated 20 observed trips, so that the final estimation data set consists of 1,726 trips.The actual estimation data set, however, has one less observation. Following initial calibration results, an outlier analysis was performed, and an observation with a miscoded income value was detected and removed from the data. Therefore, the calibration results are for a random sample of 1,725 workers.

4.2 The Estimated Models

In Table 1 we present the calibrated logit and parametrized logit captivity mode choice models. The principal observation we wish to make about the former model is the liberal use of variables in the choice utility functions designed to represent, even if in an ad hoc manner, the availability of alternatives. For example, if the household owns an automobile there will in general be alternative allocations of this resource to household members to fulfill household needs; among these is the use of the car for work travel. If there is more than one worker in the household, competition for the auto increases, so one would expect that the availability of the auto drive mode should increase with the ratio of number of cars to number of workers. Similar rationale leads to the use of other variables, such as income, household size, gender and education, in the utility functions.

Several points should be noted about the parametrized captivity model in Table 1. First, note that the utility functions now contain only alternative-specific dummies and level-of-service variables. All ad hoc availability variables have been removed from the indirect utility functions and are now used in the captivity functions.

Second, note the general increase in the scale of the parameters in the utility functions of the parametrized logit captivity model as compared to the corresponding parameters of the logit model. As indicated by our theoretical analysis in the previous section, if the captivity model is indeed the true specification, the parameters of variables in the logit model are downward biased as a function of the captivity rates. The one exception to this trend is the cost coefficient for the bus alternative; this may be due to the fact that the captivity model predicts low rates of captivity to this mode, or even to the possibility that the choice set generation model that has been postulated (captivity or choice from the full set) is inappropriate for the portion of the population which has chosen bus.

Third, the parametrized logit captivity model indicates that choice of the auto drive mode is independent of its own and competing level-of-service measures. Explanation of the choice of this mode has been wholly transferred to its captivity function. Examination of the coefficients of this function indicates the important role of auto ownership in determining the probability of utilization of auto drive.

Finally, we compare the overall statistical fit of the two model specifications, and note that the log likelihood at convergence of the parametrized captivity model is slightly smaller (and it utilizes a few more degrees of freedom) than the logit model of Table 1. This is most probably due to the a priori restrictions that removed the ad hoc availability variables from the utility functions of the parametrized captivity model. In addition, there is need to fine-tune the captivity functions. For example, distance and age variables should be included in the walk captivity function, and perhaps occupation-related dummies in the auto drive captivity function could help to increase the explanatory power of this specification. Time and monetary resources, however, prevented Swait(1984) from pursuing these possibilities, so our analysis will utilize the two models.

We now proceed to the presentation and discussion of predictions based on the two models shown above.

4.3 Model Prediction Results

Here we shall compare the predictions of the two models presented above. Predictions will be presented for two levels of uniform change of time and cost across the worker population.

Table 2 presents the predicted changes in ridership due to a 10%

travel time increase for all workers. Part (a) of the table gives the results for the logit choice (L) model, and (b) for the parametrized logit captivity (LC) formulation. One difference we note immediately is to be expected: the LC model predicts the auto drive mode to be completely insensitive not only to its own travel time, but also to that of the other modes. The logit model, however, predicts the drive own- and cross-elasticities to be significantly higher. In contrast, the logit model understates the impacts of increasing bus and train/metro travel times when compared to the LC specification. In this instance we see the increased sensitivity of the parametrized logit captivity model to bus and rail travel times offsetting the captivity effect to these modes. The walk mode is deemed less sensitive by the LC than the logit model to this low level uniform change in travel time.

In Table 3 we have corresponding predictions for a doubling in travel time, again for both model specifications. The same patterns present in Table 2 are repeated here for the motorized modes. For the walk mode, both the LC and L models predict counterintuitive results (not shown here); these are due to the quadratic travel time specification for the walk mode (see Table 1), which attributes positive utility to long travel times (>150 minutes). Within the estimation sample and under low levels of change in travel time (such as in Table 2) the specification performs well, but with high levels of change (such as a doubling of everyone's walk travel time) the counterintuitive forecasts result. This problem is a shortcoming of the polynomial specification for the variable. Due to time and monetary constraints, however, we are unable to repeat the calibration of the models to correct this variable.

If we return to the motorized modes in Table 3, we see that the differences between the LC and L models are large for the auto modes. That is, doubling automobile travel time is forecast by the LC model to not affect the drive alternative and decrease by 1.6% the ridership of the passenger mode; the L formulation, however, shows large shifts occurring from the drive mode to auto passenger, bus, train/metro, and even walk. It can certainly be argued that the LC model probably understates the true sensitivity of the population to such a large change in travel time, but it may well paint a more realistic picture of the impact of this variable on traveller behavior than is done by the standard logit model.

Tables 4 and 5 show the predictions by the two model systems for a 10% and a 100% increase in travel cost by each mode, respectively. In contrast to the travel time increases, both models predict more compatible results at both levels of cost change. The same patterns we noted above are again present in these tables.

In summary, the LC model predicts a lower response level than the logit model to uniform changes in both travel time and cost (for small and large increases in these variables). The former model, however, is quite different from the logit for the travel time dimension; the two models produce much smaller discrepancies for the cost variable.

5. Summary and Conclusions

Our theoretical and empirical analyses have shown that the biases introduced into parameter estimation and model forecasts by ignoring the existence of captivity (and more generally, that of choice set generation) can be significant. The implication of our work for transport analysts is that decisions based on demand estimates that do not consider captivity potentially result in misallocation of scarce resources.

Though our empirical work is conducted with the choice of mode to work in a developing country, the implications we have drawn should be extensible to other choice dimensions and locations.

BIBLIOGRAPHY

Adler, T. and M. Ben-Akiva(1976). "Joint Choice Model for Frequency, Destination and Travel Mode for Shopping Trips", Trans. Research Record 569, 136-150.

Ben-Akiva, M.(1973). Structure of Passenger Travel Demand Models, Unpublished PhD Thesis, Dept. of Civil Engineering, Massachusetts Institute of Technology, Cambridge, MA.

Ben-Akiva, M.(1977). "Choice Models With Simple Choice Set Generating Processes", Working Paper, Dept. of Civil Engineering, Mass. Inst. of Technology, Cambridge, MA.

Ben-Akiva, M. and J. Swait(1984), "Choice Models With Simple Probabilistic Choice Set Generation Processes", Submitted for publication to Trans. Research B.

Brog, W. and Erhard Erl(1979). "Application of a Model of Individual Behaviour (Situational Approach) To Explain Household Activity Patterns In An Urban Area and To Forecast Behavioural Changes", Presented at the Int'l Conf. on Travel Demand Analysis: Activity-Based and Other New Approaches, St. Catherine's College, Oxford, England.

CET(1979). "Projeto MUT - Modelo de Uso do Solo e Transportes de São Paulo - Volume 11 - Modelo de Transportes", published by Companhia de Engenharia de Trafego, São Paulo, Brazil.

Domencich, T. and D. McFadden(1975). Urban Travel Demand: A Behavioral Analysis, North-Holland, Amsterdam.

EMPLASA(1978a). " '77 Origin-Destination Survey - Basic Results - Bilingual Document", published by Empresa Metropolitana de Planejamento da Grande São Paulo S/A - EMPLASA, São Paulo, Brazil.

EMPLASA(1978b). " O/D 77 - Volume 1 - Referencias Basicas", published by Empresa Metropolitana de Planejamento da Grande São Paulo S/A - EMPLASA, São Paulo, Brazil.

Gaudry, M. and M. Dagenais(1979). "The Dogit Model", Trans. Research B, Vol. 13B, 105-111.

Hensher, D.(1978). "Valuation of Journey Attributes: Some Existing Empirical Evidence", in Determinants of Travel Choice, D. Hensher and Q. Dalvi, eds., Saxon House.

Heggie, I.(1983). "Valuing Savings in Non Working Travel Time: The Empirical Dilemma", Trans. Research A, Vol. 17A, No. 1, 13-23.

Kitamura, R. and Tenny Lam(1984), "A Model of Constrained Binary Choice", Proc. of the 9th Int'l Symposium on Transportation and Traffic Theory, J. Volmuller and R. Hamerslag, eds., VNU Science Press, Utrecht, The Netherlands.

Louviere, J.(1979). On the Interregional and Intercultural Transferability of Disaggregate Choice Models: Some Selected Comments, Tech. Report 111, Inst. of Urban and Regional Research, Univ. of Iowa, Iowa City.

McFadden, D.(1976). "The Multinomial Logit Model When the Population Contains 'Captive' Subpopulations", Unpublished memorandum, Sept. 30, 1976.

Meyer, R.(1979). "Theory of Destination Choice-Set Formation Under Informational Constraints", Transportation Research Record 750, 6-12.

Pirie, G.(1976). "Thoughts on Revealed Preferences and Spatial Behavior", Environment and Planning A, Vol. 8, 947-955.

Rao, C.(1973). Linear Statistical Inference and Its Applications, 2nd Edition, John Wiley & Sons, New York.

Stopher, P.(1980). "Captivity and Choice in Travel-Behavior Models", Trans. Eng. Journal of ASCE, Vol. 106, No. TE4, 427-435.

Swait, J.(1984), Probabilistic Choice Set Generation in Transportation Demand Models, Unpubl. PhD thesis, Dept. of Civil Engineering, MIT, Cambridge, MA.

Swait, J. and M. Ben-Akiva(1984a), "Constraints on Individual Travel Behavior in a Brazilian City", Forthcoming in Trans Research Record.

Swait, J. and M. Ben-Akiva(1984b), "Incorporating Random Constraints in Discrete Choice Models: An Application to Mode Choice in São Paulo, Brazil", Submitted for publication to Trans. Research B.

Westin, R. and D. Gillen(1978). "Parking Location and Transit Demand - A Case Study of Endogeneous Attributes in Disaggregate Mode Choice Models", J. of Econometrics, Vol. 8, No. 1, 75-101.

Williams, H. and J. Ortuzar(1982). "Behavioural Theories of Dispersion and the Mis-Specification of Travel Demand Models", Trans. Research B, Vol. 16B, No. 3, 167-219.

Table 1 - São Paulo Home-Based AM Peak Work Mode Choice

Choice Model Parameters	Estimated Parameters (Asymptotic t-statistics in parentheses): Logit	Estimated Parameters (Asymptotic t-statistics in parentheses): Parametrized Captivity Logit
1. Alternative Specific Constants		
- Walk	-0-	-0-
- Bus	-0.701 (-2.7)	-2.450 (-2.40)
- Auto Drive	-4.530 (-10.7)	$-\infty$
- Auto Passenger	-4.154 (-10.4)	-6.128 (-5.1)
- Rail	-2.038 (-4.7)	-5.196 (-4.6)
2. - Walk Total Travel Time, minutes, one-way	-0.0638 (-9.3)	-0.1118 (-4.1)
- Walk (Total Travel Time)^2	1.9961E-4(4.1)	3.6846E-4(2.4)
3. Motorized Modes In-Vehicle Travel Time, minutes, one-way		
- Bus	-0.0122 (-2.6)	-0.0404 (-4.3)
- Auto	-0.0322 (-2.5)	-0.0783 (-2.0)
- Rail	-0.0005 (-0.2)	-0.0081 (-1.7)
4. Cost, (Cr$1977) / Income, (Cr$1977/month); Income = { Personal income if boarder; Household income, otherwise }		
- Walk	-0-	-0-
- Bus	-655.29 (-4.2)	-497.63 (-1.8)
- Auto	-76.75 (-1.6)	-132.28 (-1.0)
- Rail	-713.02 (-3.9)	-1006.26 (-2.6)
5. Income (see above for definition)		
- Walk	-0-	-0-
- Bus	-2.11E-6(-0.2)	-0-
- Auto Drive	23.01E-6 (2.2)	-0-
- Auto Passenger	9.59E-6 (0.7)	-0-
- Rail	-9.73E-6(-0.5)	-0-
6. Number of Household Members; Members = { 0 if boarder; # persons otherwise }		
- Walk	-0-	-0-
- Bus	-0.0870 (-2.4)	-0-
- Auto Drive	-0.1338 (-2.8)	-0-
- Auto Passenger	-0.0973 (-1.7)	-0-
- Rail	-0.1459 (-2.7)	-0-
7. Availability of Auto; Availability = { 0 if boarder; # cars / # workers otherwise }		
- Walk	-0-	-0-
- Bus	-0.9289 (-3.5)	-0-
- Auto Drive	1.2071 (6.1)	-0-
- Auto Passenger	-0.3078 (-0.8)	-0-
- Rail	-0.3651 (-0.9)	-0-
8. Auto Drive Mode - dummy variables		
a) 1 if auto-owning household	2.660 (6.8)	-0-
b) 1 if auto-owning houdehold <u>and</u> head of household	0.647 (3.2)	-0-
c) 1 if female worker	-1.207 (-4.9)	-0-
9. Auto Passenger Mode - dummy variable 1 if auto-owning household	1.857 (5.4)	
10. Auto Drive/Passenger Modes - dummy variable 1 if high school or university education	0.377 (2.3)	-0-

Table 1 - Continued

<u>Captivity Function Parameters</u>

1. Walk		
- Constant	-0-	-4.628 (-4.6)
- Income (see above)	-0-	-24.79E-6 (-0.8)
- Members = { 0 if boarder; # persons otherwise	-0-	0.2004 (1.9)
- Availability = { 0 if boarder; # cars / # persons otherwise	-0-	2.8487 (6.1)
2. Bus		
- Constant	-0-	-1.470 (-2.8)
- Income	-0-	-4.80E-6 (-0.1)
- Members	-0-	-0.0612 (-0.7)
- Availability	-0-	-0.3338 (-0.4)
3. Auto Drive		
- Constant	-0-	-4.422 (-7.0)
- Income	-0-	23.68E-6 (2.3)
- Members	-0-	-0.0441 (-0.9)
- Availability	-0-	2.3449 (6.5)
- 1 if auto-owning household	-0-	2.713 (4.2)
- 1 if head of household	-0-	0.853 (1.2)
- 1 if head of houdehold and auto-owning household		-0.252 (-0.4)
- 1 if female worker	-0-	-1.137 (-4.6)
- 1 if high school or university education	-0-	0.401 (2.2)
4. Auto Passenger		
- Constant	-0-	-5.467 (-4.5)
- Income	-0-	17.62E-6 (1.2)
- Members	-0-	0.0182 (0.2)
- Availability	-0-	0.7813 (1.6)
- 1 if female worker	-0-	3.235 (3.0)
- 1 if high school or university education	-0-	0.004 (0)
5. Rail		
- Constant	-0-	-1.049 (-1.4)
- Income	-0-	-3.27E-6 (-0.1)
- Members	-0-	0.5104 (-1.9)
- Availability	-0-	1.1336 (1.3)

<u>Summary Statistics</u>

log likelihood for equal probability	-2480.5	-2480.5
log likelihoof for convergence	1437.1	-1441.6
rho-squared	0.4206	0.4188
adjusted rho-squared	0.4090	0.4035

<u>Sample Description</u>

Choosing - Walk	409
- Bus	626
- Auto Drive	466
- Auto Passenger	95
- Rail	129
	1,725

Table 2 - Predicted Impact (% Change in Ridership) of Uniform 10% Travel Time Increase

(a) Logit Choice Model

		Predicted Response in Mode				
		Bus	Auto Drive	Auto Pass.	Rail	Walk
Change	Bus	-2.5	0.9	2.3	4.3	1.0
	Auto	0.6	-1.6	1.2	0.9	0.3
	Rail	0	0	0	-0.3	0
	Walk	1.7	0.8	1.5	0.5	-4.1

(b) Parametrized Logit Captivity Model

		Bus	Auto Drive	Auto Pass.	Rail	Walk
Change	Bus	-4.6	0	5.6	12.8	1.9
	Auto	0	0	-0.3	0	0
	Rail	0.5	0	0.2	-2.9	0
	Walk	1.8	0	0.7	0.5	-3.3

Table 3 - Predicted Impact (% Change in Ridership) of Uniform 100% Travel Time Increase

(a) Logit Choice Model

		Predicted Response in Mode				
		Bus	Auto Drive	Auto Pass.	Rail	Walk
Change	Bus	-24.3	7.8	23.7	42.6	9.4
	Auto	5.6	-15.6	13.7	9.5	3.1
	Rail	0.4	0.1	0.2	-2.6	0

(b) Parametrized Logit Captivity Model

		Bus	Auto Drive	Auto Pass.	Rail	Walk
Change	Bus	-43.8	0	75.9	98.6	20.1
	Auto	0.2	0	-1.6	0.2	0
	Rail	4.6	0	2.0	-25.2	0.2

Table 4 - Predicted Impact (% Change in Ridership) of Uniform 10% Travel Cost Increase

(a) Logit Choice Model

		Predicted Response in Mode				
		Bus	Auto Drive	Auto Pass.	Rail	Walk
Change	Bus	-1.5	0.3	1.2	2.5	0.8
	Auto	0.1	-0.3	0.1	0.2	0
	Rail	0.5	0.1	0.2	-3.1	0.5

(b) Parametrized Logit Captivity Model

		Bus	Auto Drive	Auto Pass.	Rail	Walk
Change	Bus	-0.6	0	0.7	1.5	0.3
	Auto	0	0	-0.1	0	0
	Rail	0.5	0	0.2	-2.7	0

Table 5 - Predicted Impact (% Change in Ridership) of Uniform 100% Travel Cost Increase

(a) Logit Choice Model

		Predicted Response in Mode				
		Bus	Auto Drive	Auto Pass.	Rail	Walk
Change	Bus	-14.0	3.0	12.3	24.0	7.6
	Auto	1.0	-2.5	1.6	1.9	0.3
	Rail	4.0	0.9	2.1	-25.4	0.4

(b) Parametrized Logit Captivity Model

		Bus	Auto Drive	Auto Pass.	Rail	Walk
Change	Bus	-6.2	0	7.8	15.0	3.2
	Auto	0.1	0	-0.5	0.1	0
	Rail	4.1	0	1.8	-22.1	0.3

Behavioural Research for Transport Policy
pp. 135–151. © 1986 VNU Science Press.

Users' Reactions to Specific Combinations of Policy Measures

Hedwig Verron
Technical University of Berlin

Gerd Martens
Agency for Assessment and
Application of Technology, Jakarta

1. Background and Purpose of the Study

Whenever planning is taking place, a multiplicity of interests comes together being more or less heavily conflicting. This, of course, also concerns the field of transportation planning. With respect to the existing interests being in conflict it is one of the most important activities for the transportation planning community to elaborate solutions not being a-priori subordinated to a biased interest. The most practicable possibility to achieve this objective results from the development of alternative planning policies. To be able to come to an informed selection among the available alternatives, appropriate information on their impacts have to be provided. The crucial prerequisite for getting the information needed will be fulfilled by the prediction of travel demand. Insofar procedures for the estimation of travel demand have to be able to address to specific planning proposals, i.e. they, at least, have to relate to their essential characteristics. Otherwise no meaningful evaluation of the planning alternatives may be performed. Not only measures and strategies of physical nature belong to the comprehensive range of planning alternatives but also services and programs following from broader transport policy concerns.

The demand of whatever kind is resulting from human behaviour. That is why it is necessary to discuss human behaviour and its formation processes. How else an answer may be found for the question of human behaviour being a reaction to implemented planning? Only by understanding the point of view of the people affected by planning, adequate procedures for the prediction of demand will be developed.

The conceptual requirements being a "red line" within the research activities of the interdisciplinary research group "Acceptance of Local Transport Systems" may be pinpointed as following:

- The policy-sensitivity has to be developed.
- The behavioural foundation has to be elaborated.

The range of research into disaggregated behaviourally oriented considerations on passenger transport essentially meets the requirement for

---

This presentation relates to investigations of the research project "Factors of Influence on the Transport Mode Usage", which the authors performed under contract of Daimler-Benz AG, Research Group Berlin. In addition the authors refer to their results of the interdisciplinary research project "Acceptance of Local Transport Systems" under contract of the Technical University of Berlin.

policy sensitivity. Only on the basis of individual travel patterns sufficiently detailed information of attributes of the transport system may be collected. To the disaggregated behavioural mode-split models we owe important insights,like the one that the travel time components "walking" and "waiting" do influence attractiveness of the public transport mode unequally more strongly than does "in-vehicle-time". Therefore, nowadays, there is hardly any model specification, in which there is no differentiation between "in-vehicle-time" and "out-of-vehicle-time".

There is, however, no reason for assuming that time and cost are the only supply qualities effecting the transport mode choice. In the conception of behavioural models linking behaviour directly with objective attributes of the transport system consideration of further characteristics results in a huge data effort. This may be tremendously reduced by gathering first of all subjective data instead of objective ones, herewith narrowing the range of behaviourally relevant characteristics. So e.g. the attitudinal approach quickly succeeded in substantiating the importance of comfort and convenience for the transport mode choice (see e.g. Nicolaidis, 1975; Recker and Golob, 1976).

The relations between subjective attitudes and the concerning objective data will, of course, be needed as much as ever, and as e.g. Hensher and McLeod (1977) demonstrate these are possible to find even for such complex factors like convenience and comfort. In this case the advantage, of course, results from the data collection being already concentrated on those factors which have been previously identified to be important.

Already in the early stages criticism arised that the behavioural mode-split-models have not been derived from a theory on transport mode choice, but instead, a theory has been more or less adapted to the model (see Stopher and Meyburg, 1976). Therefore it may not be astonishing that the postulations on processes of behaviour underlying the model design are not correct in several points. This critical discussion does not affect the attitudinal models in the same way: The attitudinal theory, however, as theoretical basis may not be sufficient for explaining the transport mode choice, because it provides no assumptions on the perception of transport mode attributes and also no assumptions for the actual decision process. For the behavioural as well as for the attitudinal model designs three aspects of complexity are equally problematic.

- In view of explaining behaviour the first critical assumption refers to the principle of utility maximisation. Human beings, as assumed accordingly, are striving to achieve the highest possible utility when making behavioural decisions. This postulation, in recent years, has increasingly led to discussions, after Simon already in the fifties has developed his theory on a satisficing decision outcome (Simon, 1955). According to Simon's theory a behavioural alternative does not need to provide an optimal utility, instead it should satisfy some acceptable standard in order to be selected. Conjunctive and disjunctive decision rules do correspond to this thinking. But it is also contained in Tversky's elimination-by-aspects model (Tversky, 1972). In addition these ideas have already penetrated some studies of transport science (see Gensch and Svestka, 1978; Foerster 1979, 1981; Recker and Golob, 1979).

- In most behavioural travel models it is assumed, that the subjective utility of a transport mode alternative does linearly relate to the attributes of this alternative. This contradicts as well to the results of psychophysics and to the Gestalt theory of psychology as to recent knowledges on thresholds of acceptance in transport science (see Held, 1982). By applying nonlinear utility functions for time and cost Koppelman achieves a better fit of data than by linear functions (Koppelman, 1981).

- Additionally there is the crucial assumption of combining the single attributes of a transport alternative into an overall utility according to an additive pattern. Again this assumption contradicts the psychological theory by which the content of perception and memory rather combine to specific configurations than being seperated in single elements. So, there are much more configurations which determine behaviour. Systematically analyzing the combination of factors in judgement on the utility of transport mode alternatives as well as in travel behaviour, typical patterns of interactions repeatedly have been discovered (see e.g. Lewin, 1979; Louviere, Wilson and Piccolo, 1979). With the design of sequential decision models the idea of the non-additive combination of transport mode attributes has been taken into account (e.g. Hensher, 1983; Gensch and Svestka, 1978; Recker and Golob, 1979; Foerster, 1979). The following investigation also puts the emphasis on the analysis of non-additive combination rules.

2. The Acceptance-Approach

2.1 Basic Formulation and Important Terms

The approach represented by us seeks to avoid the problems of additive combination and linear relation by using no a-priori defined functions.

The ACCEPTANCE-APPROACH uses subjective ratings as a basis relating to behaviour as well as to objective qualities of transport services. In a formal way this can be expressed in the following paradigm:

$$R_i = f(C_1, C_2, C_3, \ldots C_n) \quad (1)$$

$$B = g(R_1, R_2, R_3, \ldots R_m) \quad (2)$$

where: R = subjective rating

C = objective characteristic of the transport system

B = mode choice

The alternative transport options and the situation connected to their use are considered as a complex of characteristics, i.e. they can be described by the naming with characteristics and with combinations of characteristics. This approach was initially formulated by Lancaster (see Lancaster, 1966) within the context of utility theory.

In the following there is made a differentiation between dimensions of characteristics - like the walk to the bus stop or the in-vehicle-time, etc. - and the values of these dimensions, like e.g. a walk with 10 minutes.

Ratings

While characteristics describe the objective conditions of the transport options and the situation of their use, ratings, however, describe their subjective representation through the eyes of the users and potential users. In the ratings the processes of perception and evaluation find their expression.

The entirety of the ratings related to a transport mode can be qualified as an overall attitude toward the transport mode (similar to Fishbein, 1963, 1967; Wilkie & Pessemier, 1973). Ratings are complex in different ways. At least at a superficial consideration some of them may be directly connected with objective characteristics - e.g. the rating of the walk to a bus stop - , others not - like the rating of relaxation while using the bicycle.

More detailed reflection however suggest that the more complex ratings as well as the less complex ones are built up by the same principle (see the following section). As in the case of the characteristics, here again, the dimensions of the ratings and their values will be differentiated. The values of the rating dimensions are measured with subjective scales (e.g. walks are long or short).

2.2 Subjective Ratings and Objective Characteristics

Ratings are derived from perception and evaluation of an environmental situation. They are influenced by the objective condition of the environment.

With the considerations underlying the analysis of this relationship the fundamental idea of the "Gestalt"-theory (psychology of perception) plays an essential role. According to this hypothesis the environmental situations are perceived as entireties. The entireties are structured, i.e. the single characteristics of an object perceived have determinable relations to each other. Concerning the ratings this means that a rating dimension does not only depend on its corresponding objective case, but may also be influenced by a different kind of characteristics. This holds as well for dimensions which appear to be objectively determinable, like e.g. the walk time, as for such dimensions which at the first glance do relate to more than one characteristic - like the rating of relaxation. Beyond that the hypothesis suggests non-additive relations to be expected rather than additive ones.

In this presentation the relation between subjective ratings and objective characteristics of the transport system will not be investigated. Therefore some of our results concerning this question will be summarized in advance.

- Ratings may be specified as functions on the conditions of the environment. This is not only pertaining to characteristics being easily measurable, like in-vehicle-time and walking time, but also to characteristics of transport links being quantitatively less measurable. So the rating on relaxation of riding the bicycle significantly depends on the number of cross sections to be passed and additionally on further design characteristics of the cycle path (see Wrede, 1980; Martens and Verron, 1983b).
- The relation of ratings to objective characteristics of the transport system is in most cases a non-linear one. The non-linearity is less represented by curvilinear relations than by thresholds. Thus, e.g.

concerning the walk from home to bus stop there is an obvious threshold influencing ratings at the objective time value of five minutes (see Waschke, 198o).

- Ratings are dependent on more than one objective dimension while the kind of combination of different objective dimensions into the subjective rating is usually non-additive. The results prove multiple determination not only to be effective for the more "qualitative" factors (e.g. relaxation while riding a becycle) but also for the quantitatively determinable causes of behaviour, like travel time and cost.

2.3 Subjective Ratings and Behaviour

With the interpretation of the second formal equation concerning the dependency of behaviour on the ratings, similar deliberations prevail as in equation 1. Here again, according to the theory of information integration, combination rules have to be identified, because it stands no reason that an additive combination rule has to be attributed to the joint behavioural effects of the rating dimensions.

The relating results will be presented in the second part of this paper after having considered some methodical questions at first.

3. Statistical Method of Analysis

For the data analysis a tool should be applied which is flexible in view of the theoretical considerations and which does not narrow the analysis by restrictive assumptions. Computerized procedures of the binary segmentation do meet these requirements. They facilitate an explorative approach being able to discover data patterns helping to define the structure of relations betweeen variables. The underlying principles of these procedures may be stated as following:

There is an investigation on a group of individuals or objects, each one being described by a set of variables $(y_1, y_2, \ldots, y_m; x_1, x_2, \ldots, x_p)$. The y_i will be considered as dependent variables and the x_i as independent variables or predictors. Each of these variables may be measured on any level of scale. In many studies of interest there is only one dependent y. The set of values belonging to each variable is described by Y_i or X_j.

All techniques of the binary segmentation procede as following. At first the total sample will be divided into two subgroups and to the resulting subgroups the same process will be sequentially applied. However, on each step, decisions have to be made:

- which predictor has to be included for the next split, and
- how two subsets have to be derived form the predictor values.

The fundamental strategy aims to check each predictor x_j and to choose such a split from its set of values, that the generated binary split results into two subgroups, which maximize the "distance-function" being selected. The next step decides on this variable having the highest maximum among all variables. For the y_i this may be considered as the "best explanation" having been accomplished just by one split into two groups.

This process will be repeated by sequentially splitting the subgroups resulting from the stage before.

The result of the binary segmentation is a hierarchical tree with subgroups.

The algorithm applied here presumes a dependent variable measured on an interval scale. Formalizing furthermore the following notation will be needed:

- $\bar{Y}$ is the mean value of the parent group of size n.
- $\bar{Y}_1$, $\bar{Y}_2$ are the mean values of the subgroups of a binary split - of size n_1 and n_2 - resulting from a segmentation of the values of a specific predictor x_j.
- Y_{1k} $(k = 1, 2, \ldots, n_1)$, Y_{2k} $(k = 1, 2, \ldots, n_2)$ are the observations on the dependent variable in each group.

Herewith three terms may be defined:

- $$n_1(\bar{Y}_1-\bar{Y})^2 + n_2(\bar{Y}_2-\bar{Y})^2 = \frac{n_1 n_2}{n}(\bar{Y}_1-\bar{Y}_2)^2 \qquad \text{(a)}$$
- $$\sum_{k=1}^{n_1}(Y_{1k}-\bar{Y}_1)^2 + \sum_{k=1}^{n_2}(Y_{2k}-\bar{Y}_2)^2 = \sum_{l=1}^{2}\sum_{k=1}^{nl}(Y_{lk}-\bar{Y}_l)^2 \qquad \text{(b)}$$
- $$\sum_{l=1}^{2}\sum_{k=1}^{nl}(Y_{lk}-\bar{Y})^2 \qquad \text{(c)}$$

The term (a) is the weighted sum of squares of the deviations of the subgroups means from the parent group mean, (b) is the sum of the sub-group-heterogeneities and (c) the heterogeneity of the parent group.

The heterogeneity is defined as the variation criterion of the analysis of variance: the sum of squared distances of individual Y - values from the mean value. The smaller this value is, the more homogeneous a group will be. The distance between the subgroups being maximized is the usual sum of squares between groups for two subgroups, i.e. the weighted sum of squares of the deviations of the subgroups means from the parent group mean. In this algorithm the binary split being actually generated will be this one, which maximizes (a) and minimizes (b) the distance criterion between groups - (a) and (b), of course, are equivalent in their use as a distance criterion because their sum is constant and as big as the parent group heterogeneity, i.e. as the total sum of squares

This distance criterion measures the explanatory power of a binary split on each step, i.e. it is equivalent to the sum of squares between groups ($SS_{between}$) in connexion with the analysis of variance. The division of $SS_{between}$ (a) by the total sum of squares (SS_{total}) or by (c) results into Fisher's univariate correlation ratio for one split:

$$eta^2 = \frac{SS_{between}}{SS_{total}}$$

By maximization of the distance function the portion of variation in Y being explained by the binary split will be maximized simultaneously.

4. Empirical Analysis

4.1 Data

For the data survey in two city districts of Berlin (West) - Tiergarten and Neukölln - questionnaires were handed out in 1979. About 850 people being employed were questioned concerning their work trips. Besides others, the questionaire contains questions on

- transport mode use
- car ownership
- social conditions
- qualities of service of the transport modes being used
- qualities of service of the public transport modes from the point of view of the car users
- ratings on the characteristics of all transport mode alternatives

The ratings relate to numerous characteristics, describing the single transport modes and the conditions resulting from their use. They are measured on a 7-point-scale with the poles "very negative" and "very positive". These ratings supply the basis for the analysis of behaviour.

4.2 Descriptive Results

Already simple empirical analyses of the information being at hand do provide useful support for planning activities. They are especially able to stimulate and promote the interest for planning and to direct it to relevant and more specific targets. The public opinion represented by the subjective information enables a more impressive insight into the existing state of affairs than the mere presentation of objective conditions.

45 % of the respondents use the car and 35 % public transport modes. 7,5 % are walking and only o,5 % are using the bicycle.

The statements about the quality of service already demonstrate some causes underlying transport mode choice.

For public transport users in Berlin in 1979 there are mainly two transport modes available: the bus and the subway system. Number of transfers reported by public transport users for their work trips clearly shows, that subway users transfer more frequently than bus users. But concerning the transfer conditions, measured on transfer time and transfer waiting time, they experience much more favourable conditions. Besides the unfavourable transfer conditions the headway times are considerably higher for the bus users than for the subway users. The transfer times and headway times being comparatively so unfavourable do not result from system attributes of a bus. It rather demonstrates, that the subway gets priority with respect to operational planning.

On the other side walking times to the subway station are longer than to the bus stops. This, of course, is a system attribute of the subway network not being able to be refined much more.

Car users face very favourable conditions on their car trips. Especially obvious are the short walking times from the parking lot and the low searching times for parking. Only the time regularly spent in traffic jams is relatively high. For being able to compare the public transport's quality of service for car-users and public transport users and to arrive at a conclusion for the transport mode choice, in the following "chooser" and "non-chooser" will be differentiated:

- "car-choosers" are car users not using a company car on their work trips and not needing their car for occupational purposes.
- "Public transport choosers" are public transport users having a driver's license and at least one car in their household.

In the sample there are 235 car chooser (27 %) and 77 public transport choosers (9 %). Compared to the public transport users much less "car choosers" have direct links available and much more "car choosers" have to transfer twice on their public transport links. Also the "car-choosers" notice longer headway times and longer waiting times.

In order to discover some of the reasons why the share of bicycle users is almost reduced to zero on the work trips in Berlin (West), ratings on bicycle riding in Berlin were compared to equal questionnaire items in Erlangen (1oo ooo inhabitants, 15 % bicycle users on work trips). The results demonstrate bicycle riding in Erlangen being more favourably rated on almost all dimensions. Especially obvious are the differences of ratings on accident risks, on recreation, on noise and car-pollution, on safety and on weather protection of parking facilities for bicycles.

4.3 Subjective Ratings and Behaviour

This section presents the results of analyzing the effects of subjective ratings concerning car or public transit characteristics on the use of either car or public transit for the journey to work.

The analysis of behaviour was only performed on car and public transport "choosers". On grounds of failing responses the sample reduces to n = 216. The theoretical concept as well as the empirical evidences so far suggest the assumption that only the joint effects of several relevant factors of influence allow for reasonable conclusions at the single effects involved. The transport users do evaluate the quality of service as an entity, so that the concerning single effects of subjective ratings do vary in combination with ratings on different qualities of service. The underlying combination rule does not necessarily follw an additive pattern.

At the first step a basic model will be developed to examine the joint effect of single travel time components of the public transport. Empirical results so far reveal, that an appropriate combination of single components of travel time from door to door creates a considerable contribution to the explanation of variance in the users' behaviour (see Martens, 1981; Martens and Verron 1983a, Martens 1983).

In the two following steps additionally to the ratings of the basic model some further ratings on different qualities of service will be included. In the second step it will be the weather protection at public transport stops, rather a characteristic of comfort, and in the third step it will be the rating on searching time for a parking lot exerting a strong influence in the disjoint analysis of single effects performed before.

According to former results the importance of the quality of service will be assumed to vary over different travel distances. Thus in all models ranges of travel distances differentiate "up to 5 km", "6 to 15km", and "16 to 25 km".

For the basic model besides the travel distance the subjective ratings on walking times from home to public transport stops and from stop to workplace, on waiting times, on number of transfers and on transfer waiting times have been included into the analysis. The model result (see diagram 1) explains for about 3o % of the behavioural variance. This amount is by far higher than public transport travel time not being seperated into its single components is able to explain. In principle the result makes clear, that an area oriented public transport service increases the attractiveness and the use of public transport most favourably, or on the other side, a public transport service being line oriented and thus demanding more frequent transfers will not create the expected effects. The model result provides excellent insight into the structure of the effects available. It confirms the theoretically derived assumption, that the incidence of certain effects and their importance depends on the concerning context. The appropriate combination of subjective ratings determines the entire picture of the quality of service leading to a specific user behaviour.

In the middle range of travel distances (6 to 15 km) unfavourable ratings on the egress walking time from public transport stop to workplace completely result into car use. Unfavourably rated access walking times from home to stops and favourable egress walking times allow also far a high car use of about 9o %, however not for the extremely high car use generated solely by unfavourable egress walking times. Favourable egress walking times and favourable access walking times reduce the car use down to 71 %, which is still decreasing to 5o % in the case of direct public transport links. In the case of transfers the joint effect again increases car use up to 81 %. Herewith the public transport effects resulting from favourable egress and access walking times are not only substituted but reduced to a far greater extent. In this context the transfers create an obvious car effect, of course not pushing the car use up to a level having been generated by unfavourable egress walking times or access walking times respectively. If in case of transfers, however, there are also negatively rated transfer waiting times, then again there is an exclusive car use. On the other side positively rated transfer waiting times are able to highly compensate the car effect caused by necessary transfers. In this case the car share does not exceed the amount created by the favourable combination of egress and access walking times.

While in the range of short travel distances up to 5 km the subjective ratings on egress walking times are dominating, especially the ratings on transfer waiting times do so in the range of longer travel distances over 15 km.

In order to measure the effects of policies realistically, an aggregate indicator has been developed taking into account the qualities of service being availiable (Martens, 1981). The absolute change of the probability of use ΔP derives from the following:

$$\Delta P = (P_i - P_j)\ ni/n$$

ΔP = change in the probability of use

P_i = probability of use in the situation i

P_j = probability of use in the situation j

n_i = sample size in the situation i

n = total sample size

The potentials of change being calculated demonstrate on the one hand that positively changed ratings on egress walking times and direct public transport links without transfers are able to achieve the highest substitution in favour of public transport, and on the other hand that effects of egress walking times are different according to the concerning range of travel distances.

Besides it may be recognized that combined policies, i.e. the joint improvement of several ratings, do result in higher effects than the mere summation of the single effects. Favourable ratings on egress and access walking times in a joint effect together with direct public transport links change the probability of car use by an absolute amount of 18 %. The summation of the single effect contributions achieves only 12 %.

In the second model the subjective rating on weather protection at public transport stops is included in addition to ratings of the basic model. As the model results show (see diagram 2) the effect of this rating, being very strong, pushes off some of the already known effects of the basic model. Being especially obvious, the effect due to the number of transfers is not present any more. The more specific conditions of transfer, however, being characterized by the transfer waiting time are still present in the model. By involving weather protection the proportion of explained variance raises to 35 %, which is much more than in the basic model. But still the principal conclusion from the basic model remains concerning the advantages of an area oriented service.

The importance of weather protection for explaining the behavioural variation may also be demonstrated by the potential changes. Joint effects by favourable ratings on weather protection, on access and egress walking times, and on transfer waiting times turn the absolute amount of the behavioural change to about 24 % being much more than the 13 % resulting from summation of the concerning single effects.

In the third model, instead of weather protection, the ratings on searching time for a parking lot is analyzed. As already mentioned before, this rating among all other ratings contributes most highly by 22 % to the explanation of variance according to the analyses of seperate effects in the total sample. In comparison to this result, the concerning importance decreases while being effective in combination with the subjective ratings of the basic model: the essential conclusion of which still remains firm in principle (see diagram 3).

Especially in the range of longer travel distances over 15 km the behavioural variations may be much better explained by the searching time for a parking lot. In total, however, its consideration does not account for more explanation of variance than is explained by including the subjective rating on weather protection instead. Concerning the potential behavioural changes the joint effect by unfavourable ratings of searching time for a parking lot and favourable ratings on access and egress walking time - combined in an adequate pattern - proves to be by far superior to the sum of the concerning single effects.

The last test taking into consideration the different ranges of travel distances investigates the joint effects of comfort characteristics of the public transport system. Weather protection at public transport stops is a measure being rather inexpensive, getting on and off the public transport vehicle as well - it may be modified by a reasonable standardization of the vehicles. The other characteristics, like convenient purchaseof tickets, sufficient passenger information, personal safety do need more transport personnel if these measures are required to be useful. A relatively high amount of explanation of variance by 24 % underlines the importance of these characteristics (see diagram 4).

4.4 Conclusions

Attractiveness of the public transport use may be increased by appropriate combinations of measures. Among the effective improvements those measures reducing walking times, waiting times as well as the number of transfers play a predominant role.

Insofar it may be concluded that the enlargement of a line oriented service by subway or rapid transit systems - and its resulting increases in walking times and in number of transfers - will be a failure if the attractiveness of the public transport shall be improved. An area oriented public transport service, not being essentially biased towards reduction of in-vehicle-time, does meet the needs of the transport users much more.

The result of this study in Berlin even gets higher importance, as it repeatedly demonstrates what has been discovered with different data (see Martens, 1983) and in different study areas (see Martens and Verron, 1983 b).

The importance of the single travel time components being relatively stable under different circumstances allows for the conclusion, that these are the bases for an acceptable public transport system - at least through the eyes of the users. In this context it should be underlined, that the travel time components can only be compensated among each other to a very limited amount, and moreover that the quality of the public transport system rather consists of the combination of favourable conditions.

Additionally it has been revealed that on the basis of an area-oriented public transport service further improvements may be achieved by different measures. These alternative measures have relatively similar effects, so the planners in the end may decide on cost considerations, etc...

But also, leaving the standard of an area oriented service aside, with simple and moreover low cost measures improving the users comfort, an obvious increase of attractiveness may be achieved.

References

Fishbein, M.:An investigation of the relationships between beliefs about an object and the attitude toward that object. In: Human Relations, 16, 1963

Fishbein, M.: A behavior theory approach to the relations between beliefs about an object and the attitude toward the object. In: M. Fishbein (Ed.): Readings in attitude theory and measurement. New York, 1967

Foerster, J.F.: Mode choice decision process modells: A comparison of compensatory and non-compensatory structures. In: Transportation Research, 13A, 1979, 17-28

Foerster, J.F.: Nonlinear and noncompensatory perceptual functions of evaluation and choice. In: P.R. Stopher, A.H. Meyburg, W. Brög (eds.): New horizons in travel-behaviour research. Lexington, 1981

Gensch, D.H. & Svestka, J.A.: An exact hierarchical algorithm for determining aggregate statistics from individual choice data. University of Wisconsin, Milwaukee, unpublished paper, 1978

Held, M.: Verkehrsmittelwahl der Verbraucher. Berlin, 1982

Hensher, D.A.: A sequential attribute dominance model of probabilistic choice. In: Transportation Research, 17A, 1983, 215-218

Hensher, D.A. & McLeod, P.B.: Towards an integrated approach to the identification and evaluation of the transport determinants of travel choices. In: Transportation Research,11A, 1977, 77-93

Koppelman, F.S.: Non-linear utility functions in models of travel choice behavior. In: Transportation, 1o, 1981, 127-146

Kunert, U.: Untersuchungen zum Pkw-Besitz. Diplomarbeit am Fachgebiet Integrierte Verkehrsplanung der Technischen Universität Berlin. Berlin, 1981

Lancaster, K.J.: A new approach to consumer theory. In: Journal of Political Economy, 1966, 132-157

Levin, I.P.: The development of attitudinal modelling approaches in transport research. In: D.A. Hensher & P.R. Stopher (eds.): Behavioural travel modelling. London, 1979

Louviere, J.J., Wilson, E.M., Piccolo,J.M.: Psychological modelling and measurement in travel demand: A state-of-the-art review with applications. In: D.A. Hensher & P.R. Stopher (eds.): Behavioural travel modelling. London, 1979

Martens, G.: Öffentliche Verkehrsbedienungsqualitäten und Verkehrsmittelbenutzung - Modellentwicklung mit Volkszählungsdaten. In: Verkehr und Technik, 1981

Martens, G.: Verkehrsmittelbenutzung und Bedienungsqualitäten. Ein Modell mit Volkszählungsdaten. In: Schriftenreihe des Instituts für Verkehrsplanung und Verkehrswegebau der Technischen Universität Berlin, Heft 9, Berlin, 1983

Martens, G. & Verron, H.; Einflußfaktoren auf die Verkehrsmittelbenutzung. Forschungsbericht im Auftrag der Daimler-Benz AG. Berlin, 1983 (a)

Martens, G. & Verron, H.: Akzeptanz von Nahverkehrssystemen. Maßnahmenorientierte Nachfrage-Wirkungen für den Busverkehr in Erlangen. In: G. Martens & H. Verron (Hg.): Verkehrsplanung und Verhaltensforschung. TUB-Dokumentation Kongresse und Tagungen, Heft 18, Berlin, 1983 (b)

Nicolaidis, G.C.: Quantification of the comfort variable. In: Transportation Research, 9, 1975, 55-66

Recker, W.W. & Golob, Th.F.: An attitudinal modal choice model. In: Transportation Research, 1o, 1976, 299-31o

Recker, W.W. & Golob, Th.F.: A non-compensatory model of transportation behavior based on sequential consideration of attributes. In: Transportation Research, 13B, 1979, 269-28o

Simon, H.A.: A behavioral model of rational choice. In: Quarterly Journal of Economics,69, 1955, 99-118

Stopher, P.R. & Meyburg, A.H.: Behavioral travel-demand models. In: P.R. Stopher & A.H. Meyburg (Eds.): Behavioral travel-demand models. Lexington, 1976

Tversky, A.: Elimination by aspects: A theory of choice. In: Psychological Review, 79, 1972, 281-299

Waschke, Th.: Untersuchungen zum Prozeß des Verkehrsverhaltens am Beispiel der Verkehrsmittelbenutzung in Erlangen unter Berücksichtigung der Einstellungen von Busbenutzern. Diplomarbeit am Fachgebiet Integrierte Verkehrsplanung der Technischen Universität Berlin. Berlin, 198o

Wilkie, W.L. & Pessemier, E.A.: Issues in marketing's use of multiattribute attitude models. In: Journal of Marketing Research, 1973, 428-441

Wrede, M.: Untersuchungen zum Fahrradverkehr in Erlangen. Ein Beitrag zur einstellungsorientierten Verkehrsplanung. Diplomarbeit am Fachgebiet Integrierte Verkehrsplanung der Technischen Universität Berlin. Berlin, 198o

Diagram 1
Modell 1 Results

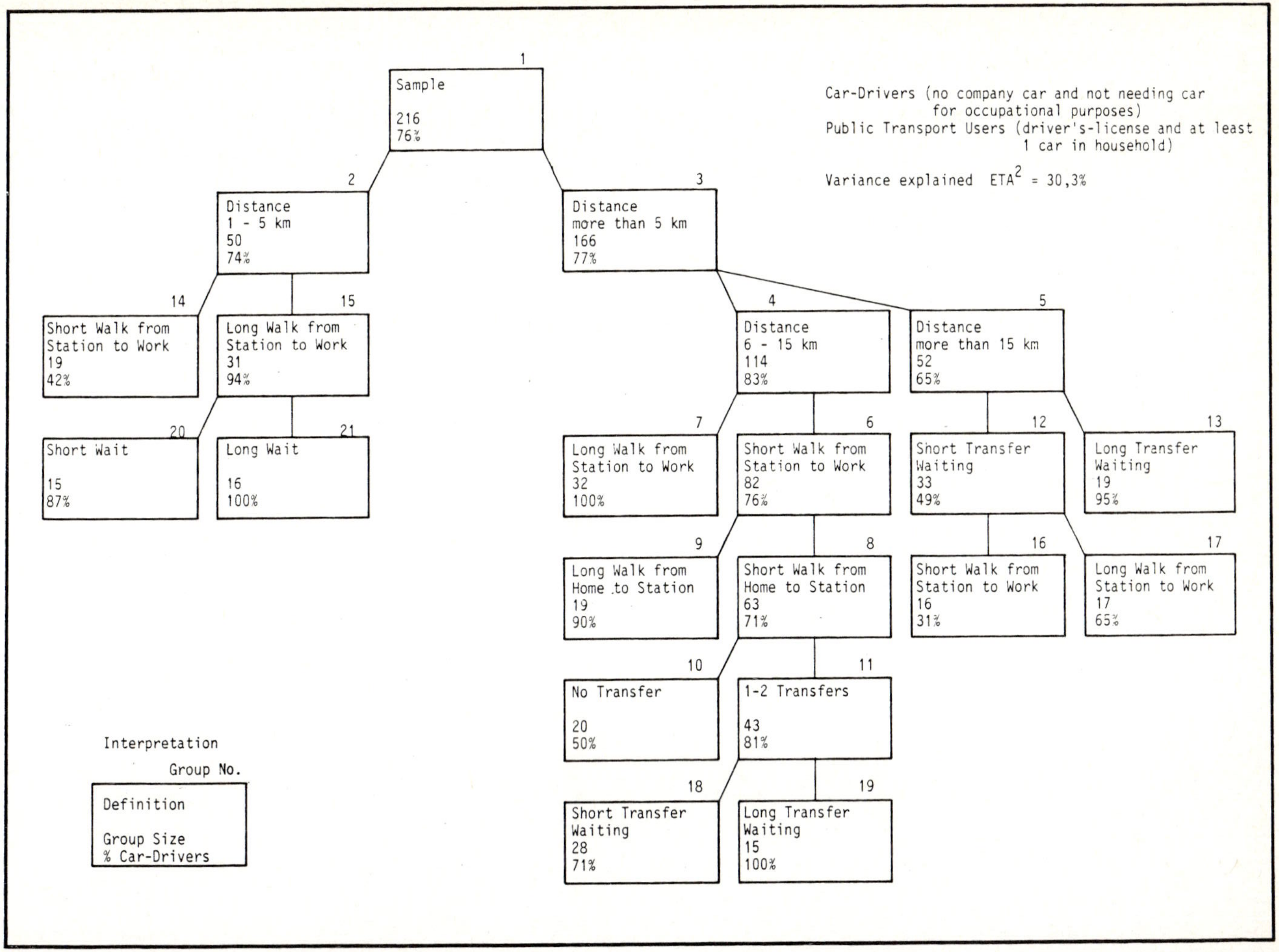

Diagram 2

Modell 2 Results

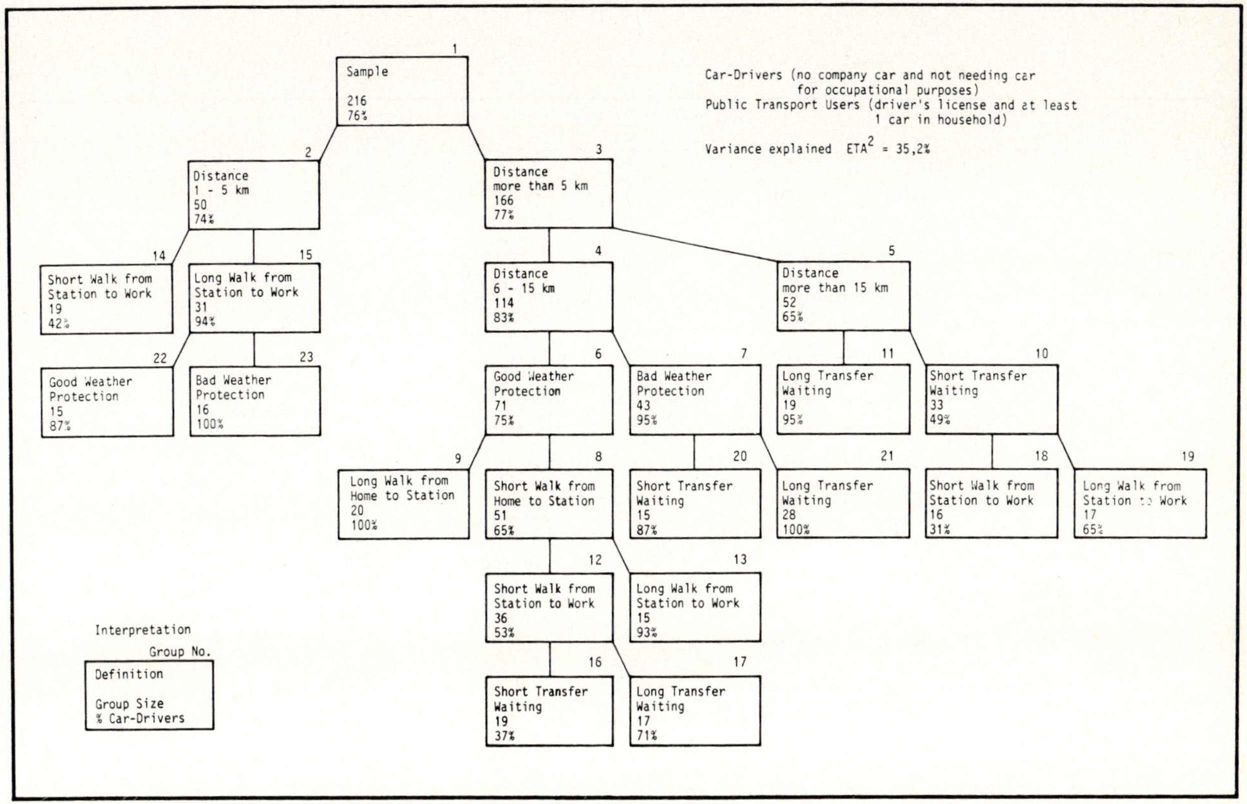

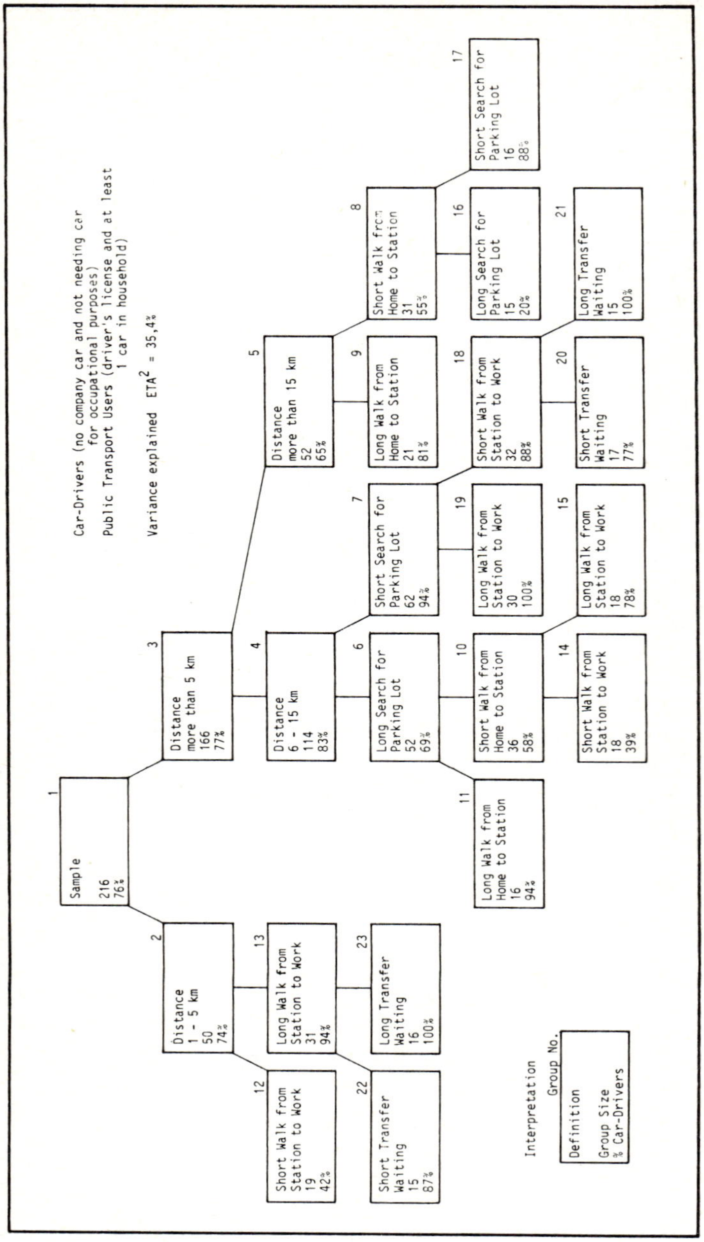
Diagram 3
Modell 3 Results
Car-Drivers (no company car and not needing car for occupational purposes)
Public Transport Users (driver's license and at least 1 car in household)
Variance explained ETA² = 35,4%
1 Sample 216 76%
2 Distance 1 - 5 km 50 74%
3 Distance more than 5 km 166 77%
4 Distance 6 - 15 km 114 83%
5 Distance more than 15 km 52 65%
6 Long Search for Parking Lot 52 69%
7 Short Search for Parking Lot 62 94%
8 Short Walk from Home to Station 31 55%
9 Long Walk from Home to Station 21 81%
10 Short Walk from Home to Station 36 58%
11 Long Walk from Home to Station 16 94%
12 Short Walk from Station to Work 19 42%
13 Long Walk from Station to Work 31 94%
14 Short Walk from Station to Work 18 39%
15 Long Walk from Station to Work 18 78%
16 Long Search for Parking Lot 15 20%
17 Short Search for Parking Lot 16 88%
18 Short Walk from Station to Work 32 88%
19 Long Walk from Station to Work 30 100%
20 Short Transfer Waiting 17 77%
21 Long Transfer Waiting 15 100%
22 Short Transfer Waiting 15 87%
23 Long Transfer Waiting 16 100%
Interpretation
Group No.
Definition
Group Size
% Car-Drivers

Diagram 4

Modell 4 Results

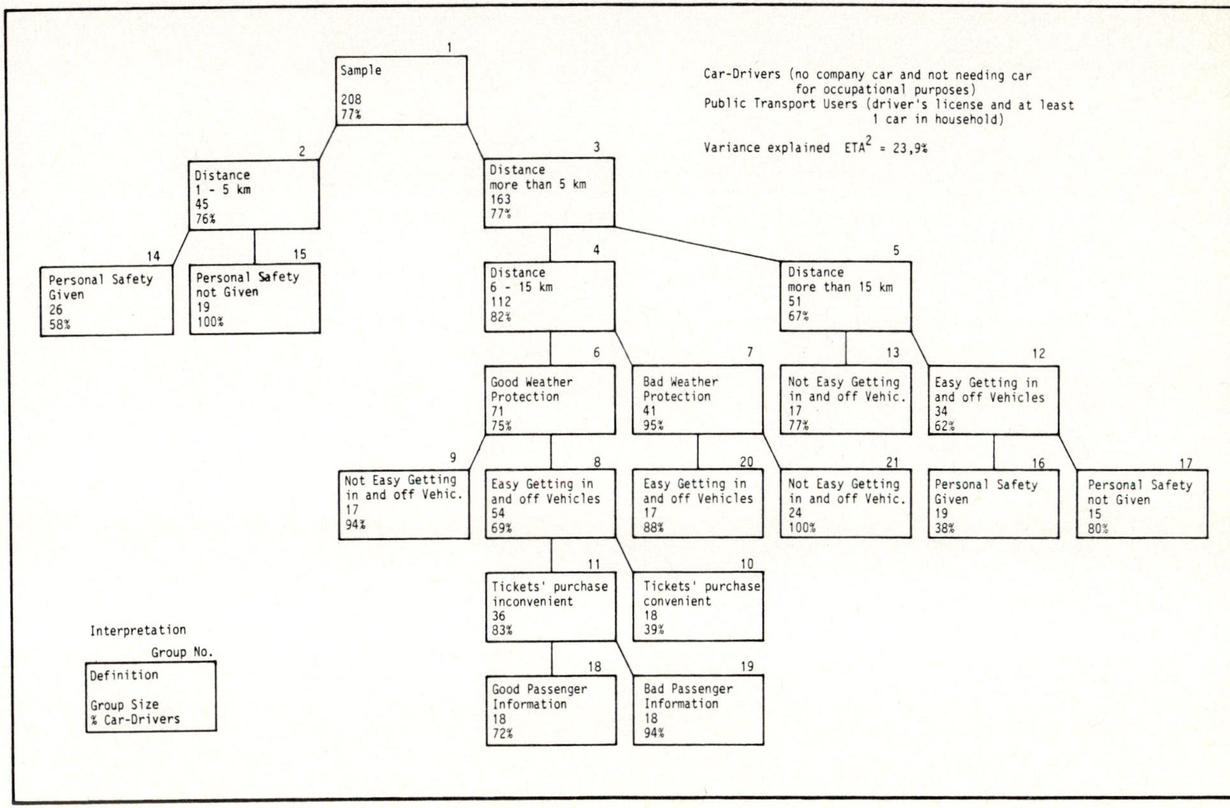

Behavioural Research for Transport Policy
pp. 153–170. © 1986 VNU Science Press.

The Role of Thresholds in Transport Choice

by

William Young

ABSTRACT

Transport policy often desires to alter the transport system to encourage or discourage the use of particular components of the system. These changes may not produce the desired effect if they are not noticed by the target population or they do not represent a significant change from the previous situation. Thresholds of awareness and acceptance have often been mentioned in the transport literature in this context. However few studies of transport related decisions have incorporated them into a modelling framework. In this paper the role of thresholds in transport choice models is discussed. The paper opens with a review of threshold types and their presence in several study areas. It then presents a discussion on the components of the transport decision process highlighting the possible roles of thresholds and their contribution to understanding decision making. The paper also outlines the historical development of the incorporation of thresholds into choice models. The breadth of application and the performance of these models, when compared with traditional choice models, is highlighted.

INTRODUCTION

Transport planners have developed a variety of statistical techniques for analysing transport decisions (Quarmby 1967; Stopher 1969; McFadden 1981; Daganzo 1980). The common feature of all these models is that choice is seen as a function of the indirect utility gained from each alternative. To calculate this utility it is assumed that an alternative is characterised by a set of attributes that contribute to a measure of total utility. Combining attribute utilities into a measure of total utility can take many forms. However most studies assume that attribute utilites are a continuous function of the satisfaction gained from each attribute. That is, every change in satisfaction no matter how small will influence the utility gained from an alternative and hence an individuals choice.

Evidence in the Psychological (Guilford 1954; Stevens 1962), Economics (Slutsky 1952) and Biology (Finney 1964) literature suggests that people may be indifferent to changes in the stimulus unless it crosses a threshold of indifference. In the transport literature this suggestion has found support in several studies on the development of transport choice models (Hensher 1972; Kavak and Demetsky 1975; Burns, Golob and Nicolaidis 1975). Particular concern was stated when it was found that models that did not incorporate indifference thresholds tended to overestimate mode shift for small changes in attribute satisfaction. It was suggested that the inclusion of thresholds of indifference may overcome this problem since they would tend to dampen the effect of small changes in attribute satisfaction. This paper investigates the incorporation of thresholds into transport choice models.

The paper is divided into three main sections. The next section describes a variety of threshold types and their application in a number of disciplines. The third section describes the areas where thresholds may be incorporated into the individuals transport choice process. The paper then describes the historical development of thresholds in transport choice models. Comparison of the performance of these models and the more traditional transport choice models are made where possible. The paper closes with suggestions for future research and some concluding remarks.

BACKGROUND

The assumption which is made most often in traditional models of transport choice is that the transformation between objective attribute levels and pervceived attribute levels is one of perfect equality. However, a long history of psychological research has indicated that very few, if any, attributes have demonstrated such a linear relationship.

Fechner (1860) in his classic work on the relationship between physical and psychological states proposed the following functional relationship:

$$R = \log S \qquad (1)$$

which has since been generalized by Luce and Galanter (1963) as:

$$R = a \log (b + S) + c \qquad (2)$$

where R = response (or perception)

S = stimulus (or objective state)

a,b,c = constants

The constant c indicates that an <u>absolute threshold</u> may exist. This threshold separates stimulus levels which elicit a response from those which do not elicit a response. This absolute threshold may be either a minimum or a maximum level.

In a transport context, the existence of lower absolute thresholds would appear to be of some relevance. For example, the perceived cost of using a car may be zero over a considerable range of actual costs. Also, trips below a certain length may not be perceived, or reported, as trips.

Similarly, the existence of upper absolute thresholds may be of significance. For instance, in residential location choice the trip of the main breadwinner may be acceptable up to a given distance. Once this distance is exceeded the trip may become onerous and hence have a strong influence on the location choice. Evidence for this hypothesis has been provided by the Highway Research board (1969), Richardson (1971) and Guest and Cluett (1976). The Highway Research Board's report on moving behaviour and residential choice concludes that the "accessibility to a number of regular, out-of-home activities, including workplace was found to be a relatively unimportant factor in household residential mobility and in a household's choice of new residence". The report does however, state that households living at more than 40 minutes distance from work have a greater tendency to move than households that lived nearer to work. Richardson (1971), contends that home to work expenses are considered only in determing an outer constraint to location choice. Within this constraint the residential environment, household quality and cost as well as the accessibility to other facilities plays a major part in influencing location choice. Guest and Cluett (1976) confirm this finding in their study of the interrelationship between home and workplace location decisions.

The second psychophysical function of importance to choice theory concerns thresholds on changes in attribute levels (as distinct from thresholds on absolute levels). The existence of such thresholds was first recognized by Weber in the 19th century and they are described by Guilford (1954) and Bock and Jones (1968). The basic theory states that changes in the level of an attribute which are smaller than the threshold will not be perceived by an individual.

It is obvious that the jnd will be a function of the magnitude of the initial stimulus. For example, a five minute change in a twelve hour trip would, most likely, go unnoticed whereas a five minute change in a ten minute trip would be noticed by a large proportion of the population. This obvious feature of jnd's can be expressed by Weber's Law as

$$\text{jnd} = KS \tag{3}$$

where K is a constant termed the Weberian fraction. Like Fechner's law (eqn. 1), Weber's Law has also been modified over the years to better match observed data. Thus Miller (1947) postulated the following equation to allow for changes in origin,

$$\text{jnd} = a + KS \tag{4}$$

whilst Guilford (1954) proposed an even more general form to include a power function,

$$\text{jnd} = a + KS^b \tag{5}$$

The just noticeable difference concept has been incorporated into a number of studies of transport choice. Foerster (1979) incorporated an absolute value of jnd in his studies. Recker and Golob (1979), Gensch and Svestka (1979) and Young (1982) combined the just noticeable difference concept and the magnitude of response to derive an acceptance threshold of the form

$$\text{Acceptable} \quad S_{ikj} \geq (1 - T_k) \max_j S_{ikj} \tag{6}$$

where S_{ikj} = satisfaction with the kth attribute of the jth alternative for the ith individual

T_k = tolerance for the kth attribute

$\max_j (S_{ikj})$ = Maximum satisfaction with the kth attribute for the ith individual over all j^{th} alternatives

One complication which must be recognised in attempting to apply Weber-Fechner relationships to transport attributes is that these psychophysical laws were derived in situations involving the presentation of stimuli in close temporal proximity. Thus in determining jnd's, for example, the changes in attribute level were made almost instantaneously whilst the experimental subject performed the test under laboratory conditions. The gathering of information on travel alternatives, however, is usually performed over long time periods. Thus the comparison of alternatives is based on information which may have been gathered some considerable time previously (and not necessarily at the same time for all alternatives). The perception of attributes is therefore subject to the decision-makers ability to recall these perceptions from the past. Whilst a considerable amount of research has been performed on the ability of subjects to recall items from long-term memory (Ebbinghaus, 1885) and short-term memory (Peterson and Peterson, 1959), the retention of information on transport system attributes has not been subject to similar scrutiny.

The existence of thresholds has therefore been acknowledged in the psychology literature for many years. There has also been discussion of thresholds in other disciplines.

For instance, economists (Slutsky 1952; Georgescu 1958) introduced the notion of thresholds of insensitivity into the theory of consumer behaviour about thirty years ago. Their application of indifference curves considered that in the choice between two commodities the decision maker will choose one or the other or be indifferent. If the decision maker is indifferent he will tend to randomise his decision. Indifference curves define all situations where the consumer is indifferent.

Similarly in management science Simon (1957) introduced the idea that people and firms are satisficers not utility maximisers. He hypothesised that individuals attempt to achieve a minimum acceptable level of utility and do not try to obtain a higher level once the minimum is realised. Simon's theory embodies the realisation that:

> "The capacity of the human mind for formulating and solving complex problems is very small compared with the size of the problems whose solution is required for objectively rational behaviour in the real world - or even for a reasonable approximation to such objective rationality"
>
> (Simon, 1957, p.198)

Finally biological experiments indicate that thresholds varied between subjects. They hypothesised that the distribution of thresholds is normal. The resulting relationship between the response and stimuli was therefore described by a probit model (Finney 1964). Finney (1964) also considered other relationships for the form of the threshold distribution, one of these was the logit model.

THRESHOLDS IN THE GENERAL CHOICE PROCESS

Before discussing the incorporation of thresholds into specific choice models it is useful to place thresholds within the overall context of the general transport choice process. This is now summarised, and the possible influence of thresholds reviewed.

The general choice process used is represented diagrammatically in Figure 1. The overall objective of the process is for the decision-maker to make a choice after consideration of the alternatives available to him. Before the final choice can be made, however, it is postulated that there are a number of discrete stages through which the choice process moves. Each decision-maker is represented in the process by the importance hierarchy. Initially each decision-maker believes that certain variables or attributes are of importance in the decision at hand and it is these importances which influence his final choice. Here in lies the first application of thresholds. Each importance level can be seen to be equal or different to another. To determine different perceived levels of importance, however, it may be necessary to incorporate thresholds. Foerster (1977) used the concept of just noticeable differences to carry out this task.

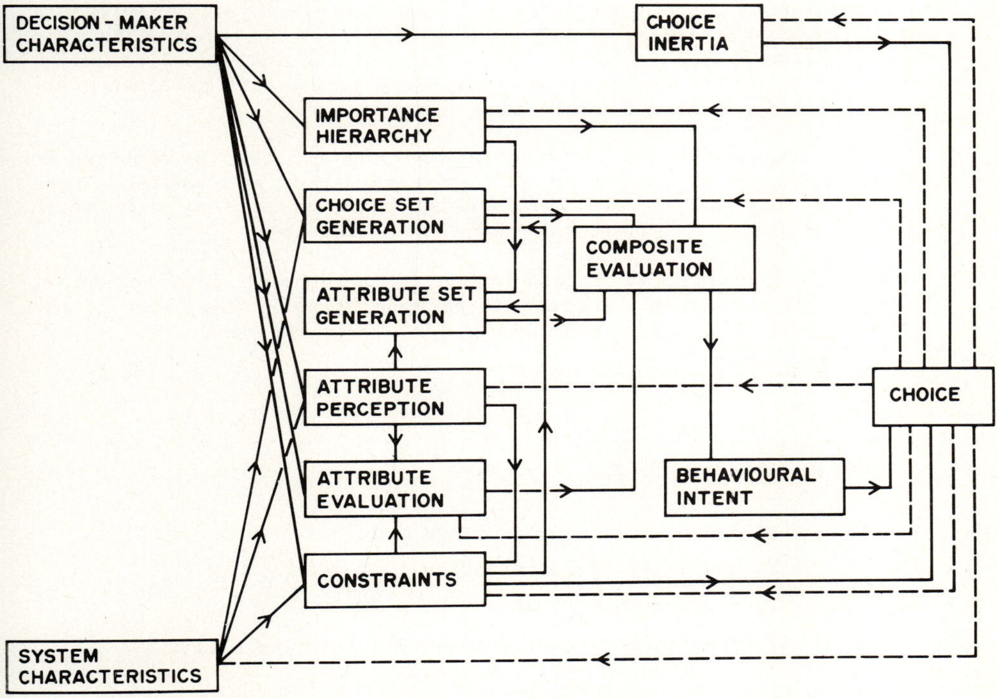

FIGURE 1 A general transport choice process

The importance hierarchy must relate to an attribute set appropriate to the decision process studied. Similarly the choice set represents those alternatives that are perceived to be available to the decision maker. Thresholds have important implications for choice set determination (Richardson 1982) since it is possible that many existing alternatives may be in the band of indifference bounded by the thresholds. They therefore will not be perceived to be real alternatives. The determination of and the alternatives in the choice set may have a strong influence on the model chosen to study the process (Williams and Ortuzar 1982a).

Before proceeding too far with the choice process it is essential to determine whether there are any constraints which significantly influence the outcome of the process. Many constraints can be considered as relating to thresholds. For instance, the financial constraints on purchasing a particular house can be overcome by a raise in household income above a threshold (Young 1982).

Given a finite choice set and a finite attribute set, the next step in the choice process is the perception of the attribute levels for each of the alternatives for each of the individuals. It must be emphasised that each individual may have a different perception concerning the physical level of an attribute, and it is this perceived level that will influence the choice. Thresholds of awareness (Hensher and Johnson 1981) may be of prime concern here since the individual may not perceive any difference between two attributes before the difference in their physical level passes some just-noticeable-difference. Further given the perceived level of an attribute each individual must then make an evaluation of the degree of satisfaction associated with that attribute. A second threshold of acceptability (Hensher and Johnson 1981) may be relevant to this stage in the choice process. Since, even if the individual perceives there to be a difference in the attribute levels, this difference may not be large enough to concern him.

The importance and satisfaction rankings can then be combined to obtain a composite evaluation of each alternative as a whole. These composite evaluations are then entered into a decision rule which gives a measure of behavioural intent or preference.

To illustrate the incorporation of thresholds into models of behavioural intention, consider three model types. The first is the logit model. The multinomial logit model takes the form

$$p(x) = \frac{e^{Ux}}{\sum_{\text{all} i} e^{Ui}} \tag{7}$$

where $p(x)$ = probability of choosing x

Ui = utility for alternative i = $f(I_i, S_{ij})$

I_i = importance of alternative i

S_{ij} = satisfaction of attribute j for alternative i.

The utility function can combine attribute importance and satisfactions in many forms. The most common form however is the linear additive form ($\sum_{\text{all } i} I_i \cdot S_{ij}$) . Attribute thresholds can be incorporated into this utility function by setting S_{ij} equal to 0 or 1 depending on whether the attribute j for alternative i is above the threshold of acceptance (Bertram and Young 1984).

The second model is the lexicographic model which states that individuals search through the attribute set in order of importance eliminating each alternative as they become unsatisfactory with respect to a particular attribute. The acceptability or not of an attribute requires the determination of an acceptance threshold (Recker and Golob 1979; Gensch and Svestka 1979). Unfortunately the lexiographic model tends to be deterministic and it is commonly thought that choice models should be probabilistic (Hensher 1972). The prime reasons for this being the apparent inconsistency of people when given the same choice and the inability of the choice model to completely replicate the choice process.

A probabilistic form of the lexiographic process was first discussed by Tversky (1972) and extended by Young and Richardson (1980). This model called the Elimination by Aspects (EBA) model takes the form

$$p(x/T) = \frac{\sum_{\text{all } s} I(x) \cdot P(x/s)}{\sum_{\text{all } R} I(x)} \qquad (8)$$

where $p(x/T)$ = probability of selecting x from total set of alternatives T

S = Any subset of attributes that are satisfactory for at least x.

R = set of attributes not acceptable for at least one alternative T

It should be noted that the EBA model can incorporate an exponential transformation of importances. The alternate form of the model would then be

$$p(x/T) = \frac{\sum_{\text{all } s} e^{I(x)} \cdot p(x/s)}{\sum_{\text{all } s} e^{I(x)}} \qquad (9)$$

This model has exactly the same functional form as the logit model for the binary choice but varies considerably as the number of alternatives increases.

Other forms of behavioural intention incorporating thresholds are the disjunctive or conjunctive models. These approaches are described in Foerster (1979) but have found little application in models of transport choice.

The measure of behavioural intent results in a choice only if various forms of choice inertia (e.g. habit or time lags) are overcome. Habits and time lags may each incorporate a threshold component (Goodwin 1977).

Having made a choice, there then exists various feedback, loops, as shown by the dashed lines in Figure 1 which may affect future choice situations.

Hence thresholds can be present in a number of stages in the choice process. Their main roles appears to be in the determination of choice sets, the perception of attributes importance and satisfaction, and in the determination of behavioural intention. Choice constraints and inertia can also be formulated in terms of thresholds.

TRANSPORT CHOICE MODELS INCORPORATING THRESHOLDS

In transport planning, thresholds have been suggested to influence choice inertia, choice set formation, perception and constraints (Richardson 1980). Empirical studies of threshold models however have not realised the full potential discussed in the previous section.

Models incorporating thresholds have tended to form themselves into two groups. The first concentrates on thresholds in individuals perception the total utility associated with each alternative. The second concentrates on attributes and considers thresholds with respect to each attribute. Both these approaches only concentrate on the behavioural intention component of the choice process.

An early example of the first category was presented by Dagenais (1969, 1975). He formulated a linear regression model, that incorporated thresholds, for the study of the purchase of automobiles. The model (Figure 2) assumes that the value of the dependent variable remains fixed until "the converting action of the independent variables and the error term induces it to overcome its reaction threshold". Maximum likelihood was used to estimate the relationship between the demand for automobiles (W) and the regression coefficients (B_i) and thresholds (N_i).

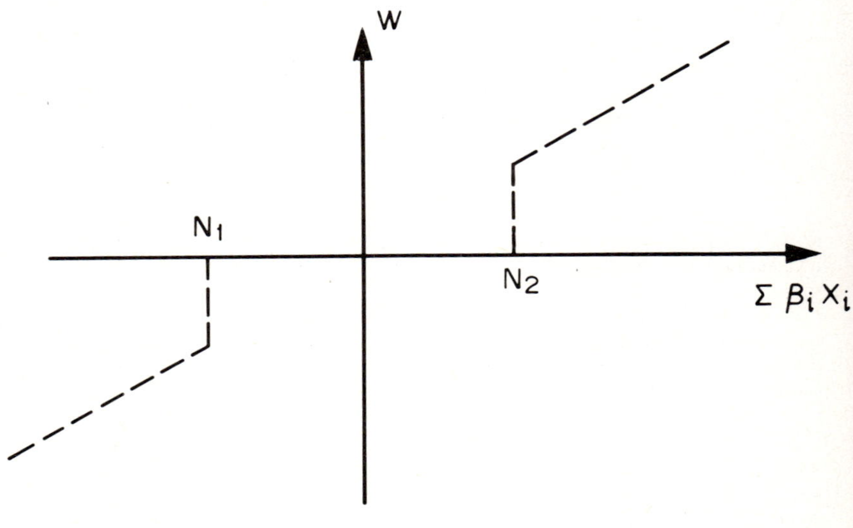

FIGURE 2 The threshold regression model

Krishnan (1977) applied this approach to mode choice using a modified logit model incorporating the concept of just-noticeable-differences, or as he referred to it, "the minimum perceived difference" (MPD). He hypothesised that an individual will prefer one alternative over another if the utility of the former should exceed that of the latter by at least a positive constant called the MPD.

If, however, the excess of utility were less than the MPD, a state of indifference would result. Krishnan applied this concept to the study of mode choice. The model combined travel time, travel cost and socioeconomic characteristics using a linear additive utility function. Estimates of the attribute coefficients and the MPD were obtained using maximum likelihood. This enabled the overall statistical fit of the MPD model to be compared with that of the logit model. Since the logit model is an MPD model, when the MPD is equal to zero, the logit model was found to have a lower overall statistical fit. This is not to say, however, that the MPD model will be a better predictor. Krishnan (1977) therefore tested the predictive ability of the model by dividing the data into two sets, calibrating a model on each set, and using these models to predict choice on the date set not used in calibration. Again the MPD model was superior. Further support for this approach can be found in Quandt (1956) and Leibenstein (1976).

The second approach used to incorporate thresholds relates the thresholds to the attributes making up the alternative. This approach has attracted a number of researchers. Some studies of attribute thresholds did not estimate the thresholds in a rigorous manner but rather used subjective rules. For instance, Foerster (1977) studied a hierarchial lexiographic mode choice model that incorporated attribute thresholds. This model considered attributes in decreasing order of importance. Alternatives were eliminated when an attribute was found to have a satisfaction below the acceptable level. In Foersters study the concept of 'just-noticeable-differences' was used to determine if the consecutive importance ratings and corresponding satisfaction ratings differed by the critical "just- noticeable difference". Foerster does not state the procedure for determining the value of the just-noticeable-difference but rather states that they "were determined by numerical procedures". Similarly, Ford and Young (1984) used heuristic approaches to determine attribute thresholds. They calibrated a logit model of residential location choice using transport survey data. During calibration difficulty was found in determining the coefficient associated with the closeness to work attribute. This was of concern given the considerable research into the relationship between workpace and homeplace (Lowry 1964). Observation of the data however indicated that most of the work trips were below 20 minutes in length. The incoporation of an absolute attribute threshold of 20 minutes was investigated. It was found to improve the performance of the model and lifted the significance of closeness to work parameters to second highest.

More sophisticated calibration processes for attribute thresholds have, however, been slow to develop. The reasons for this will become apparent in the following discussion.

One of the first attempts to estimate attribute thresholds was presented by Recker and Golob (1979). They proposed a non-compensatory model of transport behaviour based on sequential consideration of attributes. The sequential consideration of attributes was determined using interviewer stated importances. The acceptability or not of satisfaction data was determined by combining the concepts of "just-noticeable-differences" and the maximum perceived satisfaction level and estimating an acceptable tolerance. The functional form of this threshold is provided in equation (6). A constrained method of steepest descent was used to estimate the critical tolerance. The model was applied to the study of destination choice and the choice of motor car type. The output of the estimation procedure was a set of tolerances estimates for each of the choices studied. These tolerances and the split sample predictive ability of the models were compared heuristically with that of a logit model estimated using the same data sets. The logit and the sequential model provided acceptable descriptions of the variance in the data. Unfortunately the overall fit the sequential model could not be tested using existing statistical or discrete choice model tests.

Similarly, Gensch and Svestka (1979) studied a semi ordered lexiographic model of individual choice. Attribute thresholds were of the same functional form as those used by Recker and Golob (1979). To develop the model data on individuals ordering of attributes (their importance hierarchy), the satisfaction ratings for each attribute for each alternative and the choice were required. The estimation of the tolerances was provided using an uncontrained optimisation technique. The output of this model was however deterministic and comparison between the probabilistic output of the logit model was difficult.

A major problem with the previous two models related to the deterministic nature of the model output. Choice processes tend to be probabilistic in nature and an individual may, given quite similar situations, choose quite different alternatives. This may result from the individual randomising his choices to gain further information or from the models inability to fully describe the choice process. An initial step to develop estimation procedures for probabilistic threshold models was presented by Young and Richardson (1980). This model was based on Tversky's (1972) EBA model. The parameters were estimated using a least-squares approach suggested by McFadden (1979). However, like the above models this approach did not provide overall statistical measures of model performance.

Young and Richardson's (1980) model was first applied to residential location choice. Most choice models which have been applied to land use studies (Lerman, 1975; Anas, 1975) can be classed as compensatory since they use a linear additive procedure for combining the attribute satisfaction levels into an overall measure of satisfaction with each alternative. In the case of Lerman's (1975) model this results in an individual combining a large number of attributes (16) into a composite evaluation for a large number of alternatives (145). This seems beyond the mental capabilities of any individual. A more acceptable behavioural mechanism would be for the individual to rank the attributes in order of importance and then consider each attribute in order of importance. Locations are then eliminated each time an attribute describing that location is found to be unsatisfactory. This non-compensatory approach forms the basis for the EBA model discussed in this paper.

Another characteristic of many existing location choice model results from the historical development of choice models in transport planning. Most applications of choice models to transport problems have used the logit model. It was therefore not unreasonable to expect that this model would be applied to studying location choice. This was in fact the case (see Mayo, 1973; Lerman, 1975; Anas, 1975).

There is however, one major problem in applying the logit model to residential location choice and this results from the fact that the basic premise on which the model relies is violated in the case of residential location choice. This assumption is the Independence from Irrelevant Alternatives (IIA) axiom. It states that the odds of choosing one alternative over another are constant irrespective of the other alternatives in the choice set.

To illustrate the problems associated with the IIA assumption, take two suburbs, one near the centre of the city (x) and one in the outer suburbs (y). The probability of choosing x is 0.60 and the probability of choosing y is 0.40. Suburb y is now subdivided into two smaller suburbs. The second suburb becomes known as location z. If the supply of housing has no influence on choice and the new residents are indifferent to the two outer suburbs the probability of choosing y from a choice between y and z is 0.50. However the probability of choosing y from locations x, y, z given the IIA axiom holds is now 0.29. Hence the probability of choosing y or z is 0.58. Thus the subdivision of a suburban area means a developer can capture 58% of the market compared with his previous share of 40%. In fact, if a third outer suburban area was introduced, by further subdivision, the model would predict that the developer's share of the market would rise to 67% and so on.

The paradox outlined above is commonly referred to as the "red bus/blue bus" problem in the mode choice literature. It highlights a serious deficiency in applying the logit model to many transport choices. This defficiency is not present in the EBA model since it incorporates similarities between alternatives into the model structure.

An attempt to overcome the IIA problem, in the logit model, could be made by grouping the alternatives with similarities and forming a hierarchy of choices. This model is called a nested logit and was discussed by Daly and Zachary (1978) and Hensher (1983). There are however two problems with this approach.

The first problem with this approach is that there is not clear distinction between one residential area and another. Rather every residential area within an urban area is related to the others due to its spatial proximity. It is therefore necessary to adopt a complex iterative estimation procedure for determining the most appropriate groupings of areas and then determine the hierarchical order of these groupings. This is a complicated, time consuming and expensive task.

The second problem with the nested logit would result when the model is applied. Application of the nested logit would use the same hierarchy as that developed at the calibration stage. However, there can be no certainty that if a change in the urban system is made that this change does not result in new groupings and a new hierarchy of choice, hence making the old hierarchy of choice inappropriate. The building of a road bridge would for instance make two areas which were distinctly different before its construction much more similar in their spatial proximity to activities after its construction.

Both these problems cast doubt on the usefulness of the nested logit model for modelling residential location choice. A model framework that overcomes these problems is presented in Smith and Slater (1981) and Young and Richardson (1980).

Since Young and Richardson (1980) initial work development of the EBA model has gone through a number of stages and has taken a number of different directions. The first was the incorporation of a maximum likelihood estimation procedure (Young and Richardson 1983). This step enabled direct comparisons of the threshold models overall statistical performance with that of other mainstream discrete choice models. A second development (Young 1984) was for the model to determine both the critical thresholds and the importance heirarchy. This enabled physical measures of variables to be incorporated into the model (Young and Brown 1983). In turn the revealed importance model allows different transformation of the importance ratings to be incorporated into the models functional form (Bertram and Young 1984). However the estimation of importances needs to assume that the same sequencing of attributes applies to all individuals in the sampled population. This removes some of the individuality incorporated in the earlier models.

The choice between models can be made on a number of grounds. The first and most important relates to the theoretical underpinnings of the models. The EBA model may, for instance, be theoretically appropriate to residential location choice but there is no clear theoretical reason for it to be preferred in the study of mode choice.

The second factor influencing the choice of model relates to the ability of the model to replicate the data used to estimate the paramters. Various comparisons between the overall statistical performance of the EBA model and other models can be carried out. Comparison between the EBA model and the multinomial logit are presented in Table 1. It can be seen that the EBA model provides a slightly better fit than the multinomial logit in three studies, the multinomial logit is better in three studies and there is no difference in the last study. Interestingly, the logit model provides a better statistical fit than the EBA model for its application to the residential location data set (Young, 1982). The decision of which model is to be preferred cannot therefore be clearly determined by comparison of the models overall statistical fit.

Another test of the performance of the models could be to compare their predictive ability. Unfortunately research into the predictive ability of choice models incorporating thesholds is extremely rare. It is nonetheless informative to compare the influence of the thresholds, on the models response to changes in attribute levels, with that of corresponding logit models. This comparison has been carried out for each of the studies outlined in Table 1. Figure 3 presents a typical result. It can be seen, from Figure 3, that the logit model predicts a larger response to attribute changes than does the EBA. This observation is consistent with the findings in Williams and Ortuzar (1982b) and is of some concern. It is of concern since both models fit the data reasonably well yet each model would predict different responses to policy changes.

The concept of attribute thresholds, presented in the previous discussion, has been formulated in terms of acceptable/not acceptable (1-0). This aspect has been questioned by Gensch and Svestka (1979), McFadden (1981) and Williams and Ortuzar (1982b). Each have suggested that there may be a distribution of thresholds over the

TABLE 1 COMPARISON OF PSEUDO ρ^2 FOR EBA AND LOGIT MODEL

MODEL TYPE	CHOICE TYPE	EBA	LOGIT
STATED IMPORTANCE	RESIDENTIAL LOCATION (Young & Richardson 1983)	0.64	0.75
	FREIGHT FACILITY LOCATION (Young & Ogden 1983)	0.45	0.41
	FREIGHT MODE CHOICE (Young, Richardson & Kinnear 1982)	0.51	0.43
REVEALED IMPORTANCE	RESIDENTIAL LOCATION (Young 1984)	0.60	0.71
	FREIGHT FACILITY LOCATION (Ogden & Young 1984)	0.49	0.41
	URBAN MODE CHOICE (Young & Brown 1983)	0.30	0.34
EXPONENTIAL	URBAN MODE CHOICE (Bertram & Young 1984)	0.34	0.34

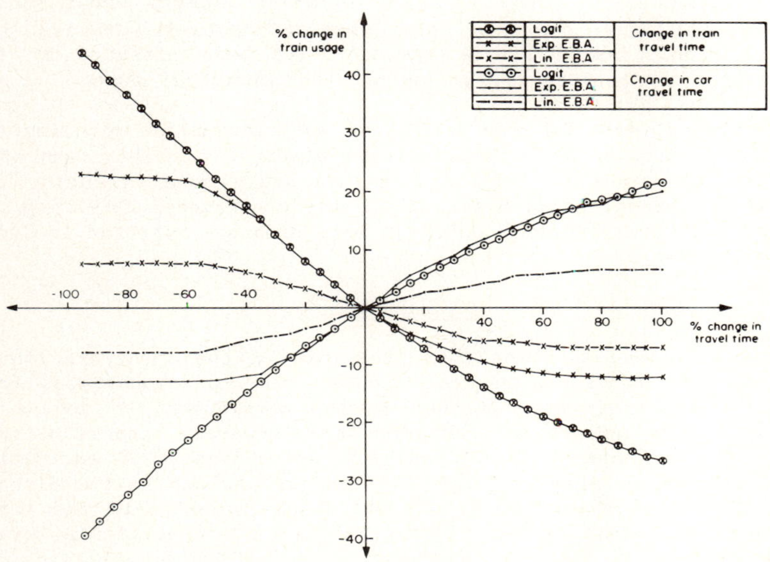

FIGURE 3 Mode choice response to changes in travel cost (Young and Brown, 1983)

population used to estimate and apply the model. To overcome this criticism McFadden (1981) and Hensher (1983) suggest that the Generalised Extreme Value model is roughly comparable in flexibility and complexity to the hierarchical EBA model and therefore could be used to replicate it. However procedures for estimating the mean and variance of the distribution of tolerances have not been studied until recently (Young 1982, 1984). Young (1982) assumes that the distribution of tolerances can be represented by a Wiebull distribution and investigates the effect of varying the variance of the tolerance distribution. It was found that the model with the best statistical fit had a distribution of tolerances consistent with the 0-1 assumption (that is the variance was equal to zero). Hence indicating that the distribution of tolerances is much narrower than previously expected.

FURTHER RESEARCH

It is evident from the previous discussion that thresholds have been incorporated into a number of choice models. These models have performed well when compared with the multinomial logit models. There is however some effort required before these models can be made available for general use. The two most important areas are the need to develop efficient calibration procedures to estimate the parameters and to then compare the predictive power of these models with that of other choice models. The only studies of the predictions given by the threshold models are simple split sample studies. These tests, although an essential stage in model testing, do not indicate if the model can predict policy changes.

Another area for development relates to the types of thresholds that have been incorporated into the existing models. Most models have combined the concepts of just-noticeable-differences and attribute magnitudes. Little research has been directed at the investigation of absolute thresholds or other threshold types.

Finally, thresholds have been seen as a method of reducing the choice set. Many choice processes may infact be a combination of a thresholds and trade-off model. The threshold model reducing the choice set to prescribed levels and the trade-off model determing the final choice. Preliminary studies in this area are reported in Young and Taylor (1984).

CONCLUSION

This paper has reviewed the literature on the incorporation of thresholds into transport choice models. It has demonstrated that there has been a growing interest in the development of models of this type. Developments in this area have, however, been relatively slow due to a number of difficulties associated with estimating critical threshold levels. These difficulties relate to the discontinuous nature of the objective function use to estimate the parameters and the incorporation of physical values of attribute levels into the models.

Attribute thresholds have, however, been estimated and in some situations improve the overall performance of the model. They were found to be particularly useful in enabling the development of models more in tune with more complex decision processes. Their main contribution to the understanding of transport choice therefore appears to be in the more complex decision processes.

REFERENCES

Anas, A. (1975). "The empirical calibration and testing of a simulation model of residential location". Environment and Planning A, Vol.7, pp.899-920.

Bock, R. and Jones, L.V. (1968). "The measurement and prediction of judgement and choice (Holden-Day: San Franscisco).

Bertram, D. and Young, W. (1984). "Incorporating attribute thresholds into mode choice models". Proc. of 12th Australian Road Research Board Conference, Vol.12, pt.6, pp.13-22.

Burns, L.D., Golob, T.F. and Nicolaids, G.C. (1975). "A theory of urban households automobile ownership decisions". Transportation Research Record 569, pp.56-72.

Daganzo, C.F. (1980). "Calibration and prediction with random utility models: Some recent advances and unresolved questions". Resource paper for 4th International Conference on Behavioural Travel Modelling, Eibsee, West Germany.

Dagenais, M.G. (1969). "A threshold regression model". Econometrica, Vol.37, No.2, pp.193-203.

Dagenais, M.G. (1975). "Application of a threshold regression model to household purchase of automobiles". Review of Economics and Statistics, Vol.57, No.3, pp.275-285.

Daly, A.J. and Zachary, S. (1978). "Improved multiple choice models". In Hensher, D.A. and Dalvi, M.O. (Eds). Determinants of travel choice. (Saxon House, Sussex).

Ebbinghaus, H. (1885). "Memory: A contribution to experimental psychology, Translated by Ruger and Bussenuis (Dover: New York).

Fechner, G. (1860). "Elements of Psychophysics". Translated by H.E. Adler, (Holt: New York).

Finney, D.J. (1964). Probit analysis, (Cambridge University Press).

Foerster, J.F. (1979). "Mode choice decision process models: A comparison of compensatory and non-compensatory structures" Transportation Research A, Vol.13A, pp.17-28.

Ford, J. and Young, W. (1984). "Using transport home interview data to model the distribution of households". Proc. of 12th Australian Road Research Board Conference, Vol.12, pt.6, pp.50-62

Guest, A.M. and Cluett, C. (1976). "Workplace and residential location: A push-pull model", Journal of Regional Science, Vol.16, pp.399-410.

Gensch, D.H. and Svestka, J.A. (1979). "An exact hierarchical algorithm for determining aggregate statistics from individual choice data". Management Science, Vol.25, No.10, pp.939-952.

Georgescue-Roegen, N. (1958). "Threshold in choice and the theory of demand", Econometrica, Vol.26, No.1, pp.157-168.

Goodwin, P.B. (1977). "Habit and hysteresis in mode choice". Urban Studies, Vol.14, pp.95-98.

Guilford, J.P. (1954). Psychometric methods. (McGraw Hill: New York).

Hensher, D.A. (1972). "The consumer's choice function: A study of traveller behaviour and values". Ph.D. dissertation, School of Economics, University of New South Wales (Australia).

Hensher, D.A. (1983). "SAD: A sequential attribute dominance model of probabilistic choice". Transportation Research A, Vol.17A, No.3, pp.215-218.

Hensher, D.A. and Johnson, L.W. (1981). "Applied Discrete Choice Modelling. (Croom-Helm; London).

Highway Research Board (1969). "Moving behaviour and residential choice: A national survey", Highway Research Board/NCHRP Report 81, Washington, D.C.

Kavak, C.F. and Demetsky, M.H. (1975). "Behavioural modelling of express bus-fringe parking decisions". Transportation Research Record 534, pp.10-23.

Krishnan, K.S. (1977). "Incorporating thresholds of indifference in probabilistic choice models". Management Science Vol.23, pp.1224-1233.

Leibenstein, H. (1976). "Beyond economic man: A new foundation in microeconomics. (Harvard University Press: Cambridge, Mass.).

Lerman, S.R. (1975). "A dissagregate behavioural model of urban mobility decisions", Ph.D. dissertation, Massachusetts Institute of Technology, Massachusetts.

Lowry, I.S. (1964). "A model of metropolis" Land Corporation Memorandum, RM-4035-RC.

Luce, R.D. and Galanter, E. (1963). "Discrimination" in Handbook of Mathematical Psychology, R.D. Luce, R.B. Bush and E. Galanter (Ed.) (Wiley, New York)

McFadden, D. (1979). "Quantitive methods for analysing travel behaviour of individuals. Some recent developments". In Behavioural Travel Modelling. D.A. Hensher and P.R. Stopher (Eds). (Croom-Helm).

McFadden D. (1981). "Econometric models of probabilistic choice". In Econometric Analysis of Discrete Data (Ed. Manski, C.F. and McFadden). M.I.T. Press, Cambridge, Massachusetts.

Mayo, S. (1973). "Local public goods and residential location: An emprical test of the tiebout hypothesis" Metropolitan Goverance Committee on Resources for the Future, Washington D.C.

Miller (1947). "Sensitivity to changes in the intensity of while noise and its relation to Mashing and loudness", Journal of the Acoustic Society of America, Vol.19, pp.609-619.

Ogden, K.W. and Young, W. (1984). "A model of manager's preferences for location of freight facilities" Transportation Planning and Technology, Vol.8, pp.283-294.

Peterson, L. and Peterson, M.J. (1959). "Short-term retention of individual verbal items" Journal of Experimental Psychology, Vol.58, pp.193-198.

Quandt, R.E. (1956). "A probabilistic theory of consumer behaviour" Quarterly Journal of Economics, Vol.LXX, pp.507-536.

Quarmby, D.A. (1967). "Choice of travel mode for the journey to work: Some findings" Journal of Transport Economics Policy, Vol.1, pp.273-314.

Recker, W.W. and Golob, T.F. (1979). "A non-compensatory model of transportation behaviour based on sequential consideration of attributes". Transportation Research. Vol.13A, pp.269-280.

Richardson, A.J. (1980). "Advanced demand modelling for advanced transort systems". Journal of Advanced Transportation, Vol.14, No.2, pp.113-131.

Richardson, A.J. (1982). "Search Models and choice set generation". Transportation Research-A, Vol.16A, No.5-6, pp.403-419.

Richardson, H.W. (1971). Urban Economics (Penguin Books, England).

Simon, H.A. (1957). Models of Man. (John Wiley and Sons, New York).

Slutsky, E.E. (1952). "On the budget of the consumer" Reprinted in Readings in Price Theory (American Economics Association) (Irwin; New York).

Smith, T.R. and Slater, P.B. (1981). "A family of spatial interaction models incorporating information flows and choice set constraints applied to U.S. interstate labour flows". International Regional Science Review, Vol.6, No.1, pp.15-31.

Stevens, S.S. (1962). "The surprising simplicity of sensory metrics". American Psychologist, Vol.17, pp.29-39.

Stopher, D.R. (1969). A probability of travel choice for the work journey". Highway Research Record 283, pp.57-65.

Tversky, A. (1972). "Elimination-by-Aspects - theory of choice" Psychological Review, Vol.79, pp.281-291.

Williams, H.C.W.L. and Ortuzar, J.D. (1982a). "Travel demand and response analysis - some integrated themes". Transportation Research-A, Vol.16A, No.5-6, pp.345-362.

Williams, H.C.W.L. and Ortuzar, J.D. (1982b). "Behavioural theories of dispersion and the mispecification of travel demand models". Transportation Research: B, Vol.16B, No.3, pp.167-219.

Young, W. (1982). "The development of an Elimination-by-Aspects model of residential location choice". Ph.D. Thesis, Monash University (unpublished).

Young, W. (1984). "A non-tradeoff decision making model of residential location choice". Transportation Research A, Vol.18A, No.1, pp.1-11.

Young, W. and Brown, H.P. (1983). "A revealed importance Elimination-by-aspects model of mode choice". Proc. of P.T.R.C. Summer Annual Meeting, Brighton, England.

Young, W. and Ogden, K.W. (1983). "Analysis of freight-facility location choice using an Elimination-by-Aspects model". Proc. 8th Australian Transport Research Forum. Vol.3, pp.265-281.

Young, W. and Richardson, A.J. (1980). "Residential location preference models: Compensatory and non-compensaoty approaches". P.T.R.C. summer conference, University of Warwick, England.

Young, W. and Richardson, A.J. (1983). "The Application of an Elimination-by-Aspects model to residential location choice". Australia Road Research, Vol.16, pt.2, pp.100-111.

Young, W. and Richardson, A.J. and Kinnear, R. (1983). "Modelling inter-urban freight mode choice". Journal of Advanced Transportation, Vol.16, No.3, pp.303-322.

Young, W. and Taylor, M.A.P. (1984). "Combining compensatory and non-compensatory models". Proc. 6th CAITRE conference, Canberra, Australia.

4
Transferability and cost-effectiveness

Behavioural Research for Transport Policy
pp. 173–192. © 1986 VNU Science Press.

APPLICABILITY AND STABILITY OF LOGIT MODELS IN SWEDEN - SOME RECENT FINDINGS WITH POLICY IMPLICATIONS

by

Staffan Algers Staffan Widlert

Stockholm County Council
Office of the Public Transport Board
Stockholm, Sweden

1. INTRODUCTION

Transportation policy consists to a large extent of a number of investment and operating decisions. Concerning road investment in Sweden, such decisions have been based on an overall economic assesement, where the consumer benefits in terms of travel time have been an important factor. Regarding investment in public transport, the number of projects explicitly evaluated in such terms has been low and restricted to the Stockholm area.

It is, however, likely that there will be an important increase in the explicit use of evaluation methods based on consumer preferences in Sweden (such as cost benefit analysis). The reason for this is that such a decision base will be required when applications for government subsidies for public transport investment are made. At present, a special task force is working on guidelines for evaluation of public transport projects in this context.

This type of evaluation is highly usable also for operating decisions. It makes it possible to find the "best" trade-off between, say, walking distances and frequencies of the public transport service. It can also be used to rank proposed bus lines, as has been done in the Stockholm County Public Transport Plan from 1983.

Obviously, these methods need information on consumer preferences - not for public transport or travel times in general, but specified to a high level of detail. Naturally, it can not be hoped to reach a level of detail that matches the perceptions of every consumer, but we think that it is feasible to achieve a level that matches the perceptions of the average consumer on the most important aspects.

The results presented here emerge from an ongoing research project concerning the applicability and stability of logit models, financed by the Swedish Council for Building Research. The aim of the project is to - based on existing data - determine the stability of logit models (in terms of time and geographical area) and to extend the range of applications. This will be done by comparing results from various studies, by estimating new models and by re-estimating earlier models. This paper draws on intermediate results from this project.

2. VALUATION OF TRAVEL TIME COMPONENTS

Several techniques for collecting information on consumer preferences exist. One of these is the logit model, which has been used on a number of occasions in Sweden. The results presented here are based on a number of Swedish studies using such models. First we discuss in some detail the results from the most recent study, conducted for the municipality of Gothenburg during the years 1982-1984. After that we compare the results from all major Swedish studies in this field.

2.1 The Gothenburg logit models

The Gothenburg logit models are developed as tools for policy evaluation at the transit operating authority, and the study is fully financed by the authority itself.

The 1982 Gothenburg travel study

The target population was defined to be households with working members. The households were selected by sampling at random from the telephone book, and interwievs were made with all working members of the household.

The response rate in the survey was 88 %. With regard to households not having a telephone, and households with secret telephone numbers, the total response rate amounts to 80 %. The study yielded 1050 observations that could be used for model estimation, that is observations having at least two alternative modes.

Apart from the data reported in the study, data on the traffic system was added on afterwards. Thus, data on trip distance, car time, public transport in-vehicle time, waiting time and transfer time were calculated on a zonal basis, whereas walk distances were measured on maps.

The mode choice model

A logit mode choice model for the journey to work was estimated. The model included the following alternatives:

- car as driver
- public transport
- walk
- bicycle

A number of different model specifications was tried, including a structured model where the walk and the bicycle modes were modelled at a lower level. The best model turned out to be number 1 in the table below:

	1	2	3
Car constant	+0.6951 (2.00)	+0.7096 (2.04)	
Bicycle constant	-0.0732 (1.92)		-1.1920 (3.39)
Walk constant	+1.2200 (3.28)		
Slow mode constant		+1.1910 (3.40)	
Expressbus constant	+1.2490 (2.94)	+1.2440 (2.92)	
In-vehicle time	-0.0132 (2.56)	-0.0126 (2.47)	
Walk to/from car	-0.0598 (3.73)	-0.0594 (3.71)	
Walk to/from public transport	-0.0315 (1.99)	-0.0308 (1.95)	
Slow mode time	-0.0452 (8.75)		-0.0509 (3.59)
Travel cost	-0.0560 (4.12)	-0.0555 (4.09)	
Waiting time	-0.0201 (2.75)	-0.0198 (2.73)	
Transfer time	-0.0286 (2.87)	-0.0285 (2.87)	
Sex - public transport	+0.6886 (3.65)	+0.6830 (3.62)	
Use of car during work - car	+1.1720 (3.58)	+1.1600 (3.55)	
Part of distance with bicycle lane - bicycle	+0.0090 (2.73)		+0.0080 (1.60)
Errands - car	+0.8333 (3.57)	+0.8272 (3.55)	
Seating possibilities - public transport	-1.1200 (5.84)	-1.1280 (5.90)	
Dummy for CBD - car	-0.8769 (3.31)	-0.8866 (3.36)	
Logsum variable		+0.8927 (8.92)	
Number of observations	1 050	1 018	138
p2	0.361	0.394	0.123

The meaning of the variables are rather self-explaining as far as the constants and the time variables are concerned.

The cost variable measures the operating cost and the parking cost for the car alternative and the fare for the public transport alternative. Regard is taken to that tax deductions are made for the journey to work. As will be pointed out later on, this is quite important.

The sex variable is attached to the public transport alternative, and the size of the parameter shows that women have a greater probability to use public transport than men, everything else beeing equal.

The variable for use of car during work captures the impact of the car being used during work on the mode choice. The variable takes the value one for the car alternative if car is used during work more than one day per month. Otherwise the variable takes the value 0.

"Part of the distance with bicycle lane" measures the percentage of the distance between the dwelling and the work place that consists of special bicycle lanes.

The variable for errands takes the value one for the car alternative if any errand (for instance leaving child at nursery) has been performed on the way to home from work.

The variable for having a seat on public transport takes the value one for the public transport alternative if the interview person has reported that he didn't have, or wouldn't have, a seat all the way.

The CBD-dummy takes the value one for the car alternative if the work place is situated in the Gothenburg CBD. Otherwise the variable is 0.

Models 2 and 3 represent a structured model were the slow modes are modelled at a lower level. As can be seen from the table, there is not much point in adopting a structured specification. The logsum variable is close to one, and the size of the other variables is not much affected.

The results of the model are discussed below, and here it will only be pointed out that the model will be used by the public transport authority of Gothenburg together with a "user - friendly" program, that has been developed for the use of the model. The program uses Sample Enumeration and it makes it possible to use the demand models without any previous experience of models or computers.

Relative weights of travel time components

The coefficients for the travel time components in the models are divided by the coefficient for the in-vehicle time. The relative weights resulting from this operation are shown in the table below:

	Choice between car and public transport	All modes	
	4	5	1
In vehicle time	1.0	1.0	1.0
Walk time to parking	-	-	4.5
Walk time to bus stop	-	-	2.4
Walk time to parking or bus stop	3.2	3.2	-
Walk time and bicycle time all the way	-	3.2	3.4
Waiting time	-	-	1.5
Transfer time	-	-	2.2
Waiting + transfer time	1.6	1.6	-

In the model for choice between car and public transport, the walk time parameter gets the weight 3.2 and the waiting and transfer time parameter gets the weight 1.6. When the model for choice between all modes is specified the same way (model 5) the same weights are obtained. This means that we can conclude that there seems to be no differences between the valuation of standard components between workers having access to car and the group of all workers. This result is important because it has earlier been argued that the limited group having full possibilities of choice between car and public transport alternatives could be special and have different valuations. The time to walk all the distance to and from work and the time to go by bicycle to and from work gets a high weight, 3.2.

When the walk time parameter is split up, then the walk time to and from bus stop gets the weight 2.4 (which is rather close to the results from other studies) and the walk time to and from parking place gets the very high weight 4.5.

The walk time to and from the parking place is normally short. The time is not an engeneering value, but reported by the interview persons. The times are also often rounded off to even five minutes. The estimated weight can therefore be rather uncertain. The walking distance to and from the bus stop is measured on a map and is therefore more accurately defined.

The subdivision into separate parameters for waiting time and transfer time results in a higher weight for transfer time than for waiting time. Waiting time is measured as half the frequency. At a lower frequency, a large part of the waiting time is so called hidden waiting time, which is spent at home or at the work place. The transfer time is in total spent at the bus stop. Therefore it is reasonable to expect that the waiting time on average is valued less negative than the transfer time.

Valuation of waiting and transfer time

Regarding the waiting and transfer time, two hypothesises can be stated. One hypothesis would be that the lower the frequency, the greater the share of the waiting time spent at home. Consequently, the longer the waiting time, the lower the waiting time weight.

The other hypothesis would be that since the transfer time is spent at the bus stop, each additional transfer time minute would be percived as more and more inconvenient. Consequently, the weight of the transfer time would increase with the transfer time.

In order to see how the lenghts of the waiting and transfer times affects the weights, a model with separate parameters depending on the lenghts of time, was estimated. The following model was obtained:

	Parameter (t-value) 6	Weight relative to in-vehicle time
Car constant	+0.6766 (1.68)	
Bicycle constant	-0.1424 (3.33)	
Walk constant	+1.1470 (2.78)	
Express bus constant	+1.2940 (2.97)	
In vehicle time	-0.0141 (2.68)	1.0
Walk time -car	-0.0662 (3.75)	
Walk time - public transport	-0.0314 (1.98)	
Walk- and bicycle time all way	-0.0450 (8.71)	
Travel cost	-0.0565 (4.16)	
Waiting time 0-20 minutes	-0.0285 (1.59)	2.0
Waiting time 21- minutes	-0.0151 (1.34)	1.1
Transfer time 0-20 minutes	-0.0193 (1.34)	1.4
" " 21- minutes	-0.0445 (1.90)	3.2
Sex - public transport	+0.7018 (3.71)	
Car during work hours - car	+1.1810 (3.61)	
Part of distance with bicycle lane - bicycle	+0.0089 (2.71)	
Errands - car	+0.8352 (3.57)	
Seating possibilities - public transport	-1.1500 (5.93)	
CBD-dummy - car	-0.8681 (3.28)	
Number of observations	1 050	
p^2	0.362	

The division of the parameters is done in the following way. The variable 0-20 takes on the value for the waiting time, respectively transfer time, for times up to 20 minutes per day and the value 20 minutes when the time exceeds 20 minutes. The variable 21- takes on the value 0 for times below 20 minutes and the value that exceeds 20 minutes for times exceeding 20 minutes. The variance of the estimated parameters is rather high, but the coeffcients are relatively stable for alternative model specifications. The sizes of the estimated parameters fit in with the hypothesis, though a statistical test at normal significance levels would reject the hypothesis that the parameters differ. Due to the coefficients, short waiting times are valued around twice as negatively as the in-vehicle time. The part of the waiting time exceeding 20 minutes per day is only slightly higher valuated than the in-vehicle time.

The transfer time is, according to the results, valued 1.4 time as negative as the in-vehicle time for short transfer times. The part of the transfer time exceeding 20 minutes per day is valued more than 3 times as negatively as the in-vehicle time.

Thus we can summarise that the achieved result is an indication that the waiting och transfer time values are different at different lenghts of times. More comprehensive studies are however needed to make it possible to get more detailed, and statistically more significant, relationships.

Bicycle lanes

The dummy variable for bicycle lanes shows that the travellers valuate those lanes highly. If, instead of the dummy variable, separate time parameters for the bicycle mode are estimated for trips with different percentage of bicycle lane, the following result is obtained:

percentage of bicycle lane between dwelling and work place	parameter for travelling time on bicycle	t-value	relative weight of travel time (bicycle lane all the way, or almost all the way=1)
0 %	-0.0621	4.47	2.0
1 -40 %	-0.0468	5.48	1.5
41-80 %	-0.0350	5.08	1.1
81- %	-0.0308	4.65	1

The table shows that the valuation of travel time on bicycle is twice as high when there is no bicycle lane as it is when there is a bicycle lane (almost) all the way.

The value of having a seat

Model 1 includes a variable for having a seat or not on the public transport alternative. The variable takes the value 1 if the person did not have, or would not have had, a seat all the way, and 0 otherwise.

The resulting coefficient shows that great importance is attached to the possibility of having a seat. However, the data set allowed for a more detailed measurement of the disutility of crowded vehicles, in that the degree of congestion could be specified. The following variables were specified (still based on reported data):

- seat available all the journey
- seat available part of journey
- stand all the journey, vehicle not very congested
- stand all the journey, vehicle congested

All but the first of the variables were defined as dummy variables and included in the model (these three variables are then relative to the first one). The following coefficients were estimated (the other variables did not change much):

seat part of the journey	- 1.0220
	(3.66)
stand all the journey, not very crowded	- 1.1870
	(3.17)
stand all the journey, crowded	- 1.1300
	(3.40)

The variance of the parameters, as indicated by the t-values, is rather low, but the coefficients are so close to each other that they do not differ significantly. Consequently, it seems as if the crucial thing is to have a seat all the way, and that otherwise the degree of congestion does not matter much.

Also, the separation into two parameters results in estimates that are very close to each other.

The parameters for the seat variable express a constant value of the disutility of standing. The values are not a function of the in-vehicle time. Of course, it would be more accurate to directly estimate separate parameters for in-vehicle time having a seat and in-vehicle time not having a seat. This has however turned out not to be possible in the Gothenburg case. (If we impose a linear relationship between the discomfort to stand and the in-vehicle time for trips where a seat is not available, than the relative weight of in-vehicle time spent standing is 2.8 times the time spent seated.)

It has, however, to some extent been possible to see how the disutility of standing varies with the time spent standing. This was done by splitting those who stand, or would have to stand, all the journey into three categories according to in-vehicle time. The following results were obtained:

	time interval in-vehicle time	mean triptime	coefficient (t-value)
stand all the journey	0-39	19	-0.8871 (3.05)
"-	40-79	56	-1.4860 (4.12)
"-	80-	99	-2.2940 (1.97)
stand part of the journey	0-	53	-1.0460 (3.97)

From the table it can be seen, that - as expected - the disutility of standing increases with time spent standing. If the coefficients are recalculated into Swedish crowns, and plotted against the mean in-vehicle trip times, the following figure is obtained:

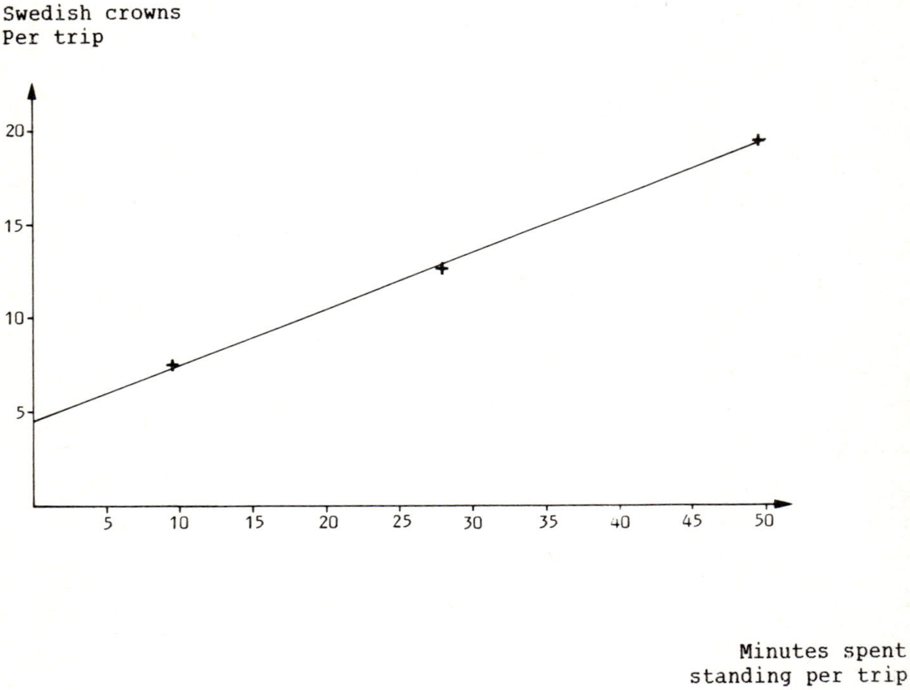

The figure shows a relationship between the disutility to stand and the time spent standing that is composed by a constant part and a part that varies linearly with time. This phenomenon, as well as the high disutility of standing presented above, may be due to the fact that the data on seat availability are reported and not measured separately. This may result in underestimating the seating availability for car choosers, and thus to a higher coefficient for the variable.

There is a constant discussion going on regarding the differences in comfort between bus and rail modes. A subdivision of the seat variable with regard to public transport mode gave the following results:

No seat on bus	-1.0880 (3.80)
No seat on rail mode	-1.0860 (4.98)

The average travel time (public transport) for those not having a seat available is equal for the bus and the rail mode alternatives (46.1 respectivly 47.7 minutes per day). Also the degree of crowding is roughly equal on the bus and the rail alternatives. The parameters are very close to each other, indicating that the time spent standing is valued very similar on buses and rail in Gothenburg. The rail alternative in Gothenburg almost solely consists of tramways.

To summarise, we can conclude that the Gothenburg data-set produces models with a very high valuation of in-vehicle time spent standing. The fact that the parameter values have been recalculated into time related weights, and are not estimated as such, makes it neccessary to treat the results with care. The disutility to stand may also be composed by a constant part and a time depending part. On the other hand, the estimated parameters have high t-values. The results are also remarkably stable when alternative subdivisions and alternative model specifications are tried.

2.2 Comparison of different Swedish studies

To get an idea of the stability of the different travel time components, relative weights from the eight Swedish studies using logit models, that have been completed up until now, are compared below. An important problem, making the comparisons more difficult, is that the obtained differences often can be the result of shortcomings in the studies that form the base for the estimation. These shortcomings are generally very difficult to discover afterwards. The comparisons that are listed below therefore do not give the answer towhether the logit model as such is stable and transferable, but instead the answer of the question how stable and transferable the so far obtained results are.

Models for choice between car and public transport

Five of the Swedish studies dealt with the choice between car and public transport for work trips.

For those five studies the parameters can be directly compared. This does not mean, however, that the parameters of the different models can be expected to be equal, since the specifications are not identical and since there are differences in the definition of the choice groupe. It is nevertheless still interesting to compare the models - will the differences be small in spite of these differences?

The first two studies in the table below were conducted in the Stockholm area. The next two studies were conducted in Gothenburg. The first four studies therefore deals with traffic in densly populated areas with good public transport. The last study was conducted in an rural area in the middle of Sweden and therefore deals with traffic in a sparesely populated area with a low level of public transport.

	AKU 1968	VBB 1980/81	SBK 1981	GS 1982	TULT 1982
Car constant	+0.9889 (5.4)		+0.6236 (2.8)	+0.7289 (1.9)	-1.6900 (3.7)
In vehicle time	-0.0073 (2.0)	-0.0220 (3.2)	-0.0234 (2.7)	-0.0242 (2.4)	-0.0162 (2.1)
Travel cost	-0.1148 (10.9)	-0.1000 (7.2)	-0.2444 (8.9)	-0.1027 (4.5)	-0.0535 (3.1)
Walk time		-0.0280 (3.2)		-0.0774 (4.4)	-0.0209 (2.6)
Waiting time	-0.0244 (2.7)				-0.0459 (3.4)
Waiting and transfer time		-0.0340 (3.8)		-0.0379 (3.8)	
Transfer time					-0.0335 (2.1)
Excess time			-0.0431 (3.5)		
Number of transfers	-0.1979 (3.0)		-0.3185 (2.6)		
Car during work	+0.6235 (3.3)			+2.8920 (2.8)	+2.2530 (4.3)
Car competition		-1.1400 (5.7)			-0.8629 (2.8)
Sex - public transport				+1.2040 (3.7)	+0.7974 (2.6)
Errands - car					+1.9020 (4.8)

It can be seen that in the three contemporary models for the densely populated areas, the in-vehicle time parameters are very close. Also the cost parameters in two of these models are close. The waiting and transfer time parameters in the VBB-, GS- and TULT-models are close. The excess time variable in the third model also seems to be consistent with the walk time and waiting and transfer time parameters in the GS- and TULT-models. The most disturbing element in this comparison is the difference in the walk time parameters in the studies. The explanation is probably that the walk-times were reported in the VBB- and TULT-studies, and measured on maps in the GS-study.

Travel time components

Apart from the five studies above that concerned the choice between car and public transport for work trips, two more mode choice studies in Sweden have used logit models.

The first is the "Västerås-model" which is a model for the choice of mode and destination for shopping trips. The second is the "Sollentuna-model" which is a model for the choice between bus and commuter rail for work trips.

The seven different studies have resulted in the following travel time components:

Area:		Stockholm		Göteborg		Sollent	Västerås	Uppsala
Study:	AIB	AKU	VBB	SBK	GS			TULT
In vehicle time	1.0	1.0	1.0	1.0	1.0	1.0	1.0	1.0
Waiting time (half frequency)		3.3			1.5	(1.5)		
Waiting time (At bus stops)	7.1	12.0			5.0-8.6			
Waiting time (at work)								2.8
Transfer time					2.2	1.5		2.1
Waiting+transfer time			1.6		1.6			
Walking time to bus stop		(2.1)	1.3		2.4	1.5	2.0	1.3
Walking time to parking place	6.3				4.5			
Waiting-, transfer and walking time				1.8				

All models except the Västerås-model concern work-trips. The above studies all show that waiting-, transfer- and walktime is valued more negative than in-vehicle time. The statistical quality of the results varies between the studies. It is interesting to note that, in spite of this, the weights for waiting- and transfer time are almost identical in three studies (the VBB, SBK and GS-studies).

One important goal with the research project about the logit models is to give recommendations about which weights to use when projects are evaluated. Since experience gathers over time, the qualities of the studies and the degree of detail of the models can be expected to grow. The latest model, which is the GS-model from 1982, also shows the most detailed and stable results. This has been born in mind when the following weights have been suggested for project evaluation.

In-vehicle time	1.0
Waiting time	1.5
Transfer time	2.0
Walk time	2.0

3. VALUE OF TIME

Another test of the stability of the models is to calculate the monetary value of travel time. For six of the studies it has been possible to calculate the value of travel time in 1982 prices.

	value of time in 1982 prices, Sw cr/hour
AKU	15
VBB	16
SBK	6
GS car/public transport	14
GS all modes	14-15
Västerås	14
TULT	18

Except from the SBK-study, all studies show a very stable value of travel time - on average the value is 20-30 % of the hourly earning, which also has been found to be the case outside Sweden.

4. CONSTRUCTION OF THE COST VARIABLE

The construction of cost variables has often been discussed. The discussion has been centred on what kind of operating costs should be included. The problem can sometimes be circumvented by estimating separate variables for out-of-pocket costs, typically parking costs, and a separate variable for distance. This is possible as long as the operating costs are linearly depending on the distance. In Sweden, however, things are not that simple. In some cases operating costs are tax deductable. Such a deduction is allowed under the following circumstances:

- when the time gain created by using car is more than two hours per day (1982 1.5 hours)
- if the car is used during work hours

In practice, this deduction results in a lumpsum which is refunded in december by the taxation authorities the year after the taxation year.

The question is of course whether the individuals are thinking in an economic way, that is if they take the possibilities of tax deductions into account when they make the choice between the car and the public transport alternatives, or if the lumpsum refund is just thought of as an early Christmas present. In the Gothenburg case, both approaches have been tried in a model for choice between car and public transport. In the table below the time and cost parameters in the two models are shown.

	travel cost after tax deduction is subtracted	travel cost without consideration of tax deduction
Travel cost	-0.0467 (3.14)	-0.0222 (2.32)
Parking cost	-0.1223 (2.93)	-0.1244 (3.03)
In-vehicle time	-0.0130 (2.52)	-0.0127 (2.50)
Walk to/from car	-0.0578 (3.59)	-0.0559 (3.51)
Walk to/from public transport	-0.0310 (1.95)	-0.0312 (1.97)
Slow mode time	-0.0448 (8.66)	-0.0444 (8.63)
Waiting time	-0.0205 (2.82)	-0.0216 (2.97)
Transfer time	-0.0269 (2.70)	-0.0264 (2.67)
Number of observations	1050	1050
p^2	0.363	0.361

As can be seen from the table, the less "rational" specification brings about a significant loss in precision for the cost parameter, as compared to the more "rational" specification. It may not be surprising, that the inclusion of a phenomenon that concerns only one alternative, and practically only when it is chosen, brings about such an improvement.

Therefore, it is also important to note, that the "rational" specification also makes the cost coefficient come closer to the parking cost coefficient.

5. THE IMPORTANCE OF THE CHOICE SET DEFINITION

To estimate a logit model, information on the chosen as well as on the non-chosen alternative is needed. Questions about the possible alternatives are complicated to have in a survey form. In most cases, several questions will be needed to make it possible to clarify the choice set. In travel surveys describing the whole travel pattern within a day, it would not be possible to ask for the full choice set for every journey. Instead, this information has to be added after the survey.

In the research project on the applicability and stability of logit models, a data set from Västerås is used. This material was collected through a postal survey in October 1975. The survey concerned shopping trips and yielded 710 persons with access to car and 250 persons without accsess to car. The data set was used to estimate models for the choice of mode and destination for home based shopping trips.

5.1 Definition of the choice set

The data set contained information on alternative modes for the last performed shopping trip, and information on the alternative destinations that the interviewee had reported as having been considered.

The data set was enlarged by adding more destination alternatives. The survey showed that the purchases mainly were performed within the neighbourhood of the individual, or in a few larger regional centres. For each dwelling area that was part of the survey, a separate destination set was defined from the observed actual travel pattern in the survey. For the selected destination alternatives, the various mode alternatives were added. The following modes were used:

- Car
- Bus
- Walk
- Bicycle

All individuals with a driving licence in a car-owning household were supposed to have the possibility to go by car. The bus alternative was supposed to be available for all destinations with a bus connection.

All individuals were supposed to be able to walk or use bicycle for their shopping trips. The survey did not contain questions on access to bicycle, handicap etc. Some individuals will therefore be assigned alternatives that they in reality can not use. The walk- and bicycle alternatives are supposed to be present for destination alternatives where walk- and bicycle trips actually have been observed in the survey, that is not for the longest regional shopping trips.

To create a relevant set of destination alternatives, information on what was actually purchased would be needed. If the purchase concerned non-durables, the relevant set of destination alternatives is easily defined. When it comes to the shopping of durables, it's more difficult to define the relevant choice set. Since we do not know anything about what kind of durables that was purchased, we have to regard all destinations with any kind of supply of durables as possible destination alternatives.

5.2 Estimated models

The average number of reported alternatives was 4.0, (chosen alternative included). When the choice set was enlarged, the average number of alternatives was 16.6, including the chosen alternative. Thus, the choice set was quadrupled, wich in this case brought about a trippled or quadrupled cost of a model estimation.

	reported choice set	enlarged choice set
Car constant	+1.0500 (4.80)	+0.4311 (2.11)
Bus constant	-0.1773 (1.08)	-0.6295 (4.65)
Bicycle constant	-0.5338 (4.00)	-1.3760 (11.32)
Weighted total time (minutes)	-0.00970 (5.86)	-0.0213 (12.76)
Travel cost (Sw cr)	-0.0755 (3.23)	-0.0974 (5.14)
Retail floor space	+0.2046 (12.14)	+0.3734 (28.31)
Sex - car	+1.2230 (5.67)	+1.4240 (6.97)
Type of purchase (regional)	-1.2370 (4.78)	-0.4170 (1.88)
Type of purchase (car)	-1.0780 (4.77)	-1.1440 (5.59)
ln L	-1039	-1893
p^2	0.234	0.335

The comparison in the table shows that the model where all alternatives are added on afterwards (enlarged choice set) is significantly better, both in terms of parameter significance and in terms of the overall fit. The difference in significances for the time and cost varables between the models with reported and coded alternatives is quite big.

5.3 Randomly restriced choice set

If the choice set is large, it could be of economic interest to reduce it. This is often the case when working with destination choice. Since the number of alternatives in our case is at most 23 alternatives per observation (16 on an average), it is not really necessary to reduce the choice set. However, this was done to study the stability of the results when the choice set is varied in size.

Since the model results are much better when the coded alternatives are used, it is also of interest to see if the improvment is to be referred to the larger amount of information wich lies in the greater choice set.

The first model in the table is estimated with all coded alternatives and differs from the previously shown models only as far as the attraction variable is concerned (in this model the size variable is defined as the log of the retail floor space).

The next three models that are shown have 3/4 of the alternatives excluded at random. The models represent three different random restrictions of the choice set.

	all	25 %	25 %	25 %	50 %
Car constant	+0.3109 (1.49)	-0.1026 (0.42)	+0.1703 (0.65)	+0.1001 (0.38)	+0.0950 (0.42)
Bus constant	-0.5813 (4.22)	-0.8959 (5.10)	-0.8683 (4.73)	-0.9498 (5.09)	-0.8092 (5.31)
Bicycle constant	-1.4230 (11.59)	-1.5320 (9.57)	-1.5550 (9.16)	-1.7950 (10.54)	-1.5070 (10.60)
Weighted total time (minutes)	-0.0225 (12.76)	-0.0195 (9.80)	-0.0233 (10.00)	-0.0220 (9.48)	-0.0221 (11.61)
Travel cost (Sw cr)	-0.1629 (7.90)	-0.1327 (5.26)	-0.1939 (6.99)	-0.1811 (6.68)	-0.1559 (6.98)
Log of retail floor space	+0.8235 (25.80)	+0.7094 (17.98)	+0.8509 (18.90)	+0.8016 (18.56)	+0.7969 (22.90)
Sex - car	+1.4160 (6.88)	+1.3800 (5.69)	+1.0930 (4.50)	+1.2180 (4.84)	+1.4520 (6.49)
Type of purchase - regional	-1.7020 (7.57)	-1.4800 (5.80)	-1.4120 (5.51)	-1.5010 (5.60)	-1.5930 (6.81)
Type of purchase - car	-1.1740 (5.70)	-0.7434 (3.06)	-0.8296 (3.33)	-0.8171 (3.20)	-0.9774 (4.38)
ln L	-1885.1	-872.5	-807.1	-801.4	-1305.7
p^2	0.338	0.4150	0.459	0.463	0.397

From the table it can be seen that the significance generally is lower when the choice set is reduced. Also the size of the parameters are affected but the differences between the models are moderate. Except for the retail floor space parameter in the second model, neither the time-, cost- nor the space parameters are significantly different from each other in the four first models.

In the last model, half of the alternatives have randomly been excluded. Of course, the parameters are now more like the parameters of the first model. The other models in the table have been estimated with the same average number of alternatives per observation as the models that have been estimated for reported alternatives (average of four alternatives per person). A comparison between the models in the table and the models with reported alternatives in chapter 5.2 shows that the most important variables, that is time, cost and floor space, have a higher degree of significance in the models with coded alternatives. The values of the coefficients are also higher in these models. Thus, the observed improvment of the estimation results when using coded alternatives is not only depending on the greater amount of information which is contained in the larger choice set, but also on the fact that the alternatives are coded and not reported.

5.4 Conclusions about the choice set

The results show that reported alternatives are inferior to coded alternatives. A likely explanation of this phenomenon is that the individuals in the survey do not report the worst alternatives, even if they are considered (or have at some time been considered). When the worst alternatives are not reported, the variance in the variables is reduced and the estimation results are less good. It may also be that the question about considered alternatives is very abstract and difficult, so that the answers are of a lower quality.

However, the coding of alternatives could be done in a more realistic way than has been the case for this data set. This could be done by asking for the type of purchase, and asking for access to bicycle and if the interviewee has some kind of handicap. It would also be possible to identify relevant alternatives with respect to the opening hours of the shops and the household constraints, like working hours.

6. CONCLUSIONS

The work presented here has produced results concerning preferences as well as methods.

The results on preferences show that it is possible to achieve detailed information on preferences, and also that this degree of detail is of great importance (such as having a seat and the existence of bicycle lanes). The stability (or variability) of such detailed results will have to be tested by doing more studies of this kind, for example the Stockholm 1985 Travel Study.

The results on the methodological side add to the knowledge that it is possible to significantly reduce modelling costs by using analyst defined alternatives instead of reported alternatives and by using techniques for sampling from alternatives. Also, it is shown that a car cost variable specification assuming economic rationality is preferable to a more simple specification.

The project will continue with estimation of mode and destination choice models on a data set for the medium-size town Jönköping from 1979. Also, the Stockholm data will be used to estimate more comparable models.

Behavioural Research for Transport Policy
pp. 193–215.

COST-EFFECTIVE METHODS FOR NATIONAL-LEVEL DEMAND FORECASTING

Andrew Daly and Hugh Gunn

Cambridge Systematics Europe b.v.

Introduction

Planning infrastructure investment for road and railway networks requires demand projections over a period of at least twenty years. To meet these needs in The Netherlands, it has recently been decided to develop a system for deriving long-term forecasts of traffic flows on the national highway and railway networks. An important part of this system is the "National Model", which will be used to predict the changes in passenger traffic flows that will result from the construction of new infrastructure, from policy decisions on the pricing and control of the transport system and from wider social and economic changes. This paper describes the design of part of the National Model.

In the current uncertain economic situation, policy interest no longer focusses on the need for large-scale investment to cope with continuously increasing traffic. Rather the interest is in efficient exploitation of existing investment to meet current needs and to anticipate those that will be generated by social and economic developments. Details of the planning context in which the National Model is to be applied are given by van Zwam and van der Valk (1984).

The design chosen for the National Model reflects this shift in emphasis, in that many data sources are expoited, and several different modelling procedures are used, specifically to allow the models to reflect the wider range of changes that are now of vital interest. In this respect the models presented in this paper differ substantially from those applied in earlier national-level studies.

The execution of the study itself also recognises the current economic climate, in that existing knowledge and data is exploited to the greatest extent possible to reduce the cost of producing the forecasts. The cost of the study and the time-scale within which it will be completed are thus significantly reduced relative to earlier national-level studies. A further advantage built into the current study is that the models will be relatively cheap to update in future years when new data becomes available.

To meet the needs of infrastructure planning the National Model must be capable of making forecasts of flows on networks defined in considerable detail. Provincial roads form essential connectors and have been incorporated into the network along with the national and international highways. The models described in this paper are concerned only with <u>weekday personal</u> traffic on this network. Weekend traffic is not used to set the highway system capacity, and freight traffic is the subject of a separate forecasting study, conducted by the Nederlands Vervoersweten-schappelijk Instituut.

Weekday passenger movements on the railway system are also to be forecast by the model. In addition to the entire passenger railway network of The Netherlands, a small number of lines of other public transport modes have been included where these form a structural part of the national network.

The main objective of the National Model is thus to produce forecasts of weekday personal traffic flows on each of the network links; this for a given set of assumptions about a future year, and for given infrastructure developments in national road and railway networks. In addition to these assignments, to evaluate the network options it will be necessary to produce overall figures, for example of traveller kilometres, and possibly to give forecasts for "slow" mode travel (i.e. bicycle and walk trips).

For network design purposes, the primary sensitivity of the model will be to predict response to travel time changes. That is, when a new link is introduced or an existing link is improved, the demand will be increased because of the benefit of the reduction in travel time. However, over the forecast period there will also be substantial and uncertain changes in a range of network and socio-economic variables, known to be important determinants of travel demand. The model will be required to forecast response to the expected changes and to indicate the sensitivity of the forecasts to uncertainty in the socio-economic variables. These uncertainties will be expressed in the form of alternative "scenarios" of coherent socio-economic developments. These scenarios are being constructed in a separate study, in which the Nederlands Economisch Instituut is playing a leading rôle.

Each scenario will contain information about the expected value of each of the key variables to which the model has to be responsive. Apart from travel time, these variables can be classified as follows:

- travel costs (e.g. fuel prices, public transport fares, parking charges);

- operating variables (e.g. public transport frequencies);

- economic indicators (e.g. employment rates, income distribution, working hours);

- social variables (e.g. household size and structure, age distribution of the population);

- planning variables (e.g. location and size of housing and employment developments).

For evaluation purposes, the impact of changes in each of these variables must be assessed for each of a number of socio-economic segmentations of the population.

Clearly, the development 'from scratch' of a model with this range of sensitivities, forecasting for the whole country with the locational detail and accuracy required for long-range network planning, would present an enormous challenge. Fortunately, work in recent years in The Netherlands has yielded a large body of knowledge concerning travel patterns. Much of this information is to be applied in the National Model, reducing substantially the cost and effort required to develop the model and at the same time increasing its relevance and effectiveness.

Three types of data are of special relevance. First, home interviews have been conducted both for a number of recent regional or urban studies (which have been exploited to set up the models used for those studies) and, more important than these regional surveys, the national home interview survey (Onderzoek Verplaatsingsgedrag, or O.V.G.) conducted by the Centraal Bureau voor de Statistiek (C.B.S.) each year since 1978, interviewing about 8500 households per year. The latter survey gives an immense amount of detail about travel patterns and their development over time. Second, large numbers of road and rail surveys have been conducted for regional and special-purpose studies and more surveys have been made specifically for the National Model. This data gives a picture of travel which is at once <u>more accurate</u> but <u>less detailed</u> than that available from home interviews. In particular long distance travel is more accurately recorded in this type of data than in home interviews. Finally, counts of traffic have been made on large numbers of links throughout the road network, and figures are also available for the traveller flows in the railway network. This data is again <u>more accurate</u> but <u>less detailed</u> than the road and rail interviews. The appropriate combination of these three types of data can give a picture of travel patterns that is both accurate and detailed. For the National Model, data collected in each of these ways is being consolidated into a "base matrix" giving the best available description of base-year traffic flows: this work is being done by Bureau Goudappel Coffeng b.v..

The Netherlands is fortunate in the richness of the data that is routinely collected. For the National Model study, background data will be exploited giving population, employment and car ownership, each of which is available in some detail for each of the municipalities in The Netherlands. Data is also available describing land uses. Road and railway networks have been prepared in computer-readable form for past studies and require little effort to bring up to date.

In addition to this data, much information is available in the form of aggregate and disaggregate models describing and forecasting the behaviour of travellers. The National Model will exploit four specific model systems that have been developed during 1982-84.

First, a system of disaggregate models developed earlier in one region (the "Zuidvleugel" models, see Daly et al., 1983) has been transferred to the Utrecht region (see Gunn et al., 1985). This model system has known transferability characteristics, and will be extended to allow forecasting for the whole country by the use of "artificial" sample enumeration (Gunn et al., 1983). The National Sample Enumeration System (NSES) resulting from this extension will be used to derive most of the personal travel response characteristics needed for the National Model.

Second, a study of long-distance travel in The Netherlands, conducted by The MVA Consultancy Ltd., gives a further possibility of estimating response characteristics for longer distances. The models developed in this Long-distance Travel Study will be exploited in the National Model for forecasting the responses of long-distance business travellers.

Third, work by van den Broecke (1982) exploits aggregate time-series data on driving licence acquisitions to develop a forecasting procedure based on age-sex cohorts. The key advantage of this procedure is that it allows time-dependent effects to be integrated into models that are otherwise purely cross-sectional.

Finally, overall national car-ownership levels are provided on the basis of key elasticities and the extrapolation of historic trends (van der Hoorn, 1984).

The key to the design of the National Model is the simultaneous use of these data sources and models. Given the varying levels of accuracy of the material, a system has been designed to exploit the more accurate, less detailed information to adjust the more detailed but less accurate sources and models. An important methodological part of the approach is the use of "pivot-point" techniques, which are described in general in the first section of the paper. The second section gives an overview of the design and structure of the whole system.

The third section of the paper describes how the aggregate car ownership and licence-holding models are applied to constrain the corresponding disaggregate models to obtain reliable forecasts for a wide variety of socio-economic groups. The fourth section goes on to describe the processes, similar in intent, in which data from the OVG home interviews, roadside and train surveys, and traffic flow data are used to constrain disaggregate model forecasts by the use of "pivot-point" methods.

In the fifth section, attention is given to the problems of calculating level-of-service measures for the transportation networks. The key issue here is the level of accuracy at which locations are represented.

1. APPLICATION OF PIVOT-POINT METHODS

Pivot-point methods are often applied in travel demand forecasting when a very accurate estimate of demand is available under one specific set of circumstances (Manheim, 1979). Relying on this estimate, the model can then be used to adapt that estimate to different circumstances by predicting only the change in demand caused by the change in the circumstances. Providing the changes are small, and the accuracy of the point estimate is good relative to the model accuracy, the forecasts obtained by "pivoting" about that point can be much more reliable than forecasts obtained by the use of the model to derive overall levels as well as changes.

The point estimate used for pivoting is most often derived from a survey of the current sitation. The model is then used to predict the changes from the current situation that will be caused by changes in policy or other variables influencing behaviour. Pivoting can also be used, however, when a forecast is available and believed to be accurate, but when further forecasts are required under different circumstances. For the National Model, pivot-point methods based both on surveys and on reliable but inappropriately-aggregate forecasts are used.

A variety of mechanisms are available for implementing the pivot-point procedure. One mechanism is to use the model to calculate demand elasticities and then apply these to estimate demand changes from the known point. An apparent drawback of this method is that the elasticities given by most models are far from constant over the range of circumstances to which the model is applicable. Errors introduced by the variation of elasticities can be reduced by segmentation of the range of each variable, but when there are several important variables (as in the National Model) the combination of segmentations will require the calculation of very many elasticities.

A further problem in the direct use of elasticities is to ensure that demands for alternatives that interact with each other (e.g. a modal split) are consistently estimated. This problem becomes most intractable when the availablity of alternatives varies between particular circumstances.

A more sophisticated approach that does not appear to suffer from any of the above problems is to exploit the reliable pivot-point estimate to introduce correction terms into the model itself. The corrected model can then be used to produce forecasts which are consistent with the point estimate but with none of the inconvenience, approximation and potential inconsistency associated with the use of elasticities. A further advantage, when it is not certain that the point estimate is absolutely reliable, is that the correction term can be examined to see whether it is reasonable. An unreasonable value for the correction term may be evidence that the point estimate itself is unreasonable. For the models described in this paper, this more sophisticated approach has been used wherever possible.

For clarity, it should be emphasised that the question of using elasticities derived from a model or of using a corrected version of the model is purely one of determining the most appropriate method for calculating forecasts, given an underlying forecasting model. The basic accuracy of the underlying model is not an issue. Even if elasticities are not employed as part of the forecasting process, it remains useful to calculate them to give an impression of the accuracy of the underlying model, and this will be done for the models imlemented in this study. In particular, it is useful to compare elasticities derived from different sources.

Pivot-point procedures are particularly useful in large systems like the National Model, where several different data sources and models of varying accuracy are being used. In this context, the pivot-point method offers the possibility of forecasting with sufficient accuracy to inspire confidence, yet sufficient sophistication to respond to all the changing variables that are of interest to policy makers.

2. SCOPE AND STRUCTURE OF THE NATIONAL MODEL

The objectives of the National Model in producing traffic assignments for network planning are broadly those of classical transport planning. It is therefore to be expected that the National Model should contain models forecasting travel frequency (generation), destination choice (distribution), mode choice and assignment, as are found in the classical four-stage forecasting models. In addition, models are incorporated for forecasting licence holding and car ownership, usually treated as exogenous in the classical models.

The component models to be employed in this system are derived from the Transferability Study (Gunn and Pol, 1985), which transferred to the Utrecht region a system of models originally developed in the Zuidvleugel area (Daly et al., 1983). These regional model systems incorporate the component sub-models required for the National Model, with the exception of the assignment procedures, which have to be added. During the Transferability Study, the models have been extensively tested and improved with regard to their transferability properties.

These regional model systems are fundamentally disaggregate: that is, the model represents the travel behaviour of an individual traveller or household. For the National Model, as for the original regional model systems, the travel of the whole population is represented as the sum of the travel of a large number of individuals: an explicit aggregation process is required. In classical studies, aggregation has typically been achieved by modelling based on zones. However, this procedure has many disadvantages of inflexibility (see, for example, Daly,

1982) relative to the more recently developed method of sample enumeration. For the regional systems, sample enumerations were used for short-term and general policy evaluations. For the National Model, the technique of sample enumeration is being extended to deal with longer-term and more location-specific forecasting, such as is necessary for making traffic assignments.

This extension of sample enumeration is achieved by two methodological developments, whose consequences are explained in detail in subsequent sections. First, the technique of "artificial" sampling is used to re-weight a sample drawn in the base year to be representative of developments that are expected for the forecast year. Second, synthetic networks and sampling of destinations are used to allow a more detailed representation of location and accessibility than is common in a study at this scale, without imposing an impossible computational burden.

Details of the design are given in the following sections. The objective in the design process is always to incorporate the effects that are most important for policy and influential on behaviour, recognising the constraints on complexity imposed by the national scale of the model. Note also that the national scale of the model reduces (but does not eliminate) the problem of modelling trips entering the study area from external areas; this characteristic is useful in achieving a representation of all the traffic with a method based on sample enumeration.

3. CAR OWNERSHIP AND LICENCE-HOLDING MODELS

Car ownership and (even more often) licence holding are generally treated as exogenous by large-scale models. In the National Model, however, these variables are treated as endogenous and will be forecast by component models. The inclusion of car ownership is particularly important because evaluation will take account of economic scenarios differing in the income levels assumed: the most significant traffic consequence of income change is in car ownership.

The car ownership and licence-holding models applied for the study are those initially developed for the Zuidvleugel area and subsequently successfully transferred to the Utrecht area. They differ from other car ownership models in incorporating an explicit treatment of licence holding and a range of other explanatory variables, primarily socio-economic in nature. However both the Zuidvleugel and Utrecht model systems were based exclusively on cross-sectional data and will therefore fail to incorporate significant time-dependent changes, e.g. in tastes, attitudes, etc.; moreover, the income dependence derived from cross-sectional data may not be appropriate for income changes over time. To correct for these changes, the models are constrained to the forecasts given by less detailed time-series models, the cross-sectional models being used effectively only to distribute a given level of ownership across the population.

The following sub-sections respectively describe the general nature of the models, the way in which "artificial" samples are constructed to be representative of base and forecast year poulations, and the differing ways in which the two models are constrained.

3.1 Overview of Car Ownership and Licence Forecasts

Recent research has concluded that both driving licence holdings and car ownership levels are significantly affected not only by prevailing levels of such background variables as household size, income and accessibility, but also by a complex of other factors unrelated to these. This latter set of factors is thought to include tastes, expectations and general 'lifestyle' effects; crucially, the combined impact of these 'lifestyle' factors has been changing over time.

The Transferability project has gone some way to establishing that the influence of the measurable background variables has remained fairly constant over recent years, and that the 'lifestyle' developments are in some sense independent from the changes in the measured variables.

The apparent importance of time trends in the development of both licence holding and car ownership argues strongly for the adoption of models which incorporate an explicit time dimension; for example, the cohort-based analyses of licence holding due to Van den Broecke (1982), or the acquisition and scrappage models of the car market due to Cramer and van der Vlis (1973) and Mogridge (1983).

However, these models operate at an aggregate level. Further, they do not distinguish clearly between the developments that are attributable to the measurable background variables (mainly socio-economic) and those which are not. For many purposes (and specifically for the National Model), there is still a need to generate forecasts capable of detailed disaggregation at a zonal level. For this reason, and given the supporting evidence of the Transferability study, a model system has been developed which uses aggregate forecasts from the time-series based models as high-level controls on the predictions of a set of disaggregate models which jointly predict the effects of the measurable background variables on both licence holding and car ownership.

The major innovation in this system is the use of so-called 'artificial samples' of households. For each zone in the country, detailed records are assembled for a sample of around 1000 households, containing both characteristics of the household and of its members. For each household, an appropriate expansion factor is calculated on the basis of a given set of key aggregate statistics for the zone (numbers of households, numbers of people in given age and sex groups, numbers of workers of each sex). These statistics are routinely collected for other purposes.

This sample is sufficient to allow the disaggregate models of household licence holding and car ownership to be applied separately for each zone in the country, and thus to augment each household record in the detailed 'artificial sample' with predictions of car-ownership and licence status.

These predictions can be expanded and summed to give aggregate predictions at both the zonal and the national level; the disaggregate models can then be adjusted to achieve consistency of the disaggregate predictions with any selected aggregate controls.

3.2 The Construction of the Artificial Samples

In the overview given above, it was stated that each zone in the system would be represented by around 1000 households, each with an appropriate expansion factor. The first major simplification to the process is to use the same set of 1000 households for more than one zone; differences between zones are thus reflected in differences in expansion factors rather than entirely different household and person records. There are arguments for and against the use of larger numbers of sets of prototypical households (these will not be discussed here). In practical terms, the problems are ones of computer storage and processing time. Several such sets could easily be used; one per zone (with several hundred zones in the system) would not be practical.

The construction of the artificial samples for the zones thus falls into two parts. Firstly, the selection of the 1000 'prototypical' households, and secondly the calculation of expansion factors. Both parts are controlled by the nature of the aggregate zonal data made available. As mentioned above, this data contains details of numbers of households, population totals by age and sex, and workforce participation by males and females separately. Thus both the prototypical sample itself, and the way in which the expansion factors are calculated, must be capable of generating zonal populations which differ in respect of these attributes.

There are of course other equally important considerations in the generation of the sample. The expanded zonal population must be realistic; it is not enough merely to match the aggregate totals. Thus, for example, a correct average household size would not be enough if the implicit distribution of the zonal population into one, two, three etc. person households was not reasonable. This requirement suggested that a 'target' or reference distribution of households be set up, based on an actual survey, and that the predicted artificial samples should not depart too far from that reference distribution.

The disaggregate models predict demand for driving licences and cars conditional on a large number of attributes of the household and of its members. These must be present in the artificial sample, and moreover they must be jointly distributed in a realistic way. The obvious source of such information is an existing survey, and in The Netherlands it transpires that the OVG data contains sufficient detail to apply the models. This data set is used to create the set of prototypical households, as follows.

First, the OVG data is used to construct a categorisation of households by the following dimensions :

1) household size : 1,2,3,4 and 5+ members
2) number of employed members : 0,1,2 and 3+ members
3) whether or not there is a female worker in the household
4) age of head of household : under 35, 35 to 65, over 65.

These dimensions lead to a 120-category classification, of which some categories are empty of necessity (one person households with more than one worker, for example) and some are empty or virtually empty despite being logically possible (one person households with age of head of household over 65 years and one female worker, for example).

In fact, in a trial using 1982 data, 96 % of the OVG households fell into just 44 of these categories.

Accordingly, it was concluded that the artificial sample could consist of households drawn from just these 44 categories, and thus that there should be only 44 distinct expansion weights per zone. Further, the base year distribution of households between these categories gives just the reference distribution that is needed to judge the credibility of the expanded samples.

There is a great deal of variation in the size of the categories. To ensure that each category was represented by a reasonable number of households in the prototypical sample, target numbers are set for each category (none less than ten households) thus defining a different sampling rate for each category, say 1 in N_i for category i. Then every N_i'th household was selected from that category in the survey data set.

For the samples drawn in each category, the average numbers of persons falling into each of eight age-sex categories is calculated, along with the average numbers of male and female workers. The zonal planning data was grouped to give totals for the same categories, and to give total household numbers in each zone. Thus there are 11 items of planning data for which target totals exist.

Using this grouping of the zonal planning data, and the reference distribution of households between the 44 categories in the base year, expansion factors are calculated for each zone for both the base and a forecast year.

In practice this is achieved by defining an objective function for each zone and each year as follows :

$$OBJ(N_i) = \sum_s \left(L_s\left(\sum_i (W_{is}.N_i) - P_s\right)^2\right) + \sum_i \left((N_i - R_i)^2\right)$$

where N_i denotes the unknown numbers of households in category i (i = 1...44)

P_s " " known totals of planning data item s (s = 1...11)

W_{is} " " known mean numbers of item s in category i

R_i " " known 'reference' target for category i

and

L_s " a weight used to increase or decrease the relative importance of the s'th item of planning data in the objective function.

The N_i are then chosen as those (non-negative) values which minimise the objective function, thus in some sense maximising the correspondence of the artificial sample with the given aggregate zonal planning items and the a priori distribution of households between categories (the 'reference' distribution). In principle, there could be a different refernce distribution for each zone. In practice, information will be available only at an area-wide or country-wide level, so that the same reference distribution will be used for different categories of zone.

The objective function being a simple sum of squares, a solution algorithm based on the Newton-Raphson technique is easily programmed.

Two further points can be mentioned here concerning the derivation of the estimates of the N_i. Firstly, it proves better to adjust the R_i in an ad-hoc way during estimation : the fit of the sample to a-priori expectation is still judged against the original reference targets, but a closer correspondence could be achieved by varying the R_i in the objective function. Secondly, the weights L_s are chosen to reflect a broad judgement on the items of planning data that were most important to fit.

Note that the procedure does not treat any of the input data items as constraints; the aim is not to achieve complete equality with all of the data items (although this will happen automatically if it is possible). In general, it will not be possible to match all planning data items whilst preserving the average reference distribution of households between categories, and an acceptable compromise must be reached. This is the function of the procedure.

The procedure described here has been employed in a recent application, providing car ownership and licence-holding forecasts for the zones of a national zoning system. These forecasts are being used in applications of a long-distance travel forecasting system (see Ashley and Gunn, 1985). In that application, few problems were encountered in adjusting an OVG sample from 1982 to fit the planning data targets; the zones of greatest difficulty proved to be the older city centre and the new areas in the polders, both of which had household structures very far from the average. It was also necessary to use different sets of reference distributions for the future year, reflecting the predicted trend towards smaller households and higher female workforce participation.

This completes the work necessary to generate the artificial samples for each zone in the system. Reasonable correspondence with aggregate zonal planning data items is ensured, whilst retaining a credible distribution of households of different sizes, number of workers and so on. The single prototypical sample of households, together with the set of zone-and-year specific expansion factors, is then available as a basis for the application of the models.

3.3 Constraining the Licence-Holding Models

Summing over the zone-specific expansion factors in base and forecast years gives an artificial sample representation of the overall national population (i.e. a national level expansion factor can be given to each category in the prototypical sample). These national-level expansion factors are also applicable for each individual in the household.

The disaggregate models of household licence holdings can then be applied to each household in the prototypical sample, with resulting predictions of the licence ownership status of the adults in the household. These predictions can be expanded at the level of the persons, and aggregated to give predictions of licence-holding by broad age and sex bands.

The age bands over which the disaggregate forecasts are aggregated can be chosen to match those used by Van den Broecke (1982). The van den Broecke forecasts contain no regional variation or socio-economic variables other than age and sex. The forecasts produced by this cohort model are however extremely robust in their treatment of time-dependent effects and are thus suitable for use as constraints for the totals of the disaggregate models, which then serve to give regional and other socio-economic variation. The procedure works by introducing correction terms into the disaggregate model to adjust its forecasts to those of the van den Broecke model for each age-sex cohort.

An important consideration is that the fashion in which the model is adjusted is compatible with the basic hypothesis that there are unmodelled trend effects which are essentially independent from the influence of the explicit variables in the model. The model consists of a set of equations giving the predicted utilities of 4 possible household 'states', being defined for the licence owning status of the first two adults in the household as follows :

1) neither adult licenced;
2) first adult licenced, second not;
3) second adult licenced, first not;
4) both adults licenced.

The model can be manipulated to give the unconditional probability that the first adult is licenced, which can be used if there is only one adult in the household. Licence holding by third or subsequent adults is not covered by the model.

The utilities are expressed as functions of measurable background variables (age, sex, working status and so on) plus three 'alternative specific constants' to adjust for the net effect of unmeasured variables. Simple theoretical reasoning, backed up by the results of the Transferability study, suggests that it is the constants that should be adjusted to achieve consistency with the van den Broecke forecasts.

This is what is done in practice. In the base year, the alternative specific constants are adjusted to bring the model predictions into line with the base year licence-holding statistics for both sexes by four age groups : 18 to 29 years, 30 to 49 years, 50 to 64 years and 65 years and older. For the forecast year, the cohort-analysis-based predictions of van den Broecke are aggregated to give figures for the same eight categories, and another adjustment made.

In fact, the procedure does not give an exact fit to the eight 'target' statistics; the models are of household licence holdings, not personal, so it is not possible to adjust differentially for specific age/sex groups. However, it is the case that the age of the 'second' adult is usually in the same category as that of the 'first', and that the 'first' adult is male and the 'second' is female. Thus by partitioning households by the age of the 'first' adult, and associating the constants for the 'first' adult with male licence holding and those for the 'second' with female licence holding, it is possible to make good approximations to the required adjustments.

This completes the licence constraint process. For each zone, we now have an artificial sample complete with predictions of probabilities for the different possible household licence states. The next stage is the application and constraint of the car-ownership models.

3.4 Constraining the Car-Ownership Models

The procedures used to adjust the car-ownership models are essentially equivalent to those used with the licence models. Once again, the mechanism for implementing the control is through the alternative-specific constants, and the adjustment is done until the required aggregate predictions are achieved.

There are two respects in which the process differs somewhat from the licence constraint. Firstly, instead of a single model addressing all possible alternatives, the household car-ownership predictions are assembled from two linked models. The first of these predicts the probabilities with which households with a single licensed driver own one or no cars. The second predicts the probabilities with which households with two or more licensed drivers own one or more than one car. A small, fixed probability is assigned to the possibility that a household has two or more licensed drivers but has no car; other states are assumed not to occur. Thus, for example, it is assumed that there are no households with cars but no licensed drivers.

The models jointly determine the proportions of households with no cars, one car or more than one car, and some thought was given to desirability of retaining 'sensible' ratios between these three states. The mechanism that was eventually adopted to achieve this was to link the adjustment of the constants between the two models in such a way as to mimic the increase or decrease of household income. Thus the ratio of the adjustments made to the models is held equal to the ratio of the marginal utilities of an increment in household income in the two models. Having done this, we shall refer to a single adjustment being made to these models rather than two independent adjustments.

The second respect in which the car-ownership constraint procedure differs from that used for licences is in respect of the level at which the constraints take place. Mean zonal household car-ownership estimates are available for the base year, and separate 'adjusted' models can be fitted for each zone, using the base year artificial samples. For the forecast year, only a single national total is available; each of the set of zonal models is then adjusted by the same amount until the application to the forecast year artificial samples returns the desired total.

Thus the constraint procedure was made consistent with the estimation of an unknown additional component of utility necessary to adjust the model predictions for factors omitted from the model specification. For the base-year zonal adjustment, these omitted factors would include local accessibility and income effects. For the forecast, the single adjustment would be associated primarily with changing tastes and income levels.

In the long-distance application of these models, mentioned previously, the zonal planning data and licence and car-ownership totals that were input to the procedure indicated continuing strong growth in licence holdings, some increase in cars owned but an even larger increase in numbers of households (with corresponding reduction in household size). The adjustments that had to be made to the models were, qualitatively, as follows :

a) in the forecast year, the constants associated with

greatly increased to achieve the predicted aggregate levels. This was especially true of licence holding amongst 'second' adults, typically the females.

b) The constants associated with the propensity to own vehicles conditional on household licence status had to be decreased as between base and forecast years.

The successful use of these procedures achieved in the long-distance application confirms the feasibility of the method. The use of the artificial samples enables a systematic account to be taken of the development of the major background variables. At the same time, the approach of re-estimating the alternative-specific constants allows the reconciliation of the overall levels with the time-series based forecasts in a fashion that is compatible with the initial hypothesis of omitted effects orthogonal to those background variables. The output from this process is, for each zone and for base and forecast years, an artificial sample of households complete with expansion factors, predicted car ownership and licence holding probabilities. There remains the task of applying the sequence of disaggregate travel demand models.

4. TRAVEL DEMAND MODELS

The application of the travel demand models falls naturally into two interconnected parts. First, for each individual and each travel purpose, the joint choice of mode and destination is modelled conditional on household car ownership and individual licence holding status, and as a function of a particular set of socio-economic and network variables. Secondly, using the probabilities of the different household car ownership and licence states from the earlier models, estimates are formed of the expected maximum utility from making the tour; this is then used with a different set of socio-economic variables to predict the number of tours actually made by household members. Note that the travel demand models operate at the personal level in contrast to the household level of the car ownership and licence models.

4.1 Computational Simplifications

There are two key simplifications which have been made to allow this process to be completed with acceptable use of computer time. First, it was noted that the only characteristics of the traveller to emerge as stable, significant influences in mode and destination choice were sex, licence holding, the numbers of cars owned by the household and the number of household members with a driving licence. Given that the models restrict household car-ownership and licence status to one of five possibilities -

1) no licences, no cars
2) one licence, no cars
3) one licence, one car

4) two licences, one car
5) two licences, more than one car -

there are thus a maximum of 2 (sex) by 2 (licence status) by 5 ('car competition' classes) different types of individual for which forecasts will be made. In fact, two of these are impossible (a licence owning member of either sex belonging to a household with no licenced members), leaving only eighteen distinct types of individual for which the models must be applied.

This reduction enables a large saving in computation to be made; instead of predicting probabilities for each member of the prototypical sample (1000 households, approximately 2750 individuals), we need only work with 18 'individuals' in the mode and destination choice models. Some of this simplification may be lost by the need to retain age distinctions in order to determine public transport fares.

The second major simplification concerns the treatment of the accessibility variable, as measured by the logsum estimate of the expected maximum utility. This is approximated by a step function over a range from very high accessibility to very low accessibility, taking only 20 distinct values.

The savings in computational effort from this simplification come in the interface between the models of mode and destination and the tour frequency models. For each of the 44 household groups in the prototypical artificial sample, the tour generation models are applied to each individual for every travel purpose, and for each of the 20 possible levels of the accessibility variable. This is done for each travel purpose (eight in this case) and for each possible state of household car-ownership and licence status. The predicted probabilities associated with these states are then used to produce the probabilities with which the individual falls into the eighteen possible 'types' of individual, and hence the expected numbers of tours generated by each of the eighteen possible person 'types'.

The reduction in computational effort due to the categorisation of the accessibility variable can now be seen; this procedure need only be performed once, retaining the results for each of the eighteen person types, 8 purposes and 20 possible accessibility levels, A 'zone' is now characterised only by an accessibility level; the computation is reduced by a factor of the number of zones divided by number of accessibility levels, a factor of about 60.

Thus for each zone the expected maximum utility from the mode-destination models for the various person types dictate the accessibility level; this enables a look-up of the expected numbers of tours generated by members of each of the 44 prototypical household groups, and then the appropriate expansion factors from the artificial sample can be applied to give the predicted numbers of tours by household members. These can be collected by purpose (say 8) and person 'type' (18 in all), and used to expand the predictions of the probabilities with which specific mode-destination pairs are selected for the different purposes.

It should be noted that the exact definitions of the purpose split and of the numbers of distinct types of individual and levels of accessibility have not yet been finalised. In particular, there may be advantage in amalgamating 'similar' travel purposes and in increasing the number of types of individual (as mentioned above) to identify those entitled to concessionary public transport fares.

4.2 Adjusting the Models to the OVG Data

The car-ownership and licence holding procedures described in Section 3 above used the idea of model adjustment to achieve agreement with exogenous aggregate totals. If the adjustment procedure takes the form of a re-estimation of alternative specific constants, it has been argued that the resulting 'adjustments' can be interpreted as estimates of omitted factors, and can be kept constant for forecasts without distorting the influence of the background variables in the models.

It is proposed that similar procedures be applied in the case of the tour generation and mode-destination models, using the OVG data in the base year as a source of aggregate statistics. It appears that sufficient data exists to define the aggregate demand by broad mode class and distance to destination, at least if purposes are grouped and the statistics are aggregated to a broad categorisation of the country to 'east', 'west', 'north' and 'south'.

4.3 Adjustment to Base Matrix

The preceding sections describe how the travel demand models will be set up and adjusted in the light of the OVG home interview data. A further important data source being used in the National Model is en route (or "screenline") interviews conducted with car drivers or train passengers. In a parallel study, this information is being processed to give a "base matrix" - a matrix of origin-destination flows based on a 345 zone system covering The Netherlands. As noted in the Introduction, this work is being carried out by Bureau Goudappel Coffeng bv.

As also noted in the Introduction, although interviews of this type give less information per interview than a home interview typically gives, the number of interviews is often much larger and the resulting traffic flow estimates are correspondingly more accurate. Clearly, data collected in this way and incorporated into the base matrix can add substantially to the description of current-year traffic.

It is proposed to exploit the screenline data by using it to estimate correction terms to introduce into the models. A range of possibilities is available extending from the introduction of a small number of terms correcting particular trip length/purpose classes to the introduction of correction terms for every cell in the matrix (analogous to a complete matrix of 'K-factors'). This last option is probably not desirable, since the matrix is not expected to be of the same good quality for short trips as for longer trips, since the short trips are much less likely to be intercepted at the screenlines. Thus at the short end, and depending on the results, correction terms may be drastically simplified or suppressed.

This proposed treatment of the base matrix information is analogous to the treatment of the OVG data and the aggregate car ownership and licence holding data described previously. That is, the model is corrected to a fixed point value believed to be more accurate than an independent use of the model, and forecasts are then made relative to that fixed point: "pivoting" around it.

4.4 Differential Assignment

A final use of the pivoting procedure is proposed in the assignment process. For many links in the national networks, good information exists for the observed flows. This information can be used to improve the quality of the forecasts by the following procedure.

An assignment of the base matrix to the base-year networks will yield a vector of flow differences or errors D_i for each link i in the networks (highway and public transport). Note that the set of links for which observed flows are available will probably not cover the entire network, but will certainly include any existing links for which forecasts are required. These flow differences are due to errors in the base matrix, in the assignment procedure or (less likely) in the link flow observations.

For a future year, a new network will be hypothesised and a new matrix will be forecast by the model. Both the base matrix and the forecast matrix will be assigned to the new network. For the links in the new network that correspond to links in the base-year network (probably the overwhelming majority) the flow differences D_i can be applied to correct the base matrix assignment, i.e. assuming that the errors D_i remain in the future year. The assignment of the forecast matrix should be corrected by the same <u>proportion</u> as the assignment of the base matrix. Symbolically,

$$F^*_i = F_i \; * \; (B_i + D_i) \; / \; B_i$$

where F^*_i is the corrected forecast flow on link i, F_i is the assignment of the forecast matrix, and B_i is the assignment of the base matrix to the new network. This formula assumes that the base matrix is correct (i.e. a complete matrix of K-factors has been incorporated) and requires adjustment if it is decided to overrule the matrix for some short-distance cells.

The method proposed takes account of the fact that assignment errors and differences produced by network changes are best treated by absolute differences, whereas the relationship of the base and forecast matrices is primarily one of proportionality.

5. NETWORK LEVEL-OF-SERVICE MEASURES

One of the most important inputs to the forecasting models, particularly for the mode and destination choice modules, is the level of service given by the networks. Level-of-service measures include travel time (in various components) and travel costs. Developing networks giving full details of travel possibilities at national level would be computationally impossible (although most of the data exists); methods were required and have been developed that enable approximations of level-of-service measures to be derived without excessive computation.

5.1 Specification of Location

Treatment of levels of service requires specification of the locations represented in the model. The base matrix, other data provided for the study and the long-distance forecasting models all operate at the level of 345 zones nationally; assignments will probably also be made using this zonal system. It would have been natural to define location for the NSES also at the level of these zones. However, for two main reasons the 345-zone system, although giving a reasonable interface with the national networks, was not acceptable for the representation of behaviour.

First, the accessibility of (national level, i.e. train) public transport usually depends on the use of "slow" modes (walk, cycle) or urban public transport (tram, bus). For these modes a difference of two or three kilometres in access can be decisive. The large zones surrounding many railway stations contain a wide range of accessibility, and cannot be used to determine how many travellers have a reasonable train option.

Second, and perhaps more important, at the 345 zone level it cannot be determined how many opportunities are within the range of slow modes. With these large zones, the overwhelming majority of trips are intrazonal, with trip lengths uncertain. Trips into directly adjacent zones can be quite short or up to 20 km in length. Reducing the zone size enables a much more realistic representation of the pattern of local opportunities and their accessibility.

It was therefore decided to represent location, for the purposes of modelling access to local opportunities only, at a level significantly more detailed than 345 zones. Approximately 1200 sub-zones have been specified (consistent with the boundaries of the 345 zones) and are used for this purpose. A number of considerations have to be kept in mind that relate to these sub-zones.

Data for base and forecast years is primarily being prepared at the 345-zone level. Therefore, applications at the sub-zone level must take account of the potential difficulties arising in splitting information from 345 to 1200 zones. It has proved relatively easy to derive base-year population figures for the sub-zones, and where no other indications exist the population figures will be used to split other data items. Although this procedure is somewhat approximate, it compares very favourably with the alternative of assuming all activity to be located at the centroid of a large zone.

The number of the sub-zones is too great to permit manual coding of connections between them and the networks. Automatic procedures for achieving these connections are described in the following section. However, the coding of connections to the network at the 345-zone level has also proved too expensive to be undertaken manually, and in the case of the larger zones it is vital to intervene manually because of the need for multiple connections of every zone.

The number of the sub-zones is also too great to permit full matrices to be forecast. It is proposed instead to sample from the matrices, selecting all origins but a carefully chosen sample of destinations. Destination sampling is now a well-established procedure, for both model estimation and application (see Ben-Akiva, Gunn and Silman, 1984). The number of destinations sampled can be chosen to balance appropriately the needs of accuracy and the computational costs. A stratified sampling scheme has been adopted in which the more important nearer destinations are sampled with higher probability (even probability 1.0 for intrazonal and adjacent zones) and lower probabilities are used for more distant zones.

5.2 Synthetic Level-of-Service Measures

Crucial to the treatment of location in detail and at national level is the development of synthetic level-of-service measures. In principle, the "Basisnetwerk" system contains more than enough detail about the road system than is necessary for making the forecasts, but it was never intended that all of this data should be used simultaneously: the volume is too great to employ in developing level-of-service measures, simply because of the computer time that would be used, but also because of the impossibility of checking so large a network. Thus a selection of the most important road links is drawn from the Basisnetwerk. Public transport information simply does not exist in the correct format apart from the selected network to be used for forecasting. Coding the detailed urban bus and tram networks that give access to the train would naturally have been far too much work.

Thus the only realistic means of completing the level-of-service data required for the study is to generate synthetic links, based on straight-line distances. This method has been employed successfully to represent walk access and complete walk journeys in previous public transport studies (Daly and Last, 1975), but for the National Model the application will be extended to apply also to car access to the highway network, bus access to the train network and to represent complete journeys from origin to destination by public transport and car modes, whenever the journey does not make use of the national networks.

An analysis of journeys observed in the surveys for the Transferability study, for which both straight-line and network distances are available, shows that the network distance for car and slow modes may be replaced by a simple function of the straight-line distance with only a small loss of accuracy. The other components of level of service, car time and the times, fare and number of interchanges for public transport journeys, can be represented in this way with a small but definite loss of accuracy, which can be reduced if the density of development of the location of the journey is taken into account.

Measuring accessibility in this synthetic manner gives the practical possibility of working with the much more accurate repesentation of location given by the 1200-zone system. Grid coordinates are relatively easily coded for zone centroids and for network nodes they are included in the data base already coded. Simple algorithms can be developed to find a reasonable set of access nodes on the networks for each of the zones (i.e. a node in each of the major directions radiating from the zone). These access nodes can be used to complete origin-destination paths by car, and to represent both slow and bus/tram access to the train network. Similarly, complete synthetic paths by car, slow and public transport modes can be represented from every zone to the nearby zones, for which use of the national networks would not be necessary.

6. CONCLUSIONS

The paper describes the methods that have been developed to provide planners of national road and railway networks with a convenient and sensitive tool for evaluating network designs. The model system being implemented features sensitivity to the key design variables of travel time and cost, but also enables designs to be evaluated in the context of uncertain economic and social futures.

The expectations for socio-economic developments in The Netherlands are for significant changes in household structures, economic rôles within the household, transportation prices and the distribution of housing and employment within the country.

These characteristics are shared with many other countries. A planning instrument is therefore required that can address the changes likely to be brought about in travel patterns, which may differ radically from present patterns. Evaluation of network options must take account of these potential developments and uncertainties.

Design of the model system has concentrated on the exploitation of existing information, in the form of both data and models, to improve the quality and reduce the development cost of the forecasting instrument. "Pivot-point" techniques play a key rôle in combining uncertain but detailed projections with more reliable aggregate figures.

Car ownership and licence-holding are forecast endogenously, using aggregate methods to forecast the totals and disaggregate models to distribute these totals among the socio-economic groups in the population.

Travel frequency, destination and mode choice are forecast by disaggregate models, adjusted to conform with the information given by the national home interview survey (OVG) and the screen-line survey information incorporated in the base matrix.

The main policy objective of the model is to evaluate potential new or improved links in the national network. To this end, the primary evaluation procedure will be the assessment of traffic flows produced by an assignment. Assignment accuracy is improved by expoiting the information currently available on the flows on important links in the network.

Because of the construction of the model, exploiting the disaggregate modelling method of sample enumeration, it will be possible to analyse the impact of policies of pricing or network change on different segments of the population. Applications of the system other than those making traffic assignments to a network will almost certainly be made. These applications will enable the consequences of a wide range of policy and other developments to be evaluated at national level and exploiting the convenience, quickness and cheapness of sample enumeration.

References

Ashley, D. A. and H.F. Gunn (1985), "The Use of Artificial Samples in Car-Ownership Forecasting", forthcoming.

Ben-Akiva, M., H.F. Gunn and L.A. Silman (1984), "Disaggregate Trip Distribution Models", Proceedings Japan Society of Civil Engineers, No. 347.

Broecke, van den (1982), "Rijbewijsbezit (Driving Licence Holding) in Nederland", Van den Broecke/Social research bv.

Cramer, J. S. and M.B. van der Vlis (1973), "Het aantal personenauto's in Nederland" (The Number of Passenger Cars in The Netherlands), Report to Projectbureau I.V.V.S., Ministerie van Verkeer en Waterstaat.

Daly, A. J. (1982), "Policy Analysis Using Sample Enumeration - An Application to Car Ownership Forecasting" P.T.R.C. Summer Meeting.

Daly, A. J. and A. Last (1975), "Modelling the Walk Mode", P.T.R.C. Summer Meeting.

Daly, A. J., H.H.P. van Zwam and J. van der Valk (1982), "Application of Disaggregate Models for a Regional Transport Study in The Netherlands", presented to World Conference on Transport Research, Hamburg.

Gunn, H. F., M.E. Ben-Akiva and M.A. Bradley (1985), "Tests of the Scaling Approach to Transferring Travel Demand Models", Transportation Research Board Annual Meeting.

Gunn, H. F., P.D. Fisher, H.D.P. Pol and A.J. Daly (1983), "Synthetic Samples as a Basis for Enumerating Disaggregate Models", P.T.R.C. Summer Meeting.

Gunn, H. F. and H.D.P. Pol (1985), "Model Transferability: the Potential for Increasing Cost Effectiveness", presented to International Conference on Travel Behaviour.

Hoorn, A.I.J.M. van der (1984), "Car-Ownership Forecasting for the Zuiderzee Line", Internal Memorandum, Dienst Verkeerskunde, Rijkswaterstaat.

Manheim, M. L. (1979), "Fundamentals of Transportation Systems Analysis", Vol. 1, M.I.T. Press.

Mogridge, M. J. H. (1983), "The Car Market", Pion.

Zwam, H. H. P. van and J. van der Valk (1984), "Long Term Policy and The National Model", P.T.R.C. Summer Meeting.

Behavioural Research for Transport Policy
pp. 217–243. © 1986 VNU Science Press.

Model Transferability : the potential for increasing cost-effectiveness

Hugh F. Gunn (Cambridge Systematics Europe bv)

Heero Pol (Dienst Verkeerskunde, Rijkswaterstaat)

Introduction

In recent years, much research effort has been devoted to the study of the properties of transferability of mathematical models of travel demand. Tests of performance in new contexts offer possibilities of fresh insights into the validity of the models themselves, and indeed perhaps into the behavioural content of the underlying theories from which the models have been derived.

For the practitioner, there is another equally valid, if more mundane, consideration. Transferable models could cut costs.

Despite the constraints on public spending that have accompanied the current economic recession, many important planning functions still rely on detailed long and medium term forecasts of national or regional travel demand. Substantial advances have now been made in establishing working model systems which can predict aggregate flows within a framework which is theoretically defensible in terms of simplistic individual behavioural rules (van der Valk and Feijen, 1983).

This has much improved the credibility of the models, and in particular has removed the potential that existed for the pathological misspecification which had in the past led to absurd forecasts (Williams, 1977, Daly and Zachary,1977). However, little progress has been made against two major practical problems that have always bedevilled the large scale model, namely unacceptably high costs and long production times.

This paper presents some results from the 'Overdraagbaarheid' study, an experiment conducted recently in the Netherlands in which an entire regional travel demand model system was transferred from one region to another. The aim of the experiment was to explore the extent to which results obtained in the first study were replicated in the second. The conclusions, which are summarised below, were sufficiently encouraging to proceed to the stage of developing techniques to implement transfers of such model systems without the costly and time-consuming purpose-designed transportation survey which usually precedes such modelling effort.

Rather than an abstract or general treatment of the subject, the focus of the paper will thus be on the practical significance of model transferability, and that in the context of one particular set of travel demand models. However, emphasis will be given to the interpretation of the concept and the transfer procedures in terms of an underlying behavioural model, namely individual random utility maximisation subject to constraints.

An outline of the model system.

The degree of validity (and potentially of transferability) of any model system is inevitably closely connected to the quality of both the input data and of the technical specification of the system itself. In this section we give a brief account of the structure of the model system under investigatation, and a description of the data bases that have been assembled for initial estimation and to test transferability.

The basic regional travel demand model system that is involved in the experiment was designed during the Zuidvleugel study (1977 to 1981) and is fully described in the final reports from that study. The general structure differs little from the standard four-stage transportation planning models of the sixties and seventies, in that travel demand is characterised as a function of household decisions on mobility (primarily car-ownership) together with individual decisions on the numbers of trips to undertake for different purposes, and the exact destinations to be visited and the travel modes to be used. It is assumed that only minimum generalised cost paths are relevant.

Three aspects of the system set it apart from the earlier standard models.

Firstly, the unit of travel is taken as the home-based tour, rather than the trip. The advantages of this are that demand can be made conditional on network conditions of both outward and inward legs of the tour, and that consistency of mode used may be imposed. The major disadvantage is that some non-home-based travel is lost from the system, and must be treated in some other way.

Secondly, the models and the data are defined at a disaggregate level. The practical advantage of this is that very many more variables may be included in the specification, and more levels of stratification by purpose can be made, thus making for better levels of statistical precision and more policy-relevant models. The major disadvantage is that more data must be collected and processed, both to estimate the models in the first place and then to apply the models to generate forecasts.

Thirdly, the models are interlinked in a logical and systematic way, allowing the decisions on numbers of tours to be undertaken to be made conditional on the expectation of the maximum utility to be derived from the available destinations and travel modes, and allowing household decisions on mobility to be influenced by relative accessibility.

Various simplifications are introduced at each level, to reduce the number of possible alternatives to an analytically tractable number and thus to allow the entire system to be constructed from simple discrete choice models.

Thus household mobility decisions are characterised only by the licence-ownership status of (up to) two adults in the household and by the categorisation of the household into one of three car-ownership groups (no car, one car or more than one car).

For each activity purpose, estimates of the numbers of tours undertaken are generated by two sequential binary models. The first gives the probability that a first tour will be undertaken, the other the probability that another tour is made, conditional on at least one having already been made. This latter probability is assumed not to be conditional on the actual number of tours previously made.

Travel modes and activity purposes are classified into a small number of categories, and potential elemental destinations (addresses) are grouped into geographical zones. For both estimation and application, only a sample of possible destination zones is made available for each tour.

The data bases involved in the transferability experiment (the Overdraagbaarheidstudie) were assembled from two household surveys conducted in different regions of the Netherlands. The first survey took place in late 1977, and involved 3000 households drawn from the region around Rotterdam and The Hague. The second took place in the spring of 1982, drawing 1000 households from the region around Utrecht. In each case, an initial interviewer-administered household interview was followed by a self-completed one day travel diary.

Detailed comparisons of the two data sources are given elsewhere, (Gunn et al, 1985) along with supporting evidence from a control data set drawn from an annual national travel demand survey. In most respects, the two areas and the two data sets were quite similar, except that levels of licence ownership and car ownership were very much higher in the second survey, as were tour rates by all modes and for all purposes.

Common sources and definitions were used for all location coding, for the extraction of level-of-service measures and in the assembly of zonal land-use statistics.

Model 'Transferability'

Ideas about model transferability have actually changed very little in ten years, although the body of empirical evidence that is now available is considerable, thanks to painstaking efforts by many researchers (see Koppelman and Wilmot, 1983, Ben Akiva,1981, and the references therein).

Whilst it is possible to consider the phenomenon of model transferability (or lack of transferability) on a case-by-case, empirical basis, it is rather more enlightening to start from the simple rationale for transfer that can be developed within the framework from which the model systems have themselves been derived. In particular, this means that ideas about transferability will be compatible with random utility maximisation.

The reasoning focusses attention on the interpretation of the coefficients involved in the model; we can outline the basic arguments as follows.

Given that we have approximated travel demand as a discrete choice problem and given the initial assumption of random utility maximisation, and further given that the utility of an alternative will be approximated by a linear function of measurable variables, the co-efficients attached to each measurable variable can be interpreted as marginal utilities.

The basic forecasting assumption is that these marginal utilities are determined by the individual's tastes, constraints,etc, (Train and McFadden, 1978) and that neither these nor the marginal utilities will change between base and forecast.

In fact, the estimated coefficients returned from conventional logit programs measure the marginal utilities of the various measurable variables scaled by the standard deviation of the unobservable elements. Forecasting with such models thus involves the assumption that the ratio of the marginal utilities to the standard deviation of these unobservable elements will remain at the same level as in the base period. Alternative specific constants are introduced into the functional specification to allow for the unknown mean utility of the unobservables for each alternative, and for the forecast, these too are assumed to remain at the levels estimated on the base data.

This widely used and highly simplified model of behaviour can be pursued further to generate a number of hypotheses about possible 'model transferability'. Suppose that we have two different populations facing two different sets of travel demand options, and that we fit a model to each using a selected set of measurable variables. Suppose that there is indeed a single 'true' process governing choice in each of the two cases, but that the utility functions also contain contributions from unmeasured variables.

If the unmeasured variables happen to be distributed identically over the two populations, then the model estimated on one should apply without any adjustment to the other. Elsewhere we have called this a 'naive' transfer to emphasise the improbability of the hypothesis.

If the unmeasured variables differ between the two populations in respect of both variance and mean values over alternatives, then the model estimated on one population would require adjustment in respect of both overall utility scale and alternative specific constants before it could be applied to the second. We have called this a 'complete' transfer, viewing this degree of transferability as the very most that is reasonable to expect of a well specified model, at least on a priori grounds.

Further structural forms for transfer transformations have been generated on the hypothesis that the (measurable) variables in the models can be collected into subgroups on the basis of the function that they fulfil, and that that function may be more or less important in different contexts (Gunn et al 1985). Such hypotheses suggest that variables should be transferred in blocks, each attracting a different scale factor.

Transferability Tests.

Many reported studies of model transferability have approached the subject as an investigation of the relative support for one or more of these various hypotheses against a 'null' hypothesis that models cannot be transferred. An armoury of classical statistical tests is available to test whether or not any one of the transfer models is the underlying 'true' model, i.e. that it is as well supported by the evidence as a completely re-estimated model.

In reality, of course, there is no underlying single 'true' model for both populations ; indeed, there is no 'true' model for either population taken separately. The discrete choice models are merely convenient approximations, and given sufficient data any such hypothesis about model transferability will always be rejected. This is not to say that the test statistics are uninformative; on the contrary, they are extremely useful measures of relative model performance. However it does suggest that experiments accepting or rejecting model transferability on the basis of such tests are somewhat irrelevant. They will always accept the hypothesis with small enough data sets, and always reject it with large enough data sets.

This problem has been noted many times before, both in connection with model transfers and indeed with the validation of individual models against independent data. The difficulty is that the nature of the data makes it rather hard to deal with residual or specification error. The evaluation of transferred, or approximate models requires just such information.

In the course of the Overdraagbaarheid project, we have followed other researchers in trying to develop a pragmatic approach to the appraisal of model transferability (e.g. McCarthy, 1981, Koppleman and Wilmot,1982). In particular, we have tried to view the problem from the standpoint of the practitioner, to whom two models which give substantially the same results are effectively 'transferable' regardless of statistical evidence that they are not completely interchangeable.

It must be said that this approach is not without its problems. A basic difficulty is that there are rather a large number of standpoints that the practitioner may have to take, and rather a wide range of questions that may be put to the model. A transferred model may be perfectly adequate to answer some of these, but entirely misleading for others.

To focus our investigation, we have tried to appraise the performances of the various transfer models under three basic headings, which we have entitled 'relevance','adequacy' and 'validity'.

Where a group of variables has been transferred from one model to another, we would judge the process 'irrelevant' if the scale coefficient attracted to the group were statistically insignificantly different from zero. We would deem the transfer to be 'inadequate' if the resulting model could be greatly improved by a more detailed specification, and finally we would judge the 'validity' of the model in terms of its ability to predict with reasonable accuracy the patterns of observed demand.

Simple statistical tests were associated with each of these aspects of the overall appraisal. Estimated 't'-statistics for the scale factors were used to judge relevance; likelihood ratio tests against more detailed specifications were used to judge adequacy, and finally sets of 'prediction tables' were generated to appraise validity.

The basis for the prediction table is an a-priori categorisation of the data, and a comparison of the model predictions for each category with the actual observations. The similarity of the prediction and the observation is then judged in the context of an expectation of the size of random sampling error which should be associated with that category. For ease of presentation, each category has been sorted into one of three groups, according as the prediction is within one standard error of the data ('given random sampling error, in 70% of cases the difference would be this great were the model true'), between one and two standard errors away ('this difference would only arise in 25% of cases were the model true'), or more than two standard errors away ('were the model true, this sort of discrepancy would arise only in 5% of cases).

More detail of these tests is given in Gunn et al.(1985), and a comprehensive review of these and other tests may be found in Koppelman and Wilmot (1982). More exact tests on the basis of validation tables are available in Horowitz (1983).

An overview of the results of Overdraagbaarheidstudie.

In this section, we shall give some summary statistics for all of the stages in the transferability experiment, starting with the models of mode/destination choice and progressing to tour frequency and household car and licence ownership. In view of the exceedingly large numbers of models involved, and indeed the large numbers of variables in each, we shall not give full details of the models themselves, but rather concentrate on the estimated transfer functions and the comparative measures of model adequacy and validity.

An overview of the results of the Overdraagbaarheidstudie

In this section we shall give some summary statistics from all the stages in the Transferability experiment, starting with the models of mode and destination choice, progressing to tour frequency and then to household car and licence holdings.

In view of the exceedingly large numbers of models involved, and indeed the large numbers of variables in each, we shall restrict attention to the estimated transfer functions and the comparative measures of model adequacy and validity. Full details of the specification of each component model may be had from the project reports of the Overdraagbaarheidstudie.

Firstly, for the joint mode and destination choice models, we shall consider three different levels of transferred model, and attempt to summarise the performance of these models in relation to the 'zero coefficient' model in which all possible outcomes are deemed equally likely, and in relation to a best available model. This last represents a re-estimation of all coefficients on the second survey data set, and in fact also includes some area-specific effects introduced to improve the specification.

The three transfer models can be described as follows :

a) the 'naive' transfer - in which the base model has been applied without any correction to the transfer area data;

b) the 'complete' transfer - in which a single scale factor has been estimated for all of the variables in the utility functions, and alternative specific constants have been re-estimated;

c) a 'partial' transfer, in which the level-of-service variables have been scaled separately from any others.

The 'partial' transfer involves two separate scales in most cases; the exceptions are for the purposes 'work' and 'education', for which the base models had separated mode choice from destination choice. The transferred models are of simultaneous mode and destination choice for all purposes. For 'work' and 'education' three separate scales have been estimated, one for the level-of-service variables, one for all other variables transferred from the base mode-choice model, and one for all other variables transferred from the base destination choice models.

Table 1 sets out the respective log-likelihoods of the five models for seven travel purposes.

The overall implications of the table are very clear. The base models would go some way to explaining the variability in the transfer survey, but each successive refinement of the specification is clearly justified in statistical terms. For example, the move from the 'naive' transfer to the 'complete'

Table 1 : Joint mode/destination models - log-likelihoods

tour purpose	zero coefficients	naive transfer	complete transfer	partial transfer	final model
work	-2736	-2408	-2226	-2221	-2169
education	-3345	-1717	-1504	-1483	-1398
social	-2229	-1523	-1499	-1481	-1354
recreation	-2626	- 949	- 919	- 910	- 765
shopping	-2168	- 858	- 824	- 824	- 805
personal	-1554	- 828	- 804	- 803	- 770
other	-2471	-1566	-1413	-1392	-1336

note : all models include "size" variables - see Daly (1982) for an account of the use of these attraction variables.

Table 2 : Joint mode/destination models : complete transfers - scaling factors and statistics

tour purpose	scale factor	t(0)	t(1)	ASC1-factor	ASC2-factor
work	0.70	41	18	1.32	-2.65
education	0.60	36	24	0.87	1.20
social	0.91	36	4	1.19	1.02
recreation	0.76	28	9	0.94	-1.69
personal	0.83	29	6	0.92	0.60
shopping	1.09	27	2	0.89	0.84
other	1.22	41	7	1.43	4.38

notes:
(a) the columns headed t(0) and t(1) give the t-statistics for the hypotheses that the scale factor is 0 or 1 respectively. Values in excess of 2 indicate that the corresponding hypothesis can be rejected at (approximately) the 95% level of confidence.
(b) the first alternative specific constant refers to the car mode, the second to public transport.

transfer involves the estimation of three coefficients. This extension of the specification would be strongly justified terms of the likelihood ratio test if the log-likelihood were increased by 5.65 units, and the average increase is over 94 units.

By 'strongly justified' here we mean that the hypothesis that the 'true' scale factor is 1.00, and the 'true' alternative specific constants are those from the base model, can be rejected at the 99% level of confidence. The test involves the comparison of the difference in log-likelihoods with a critical value. Given that the models are all 'nested', this critical value is calculated as one-half of the value of the chi-squared statistic with as many degrees of freedom as there are extra parameters in the more detailed model.

Similarly, the extension of the specification to isolate the level-of-service scales from the others involves an extra two coefficients for the 'work' and 'education' purposes, and would be equally strongly justified if log-likelihoods were increased by 4.6 units - true in both cases. For the other purposes, the critical level would be an increase of 3.3 units, which is exceeded for all except the shopping and personal business models. We can infer that the 'partial' transfer is better than the 'complete' transfer.

Finally, the transition from the 'partial' transfer to the final model involves some 20 extra coefficients in each case; this refinement would be strongly justified if the log-likelihood could be increased by 19 or 20 units; the average is around 80.

It may be concluded that the base models can be greatly improved by the various transfer transformations we have discussed, but that the transferred models are themselves only approximations which can also be improved if an entire new survey is available.

Tables 2 and 3 sets out the actual estimated values of the transfer scale factors for 'complete' and 'partial' transfers respectively, along with approximate t-statistics conveying estimation accuracy. Scaling factors are also given for the alternative specific constants; these particular models address choice between car, public transport and the 'slow' modes (walk bicycle or moped), and thus incorporate only two constants.

The main points to be made about these tables are as follows :

a) all of the transfer scale factors on utilities (or components of utilities) are markedly statistically significantly different from zero and, in most cases, are also different from unity;

b) excluding work and education, for which the structural differences between base and transfer models make comparisons difficult, the factors for the utility components range between 0.76 and 1.22 for the 'complete' transfer, and between 0.75 and 1.38 for the level-of-service variables ;the range for the other variables for the 'partial' transfer is between 0.86 and 1.42;

Table 3 : Joint mode/destination models
partial transfers - scaling factors and statistics

tour purpose	S-F1	t(0)	t(1)	S-F2	t(0)	t(1)	S-F3	t(0)	t(1)	ASC1	ASC2
work	0.60	10	6	0.89	11	1	0.74	18	6	1.36	-2.89
educ	0.29	5	12	0.35	4	7	0.84	18	3	0.65	1.34
social	0.85	33	6	1.37	16	4				1.41	1.06
recr'n	0.75	28	9	1.42	9	3				1.15	-2.46
pers'nl	0.82	26	6	0.93	8	1				0.99	0.64
shop	1.11	26	3	0.94	8	1				0.83	0.83
other	1.38	32	9	0.86	14	2				1.12	4.06

notes : as for table 2, plus:
a) S-F1 refers to level-of-service variables
b) S-F2 and S-F3 refer to variables from base mode split and distribution sub-models respectively, for work and education
c) S-F2 refers to all non-level-of-service variables for other models

c) the factors applying to the alternative specific constants are much more variable, taking large positive and negative values.

It seems that the alternative-specific constants from the base model are not relevant in the new context, but that the composite utility estimated on the basis of the measurable variables remains a useful predictor. The scale of this composite utility appears to vary from model to model, but not in any systematic way. The average remains around 1.0, whether or not level-of-service variables are split from the others.

Finally, tables 4 and 5 illustrate the performance of the various models in predicting response in a pre-defined set of validation tables. We have discussed above the criterion used to assign a prediction of demand to one of three categories, according as the difference between the observed demand and the predicted demand is less than one standard error, between one and two standard errors, or greater than two standard errors in size. The categorisation is an approximate one, but in general the more cells that fall in the first category than in the other two, and the more that fall in the second rather than the third, the better the model.

We have used only three stratifying variables to indicate the

Table 4 : Joint mode/destination - summary of validation tables

MODEL category	NAIVE A	NAIVE B	NAIVE C	COMPLETE A	COMPLETE B	COMPLETE C	PARTIAL A	PARTIAL B	PARTIAL C	FINAL A	FINAL B	FINAL C
WORK												
distance	23	10	31	23	18	23	23	17	24	32	17	15
profession	35	13	26	45	25	4	49	21	4	53	15	6
pop.density	14	10	21	26	12	7	26	11	8	23	12	10
EDUCATION												
distance	25	6	28	31	10	18	34	9	16	34	17	12
profession	30	5	9	27	11	6	27	7	10	32	9	3
pop.density	19	6	17	29	9	4	33	6	3	30	11	1
SOCIAL												
distance	23	15	21	28	15	16	32	14	13	30	16	13
profession	43	19	8	43	19	8	46	18	6	52	13	5
pop.density	23	15	8	22	14	10	22	11	13	28	15	3
RECREATION												
distance	24	7	20	28	7	16	28	6	17	30	8	13
profession	50	11	3	45	14	5	47	14	3	53	9	2
pop.density	23	9	9	26	7	8	26	7	8	28	10	3
SHOPPING												
distance	23	8	8	24	5	10	23	5	11	24	5	10
profession	33	6	13	41	9	2	41	8	3	41	9	2
pop.density	25	5	9	29	3	7	29	3	7	29	3	7
PERSONAL												
distance	26	9	16	31	10	10	30	12	9	34	12	5
profession	41	9	9	48	3	8	48	3	8	48	7	4
pop.density	25	10	6	28	9	4	29	8	4	29	10	2
OTHER												
distance	24	17	11	33	10	9	29	10	13	33	10	9
profession	42	12	10	48	11	5	44	16	4	48	14	2
pop.density	15	12	15	24	12	6	27	11	4	29	11	2

notes:

categories A,B and C refer to the number of cells in which the model and the data are within one standard error, between one and two standard errors and more than two standard errors apart, repectively.

Table 5 Joint mode/destination - overall measures of performance

MODEL	TOTALS IN CATEGORIES ..A..	..B..	..C..	RHO=SQUARE(0)	'SCORE'
WORK					
naive	72	33	78	0.12	1.03
complete	94	55	34	0.19	0.67
partial	98	49	36	0.19	0.66
final	108	44	31	0.21	0.58
EDUCATION					
naive	74	17	54	0.49	0.86
complete	87	30	28	0.55	0.59
partial	94	22	29	0.56	0.55
final	96	37	16	0.58	0.48
SOCIAL					
naive	89	49	37	0.32	0.70
complete	93	48	34	0.33	0.66
partial	100	43	32	0.34	0.61
final	110	44	21	0.39	0.49
RECREATION					
naive	97	27	32	0.64	0.58
complete	99	28	29	0.65	0.55
partial	101	27	28	0.65	0.54
final	111	27	18	0.71	0.40
SHOPPING					
naive	81	19	30	0.60	0.61
complete	94	17	19	0.62	0.42
partial	93	16	21	0.62	0.45
final	94	17	19	0.63	0.42
PERSONAL					
naive	92	28	31	0.47	0.60
complete	107	22	22	0.48	0.44
partial	107	23	21	0.48	0.43
final	111	29	11	0.50	0.34
OTHER					
naive	81	41	36	0.37	0.72
complete	105	33	20	0.43	0.46
partial	100	37	21	0.44	0.50
final	110	35	13	0.46	0.39

notes : as for previous table

performance of these models (being the only three common to all validation runs). Mode/destination alternatives in the model have been classified by distance, the occupation of the traveller and origin population density in turn, and for each classifiction two validation tables have been constructed. The first of these compared observation with forecast for the mode chosen to access the destination actually visited. The second compares observation with prediction for access to destinations, aggregating over modes. These tables were replicated from those used in the course of the Zuidvleugel study, and are designed to generate insight into the effectiveness of the model both as a predictor of mode and as a predictor of destination choice. The numbers in the tables are simply accumulated over both of these tables.

The figures in table 4 corroborate that successive extensions to the model specification do result in appreciable improvement to the model, when judged in terms of decreasing numbers of cells in categories C and B , but suggest that the biggest single improvement is the upgrading of the 'naive' model to the 'complete transfer'.

Table 5 summarises the information into overall totals for each model, and further collapses these into single 'scores' designed to reflect model fit. The definition of the 'score' is entirely arbitrary; weights of zero have been assigned to cells in category A, one for those in category B and two for those in C. The 'score' is calculated as the average weight, so that in some sense the lower the score, the better the fit. The measure is offered as a convenient summary statistic to compare with more conventional measures of fit such as the rho-square index (also given on the table). However, it has no strong interpretation.

The overall scores for the 'naive','complete', 'partial' and 'final' models average out at 0.73, 0.54, 0.53 and 0.44 respectively. On this scale, the 'complete' transfer generates around 65% of the total possible improvement over the 'naive' transfer.

In terms of rho-square index, the corresponding figures are 0.430, 0.464, 0.469 and 0.497. Thus the increase in rho-square associated with the 'complete' transfer is around 50% of the total possible improvement.

Both indicators of fit suggest that the 'partial' transfer improves the model, but only very marginally.

Turning now to the tour generation models, table 6 sets out the log-likelihoods for the 'zero coefficient' model, a single 'complete' transfer and a final, re-estimated model, and Table 7 gives scale factors for both composite utilities and alternative specific constants.

The following points may be made :

a) the base survey work models were not transferred; they were appropriate only to employment levels at the time of that survey. Subsequent modelling work for forecasting has made workforce

Table 6 : Tour generation models : log-likelihoods

tour purpose	zero coefficients	complete transfer	final model	extra parameters
educ1	-1187	- 249	- 245	3
educ2	- 503	- 286	- 280	6
recr'n	-1691	-1298	-1295	3
social	-1691	-1197	-1195	3
personal	-1691	- 978	- 974	3
shopping	-1691	-1086	-1084	4
other	-1691	-1138	-1137	3

note: educ1 refers to education tours by non-students, educ2 to education tours by students;

Table 7 : Tour generation models
complete transfers - scaling factors and statistics

tour purpose	scale factor	t(0)	t(1)	ASC	%rate change
educ1	0.67	4	2	1.10	0
educ2	0.88	7	1	0.97	0
social	0.87	10	1	1.10	+31
recreation	0.91	8	1	0.89	+71
personal	0.68	7	3	0.99	+27
shopping	0.86	16	3	1.14	+11
other	0.83	10	2	1.40	+36

notes : see earlier tables, plus
a) %rate change gives the difference in tour rate between base and transfer surveys.

Table 8 : Tour generation - summary of validation tables

MODEL	COMPLETE			FINAL			COMPLETE			FINAL		
category	A	B	C	A	B	C	A	B	C	A	B	C
	NON-STUDENT EDUCATION						STUDENT EDUCATION					
age	5	2	1	6	2	0	1	4	0	4	1	0
sex/car	1	2	1	1	2	1	3	2	0	4	1	0
education	3	3	1	5	2	0	8	0	0	7	1	0
occupation	4	3	1	4	2	2	1	0	0	1	0	0
household size	4	2	1	4	2	1	4	4	0	4	4	0
income	8	0	1	8	0	1	8	0	1	8	0	1
cars/licence	4	2	1	4	2	1	3	4	0	4	3	0
	RECREATIONAL						SOCIAL					
age	2	6	1	5	3	1	6	3	0	8	1	0
sex/car	2	3	0	4	1	0	3	2	0	3	1	1
education	6	1	2	6	1	2	4	3	2	4	3	2
occupation	2	7	0	6	3	0	6	3	0	6	3	0
household size	3	1	4	3	1	4	6	1	1	6	1	1
income	5	3	1	5	3	1	7	2	0	5	4	0
cars/licence	6	1	0	6	1	0	4	2	1	5	1	1
	PERSONAL BUSINESS						SHOPPING					
age	7	1	1	7	1	1	3	3	3	4	3	2
sex/car	3	2	0	4	1	0	5	0	0	4	1	0
education	7	2	0	5	4	0	5	4	0	7	2	0
occupation	5	4	0	6	3	0	6	3	0	6	3	0
household size	4	4	0	4	4	0	4	3	1	4	3	1
income	8	1	0	7	2	0	6	2	1	6	2	1
cars/licence	6	1	0	6	1	0	4	3	0	4	3	0
	OTHER PURPOSES											
age	4	4	1	5	3	1						
sex/car	5	0	0	5	0	0						
education	5	2	2	5	2	2						
occupation	7	1	1	7	1	1						
household size	6	2	0	6	2	0						
income	7	2	0	7	2	0						
cars/licence	4	2	1	4	2	1						

notes :

a) sex/car denotes a categorisation by sex and car-availability

b) cars/licence denotes a categorisation by the number of cars and the number of driving licences in the household

c) category definitions as on Table 4

participation an exogenous variable, leading to rather different models;

b) only the first of the binary models is discussed, i.e. that which predicts the probability that at least one tour be made for that purpose. The transfer survey data base was not sufficiently large to provide sufficient evidence on multiple tour makers to transfer the second models;

c) differences in the joint mode and destination models as between base and transfer will inevitably change the measurement of the 'logsum' variable which is used to reflect accessibility differences. The logsum variables were estimated separately, but in no case did they attract coefficients which were statistically different from 0.0 at the 95% confidence level. They are not reported in the table.

The conclusions from this table are slightly different from those drawn in connection with the mode/destination modelling. In this case the transfer models are almost indistinguishable from the completely re-estimated final models. The final models introduce at least 3 extra coefficients in each case, and would need to increase log-likelihoods by more than 5.6 to demonstrate improvement at the 99% level. In fact, even at the 95% level, the models are not improved by the move from 'complete' transfer to re-estimation.

This impression is supported by the validation tables, as Tables 8 and 9 testify. The 'complete' and 'final' models are seen to be almost identical in terms of their explanatory power as measured by the rho-square index, and only slightly worse in terms of the constructed 'score'.

Finally, we come to the models of household driving licence status and car-ownership level. Tables 10 and 11 set out the various log-likelihoods and scale factor values for the models. No results are given for the model predicting car-ownership levels in multiple licence owning households; lack of data and poor income reponse has prevented any inference on the basis of the transfer survey.

The transferred licence ownership model exchanges 8 parameters for 11 units of likelihood; the likelihood ratio test would once again confirm that the fully reestimated model was statistically superior to the transferred model (at the 95% level of confidence) . Other points to note are that

a) the transfer scale factor is significantly different from zero, but not from 1.0 (at the 95% level of confidence), and

b) the alternative specific constants have changed in a systematic way; it turns out that this is such as to increase licence holding overall, and especially amongst women.

The car-ownership models show the complete transfer exchanging 6 units of likelihood for 4 co-efficients, suggesting that in this case the statistical evidence also supports the extra detail of

Table 9 : Tour generation - overall measures of performance

MODEL	TOTALS IN CATEGORIES ..A..	..B..	..C..	RHO-SQUARE(O)	'SCORE'
NON-STUDENT EDUC					
complete	29	14	7	0.79	0.56
final	32	12	6	0.79	0.48
STUDENT EDUC					
complete	28	14	1	0.43	0.37
final	32	10	1	0.44	0.28
RECREATION					
complete	26	22	8	0.23	0.68
final	35	13	8	0.23	0.52
SOCIAL					
complete	36	16	4	0.29	0.43
final	37	14	5	0.29	0.43
PERSONAL					
complete	40	15	1	0.42	0.30
final	39	16	1	0.42	0.32
SHOPPING					
complete	33	18	5	0.36	0.50
final	35	17	4	0.36	0.45
OTHER					
complete	38	13	5	0.33	0.41
final	39	12	5	0.33	0.39

notes : see earlier tables.

the complete reestimation. The transfer scale factor is once again statistically significantly different from zero (at the 95% level of confidence) but not significantly different from 1.0. The alternative specific constant has not changed very much.

Thus the most striking difference as between base and transferred licence and car-ownership models concerns the alternative specific constants in the licence model. It has been the case that over the five year period between the two surveys, driving licence acquisition rates have increased sharply, especially amongst women. The inference is quite strong that the adjustment of the constants has been prompted not by area-to-area differences, but by changes over time that the separate cross-sectional models cannot identify.

This suggested that a rather different validation test should be carried out on the base models. Using the control data set taken from the National Travel Survey (the O.V.G. data set) for the years 1979, 1980 and 1981, the base models were run to predict car-ownership and licence holding in each year.

The results are set out in Table 12. Distinguished separately are results from a 'full' forecast, in which driving licence holding were predicted first, and then car-ownership conditional on these holdings, and a 'limited' forecast, in which car-ownership was predicted conditional on actual observed driving licence status.

Table 10 : Licence and car-ownership models - log-likelihoods

model	zero coefficients	complete transfer	final model	extra parameters
licences	-923	-616	-605	8
cars	-198	-113	-107	4

notes : as for earlier tables

Table 11 : licence and car-ownership models : factors

model	scale factor	t(0)	t(1)	ASC1	ASC2	ASC3
licences	0.91	14	1	0.65	0.61	0.61
cars	0.94	5	0	1.07		

Table 12 : licence and car-ownership predictions with base models

	1979	1980	1981
driving licences per household			
observed	0.93	1.04	1.06
predicted	1.00	1.03	1.01
cars per household			
observed	0.72	0.82	0.81
predicted 1	0.71	0.77	0.77
predicted 2	0.75	0.77	0.75

notes : two predictions of cars per household are given. The first is conditional on observed licence holding, the second on predicted licence holding.

Detailed analyses indicated that the models were performing well at predicting the marginal effects of the various background variables on licence and car-ownership levels, although the absolute levels were somewhat different.

The conclusions drawn from this work were that there were secular trends at work which could not be identified from the cross-sectional analyses. Adjustment by means of altering the alternative specific constants seemed adequate to allow for these trends, but the information required to make the adjustment had to be taken from data or analyses which explicitly recognised these trends.

Summary of conclusions from the transferability experiment.

The major conclusions we have drawn from this work are as follows.

Firstly, the 'complete' transfer procedures quite clearly bring 'relevant' explanatory power to the models. The t-statistics show that in all cases the scale factors were significantly different from zero. They were also significantly different from unity in most of the joint mode/destination models and in most of the tour generation models, although not for the licence or car-ownership models.

The alternative specific constants are rather less similar between base and transfer models. For the licence holding models, the constants had changed quite markedly, apparently in response to factors not in the model specification, and which had been changing systematically and quite rapidly over time.

The 'partial' transfer procedures improve the models somewhat, although the simple partitioning that we have discussed, separating level-of-service measures from the rest, does not seem to have had great practical impact on model performance.

Finally, the extra parameters involved in a complete reestimation of the models on the transfer survey data have almost always been strongly statistically supported in terms of the resulting improvement in fit.

In general the transferred models were not 'adequate' to replace the fully re-estimated model.

The figures on Table 1 illustrate the position rather clearly. Taking simple averages over all the models, the gain in likelihood units per parameter fitted is 31 for the move from the 'naive' model to the 'complete' transfer, 9 for the move from the 'complete' to 'partial' transfer, and 4 for the move from 'partial' to 'final' model. Each stage is soundly justified in statistical terms, but the trend of diminishing returns per extra parameter is striking.

Figure 1 : 'naive' prediction plotted against 'final' model

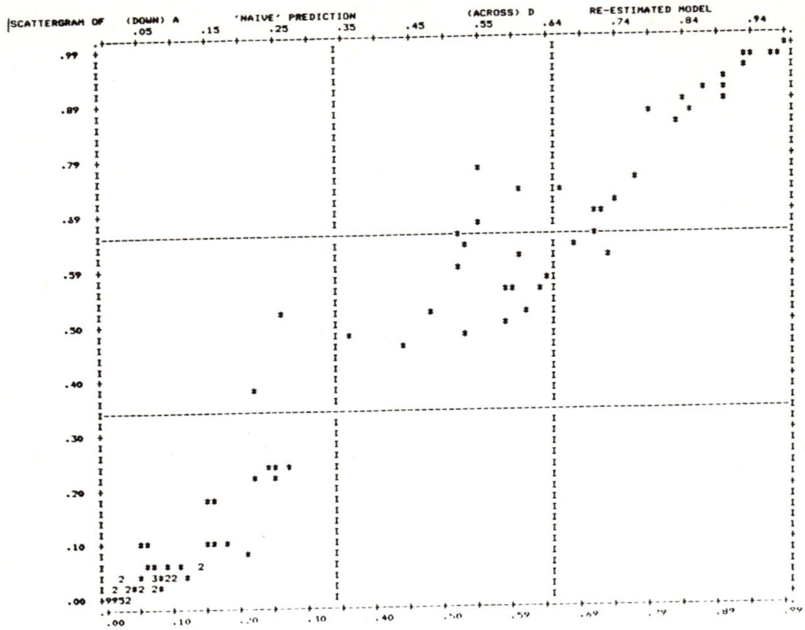

Figures 1, 2 and 3 provide a visual reinforcement for this observation. Some 177 cases have been selected at random from the data set for the joint choice of mode and destination for shopping trips, and the probabilities accorded to the selection of the three available travel modes have been extracted. Plotting the predictions of the 'naive' model against that of the 'final' model (figure 1) shows a high degree of correlation, but more scatter than in the plot of the 'complete' transfer against the 'final' model (figure 2). The plot of the 'partial' transfer against the 'final' model shows some slight further decrease in scatter.

The visual evidence is backed up by the descriptive statistics giving the usual 'R-square' measure and the standard error of estimation. The 'R-square' values are respectively 0.984, 0.996 and 0.996 for 'naive','complete transfer' and 'partial transfer' comparisons with the 'final' model. The corresponding standard errors of estimate are .043, .021 and .020 for the same three comparisons. These statistics should be interpreted with some caution, given the nature of the data, but the overall indication for the relative performance of the models is clear.

In conclusion, the best models for the base and transfer surveys

Figure 2 : 'complete transfer' plotted againts 'final' model

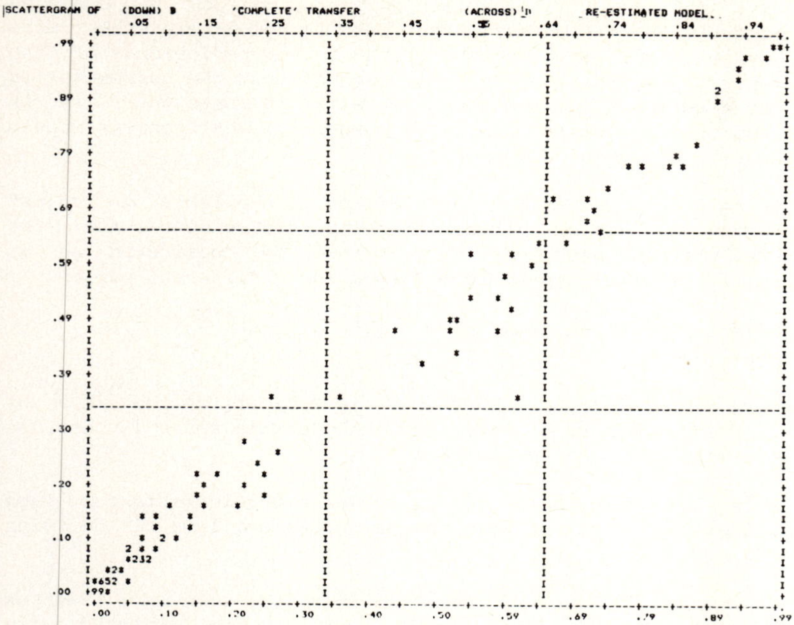

Figure 3 : 'partial' transfer plotted against 'final' model

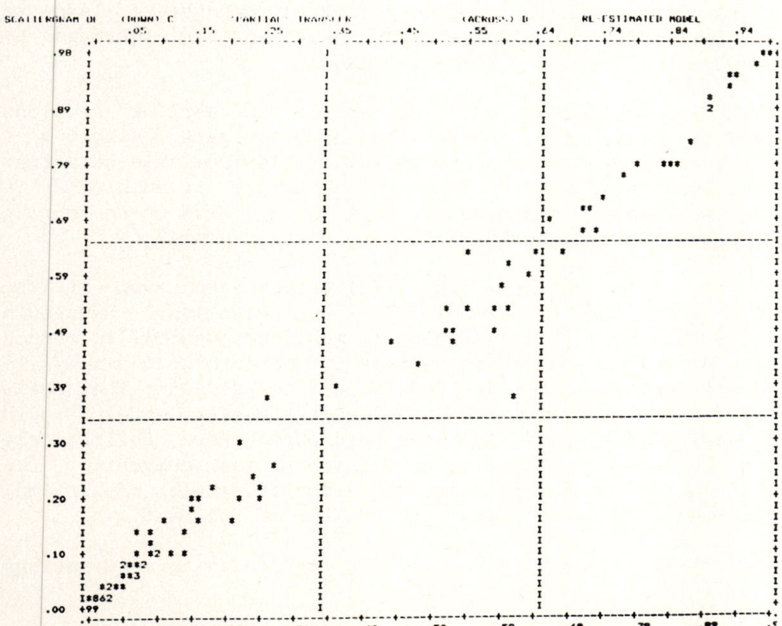

are different. A simple scaling along the lines of the 'complete' transfer will go a long way towards adjusting the base to the new context. The result will remain an approximation, and could be improved if there were sufficient data to completely reestimate the system. However, we would point to the performance of the models in the validation tables to suggest that the approximation may be a workable one and could be useful in cases where it is not possible to collect new data. Overall, the transfer models retain a high degree of validity.

Providing that sufficient information is available to adjust constants and scales, then, there is some evidence that the transfer of the base model system should be considered as an option if a similar system is required for a different region.

Transferring model systems.

We have concluded above that there is support for the idea of transferring the models, provided that certain scale factors can be estimated for the new context.

In this section we shall discuss the data requirements for such an estimation, together with the practical problems of deriving the estimates.

Before doing so, however, we shall introduce some more ideas on the potential for cost saving that can be achieved with transferable model systems.

The application of a disaggregate model system to predict regional demand requires that a representative sample of households be available, together with an inventory of land-use and estimates of level-of-service measures for journeys by different modes to potential destinations.

The experiment that we have summarised in earlier sections involved the collection of household interview data, albeit on a smaller scale than for the base survey. It is from this data that we have estimated transfer factors. However, if such data is already available, the marginal cost saving from transferring models rather than re-estimating is relatively small.

Of course, cost savings can still be achieved even if the transfer data must be assembled from a conventional household survey of this sort ; for example, it would be possible to reduce the size of the survey if the purpose of the data were simply to provide a data base for application and to estimate the scale factors. Another circumstance in which transfer models could prove useful would be if the household survey were deliberately biassed to focus on a particular section of the community. The basic transfer model could be fitted for the whole population, and enriched with detail about the section of interest.

However, the most exciting possibility for achieving cost savings with transferable models comes in conjunction with purpose-designed artificial samples.

This potential development has been kept in mind throughout the Overdraagbaarheidstudie, and a parallel programme of research has been conducted into the construction and use of artificial samples.

The essence of the idea is that a data base is assembled with all the required detail for the application of the models. It contains a sample of households, together with expansion factors, which is representative of the study area population in all key respects. It contains a detailed inventory of land-use. Lastly, for a sample of locations across the study region, it contains information on the level-of-service measures by different modes for journeys to samples of possible destinations.

The cost saving (and the artificiality) arises because the locations are not specific to the actual individual households, as would be the case were a conventional survey mounted. Existing sources of data (regular surveys or censuses) can be used to supply the household data, leaving the analyst with the problem of selecting some representative locations and then processing networks as usual, but without the expense and delays that accompany the design and conduct of a household survey and the coding and processing of the data.

A series of tests have been undertaken to explore the extra errors that this approach introduces (Gunn et al, 1982) but indications are that the model systems are quite robust in the face of such an approximation.

For the purposes of this paper, we note only that if an artificial sample is available, then the estimation of the various scale factors required for the models can be accomplished using aggregate statistics, most or all of which may be taken from existing sources, such as routinely collected travel surveys.

Estimation of scale factors from aggregate statistics.

Given a set of estimated models together with the requisite input data, there is in principle no difficulty in adjusting constants and scale factors to maximise the correspondence of the prediction with a given set of aggregate statistics.

In practice, of course, It is not so easy to calculate the appropriate adjustments. Even in the simplest case of the adjustment of alternative specific constants in a single model to achieve equality of aggregate shares, the problem is non-linear and requires an iterative search during which the model is applied at each step. Dehghani and Talvittie (1983) give the necessary equations to generate a Newton-Raphson algorithm for this problem. We shall suggest a rather simpler approach below.

First, it should be noted that the number and form of the

aggregate controls that are needed to transfer an entire regional travel demand model make for a much more difficult technical problem. The transfer procedures that have been devised for the application of the Overdraagbaarheidstudie models to forecast national travel demand involve the following :

a) adjustment of two constants in the household driving licence model to achieve consistency with externally supplied estimates of national average licence holdings per person, by sex and age group;

b) adjustment of a constant in each of two car-ownership models to match externally supplied estimates of average zonal household car-ownership;

c) adjustment of constants in tour generation models to match externally supplied estimates of regional travel demand, and

d) adjustment of mode-specific constants and overall scale factor in the joint mode/destination models to match externally supplied estimates of numbers of tours by each mode in given distance bands.

The potential costs of estimating all of these constants is daunting. Simple methods, involving a degree of approximation, have been adopted.

Firstly, it is easier (and cheaper) in many cases to conduct a grid search rather than a directed search. By this we mean that if values are sought for one or more variables to maximise some objective function (correspondence to exogenous totals) one can limit each variable to a set of values over a preselected range, (given reasonable guesses about maximum and minimum values), and evaluate the models at each of these values (thus two variables each taking 10 possible values would define 100 points at which to evaluate the model). The maximum within this restricted set may then be used to approximate the true maximum - or a further grid search conducted with tighter ranges if more precision is desired.

Secondly, when logit estimation software is available with the option to constrain parameters, a simple trick will allow the estimation of new alternative specific constants, or indeed of dummies for any selected group of alternatives.

The method proceeds as follows : first, the un-transferred model is applied to each case in the (artificial) sample, and predictions made of the probability with which each of the alternatives (or groups of alternatives) would be selected. The natural logarithm of this probability is then entered as the sole variable in a dummy 'attribute' file for each case.

Next, a dummy 'data' file is generated for each case, in which the various alternatives are selected in the desired proportion of cases. The order in which these are entered is immaterial; thus if we wish to estimate new constants in a binary choice model in order to ensure that alternative 1 is chosen in 28% of cases, and we have 100 cases in our (artificial) sample, then we can set up the dummy data file so that 28 cases 'choose' alternative 1, and 72 'choose' alternative 2.

If a logit estimation is carried out on the dummy 'data', with the parameter of the logged probability variable constrained to unity and alternative specific constants included in the specification, then the returned values of the constants give the required adjustment to the model.

If a scale factor is sought in addition to such constants, and sufficient data is available to estimate both, then a grid search on the scale factor performed jointly with this pseudo-logit estimation of the adjustments to the constants will return the required estimates.

In conclusion, the transfer procedures we have recommended involve the estimation of a rather large number of factors; however, practical solutions have been found and implementable methods now exist.

More detail about these techniques, and about the artificial sample method, is given in Daly and Gunn (1985), in which a methodology is outlined for the transfer of the Overdraagbaarheid system to the whole of the Netherlands (the 'National Model' study).

Conclusions.

We have argued above the 'perfect transferability' is both an impossibly and unnecessarily high hurdle for a practical forecasting model to have to leap. All models are approximations. In comparing a transfer model with a completely reestimated model, it seems to us to be more constructive to enquire how similar are the models, rather than whether or not they have been measured with sufficient statistical precision to establish that they are different.

On the whole, we have been impressed by the degree to which each separate component model, and indeed the Zuidvleugel system as a whole, has transferred. It seems clear that some transfer adjustment is necessary, and that the simple scaling approach we have described will not completely provide this adjustment. However, it will go a long way towards it, and simple methods based on available statistics have been devised to implement the correction.

A major discovery was the degree to which the development of driving licence holdings was underpredicted by the cross-sectional models, despite the relationship with the variables explicit in the model remaining quite stable. The need to adjust the models for the base year is obvious; equally, it seems that an explicit recognition of potential developments is needed to make a further adjustment for the forecast years. In practice, this means that the disaggregate models will be adjusted to be compatible at an overall level with aggregate time series forecasts based on cohort analyses. The transfer factor procedures may be seen as ways in which to operate the models in a constrained environment, and to input into the system information which cannot properly be measured from cross-sectional data.

Finally, our conclusions have been limited to the two contexts of the Overdraagbaarheid experiment. Much more information will be generated in the course of the application of the models in the Landelijk Model study. These results will be published in due course.

Bibliography

Atherton, T.J. and M.Ben-Akiva (1976) Transferability and Updating of Disaggregate Travel Demand Models, Transportation Research Record 610.

Ben-Akiva, M. (1981) Issues in Transferring and Updating travel Behaviour Models. in 'New Horizons in Travel Behaviour', P.R.Stopher et al (eds) Lexington Books, Lexington.

Daly,A.J. (1983) Estimating choice models containing attraction variables. Transportation Research B, vol 16b.

Daly,A.J. and H.F.Gunn (1985) Cost-Effective Methods for National Level Demand Forecasting. The 1985 International Conference on Travel Behaviour, Noordwijk.

Daly, A.J. and S. Zachary (1977) Improved Multiple Choice Models. P.T.R.C. Summer Annual Meeting

Deghani, Y. and A. Talvittie (1983) Forecasting accuracy, transferability and updating of modal constants in disaggregate mode choice models . P.T.R.C. Summer Annual Meeting.

Gunn,H.F., M.Ben-Akiva and M.Bradley (1985) Tests of the Scaling Approach to Transferring Disaggregate Travel Demand Models. Transportation Research Board.

Gunn, H.F. , P.Fisher, A.J.Daly and H.Pol (1982) Synthetic Samples as a Basis for Enumerating Discrete Choice Models. Proc. of the P.T.R.C. Annual Summer Meeting.

Horowitz, J.L. (1983) Testing Probabilistic Discrete Choice Models of Travel demand by Comparing Predicted and Observed Aggregate Choice Shares. University of Iowa.

Koppelman,F.S. and C.G.Wilmot (1982) Transferability Analyses of Disaggregate Choice Models. Transportation research Record 895.

Koppelman, F.S. and C.G.Wilmot (1983) Identification of conditions for effective transferability of disaggregate choice models. Report no. UMTA 429-I, Northwestern University.

McCarthy, P.S.(1981) Further evidence on the temporal stability of disaggregate travel demand models. Mimeo, Department of Economics, Purdue University.

Train, K. and D. McFadden (1978) The goods/leisure trade-off and disaggregate work trip mode choice models. Transpn Research, vol

v.d.Valk, J. and B. Feijen (1983) The 'Zuidvleugel' Study. P.T.R.C. Annual Meeting.

Williams, H.C.W.L. and M.L. Senior (1977) Model Based Transportation Policy Assessment. I.T.S.,University of Leeds.

Behavioural Research for Transport Policy
pp. 245–265.

Temporal Stability of Transport Demand Models [1/] in a Colombian City

Valerie Jean Kozel
The World Bank, 1818 H Street, NW, Washington, DC 20433

Travel demand models are primarily estimated for forecasting traveller response to policy actions and transport investments. Accurate forecasts require not only that the model be specified correctly (e.g., have an appropriate functional form and included variables), but also that model coefficients are temporally stable and that accurate forecasts of necessary explanatory variables can be made. A variety of factors could cause coefficient values to change over time (for example, problems caused by excluded variables, errors associated with incorrect functional form, changes in the constitution or underlying behavior of the specific urban population) which are difficult to identify or correct a priori. Likewise, it is extremely difficult to make accruate forecasts of explanatory variables, particularly over a broad time frame, due to the complex structure of the urban travel environment. Errors in variables and changes in base parameters introduce potential errors in model forecasts.

This paper addresses the latter source of error, namely, the effects of shifts in base year coefficients over a forecasting period. Methodologically, it draws on a body of work pertaining to so-called temporal and spatial transferability of transport demand models. Initial studies were done by Watson and Westin (1975), Atherton and Ben-Akiva (1976), Hensher and Johnson (1977), Talvitie and Kirshner (1978), and Train (1978). Later research includes applied and theoretical studies by Silman (1982), Koppelman and Wilmot (1982), McCarthy (1981), Gunn, Ben-Akiva, and Bradley (1984) and McCoomb (1983).

Errors in forecasting explanatory variables may also be an important source of prediction error in demand evaluations. Numerous past planning studies have found a high level of temporal stability for certain demand statistics (for example, household trip generation rates) within socio-economic groups, which suggests that predicting changes in the number of individuals belonging to a group is a critical factor in forecasting future levels of trip making. An analyst must make accurate forecasts across a population of the variables that dimension a group (typically income, automobile

---

1/ Support for this work was provided by the City Study research project (RPO 671-47) funded by the World Bank. The views reported here are those of the author and should not be interpreted as reflecting the views of the World Bank or its affiliated organizations. The research benefited greatly from discussions with Richard Westin and Alvaro Pachon, my coworkers in the transport sector, and Gregory Ingram, the project supervisor. In addition, I am grateful to Moshe Ben-Akiva and Joffre Swait for numerous comments and suggestions.

ownership, and household size and composition) in order to predict changes in group composition. For purposes of the present study, however, we assume that changes in the underlying explanatory variables are known with certainty, and we can obtain a sample drawn from the "true" population both in the base (or model estimation) year and in some later year which demarcates an analysis period.

Only a few of the above cited studies specifically examine questions of temporal stability of model parameters (viz. Hensher and Johnson, Train, Silman, and McCarthy) with a view to assessing the impact of changes in these estimates on model predictions. Data requirements for such an excercise are high, for, as Train points out, 'how well disaggregate (mode choice) models predict is an empirical question rather than an analytic one' (Train, 1978, p.174). A stringent test of temporal transferability requires full household-level data from two distinct points in time. Furthermore, the two data sets must be sufficiently similar so that changes in traveller behavior can be distinguished from differences caused by dissimilar sampling frames, measurement error, or differences in the way basic variables are defined (for example, estimation of travel times based on networks and reported travel times). The potential value of such studies is high, however; it is not only possible but likely that the impact of various factors on travel decisions will change over time, and, that these changes will be systematic and comprehensible. The better such changes are understood, the more likely it is that they can be anticipated and corrected for. Such corrections necessarily improve the accuracy of model predictions.

We are in a fortunate position to evaluate the temporal stability of a specific set of mode choice models estimated for Bogota, Colombia in having two seperate travel surveys for the city -- one performed in 1972, and another performed six years later in 1978. Although the survey instruments were rather different and we suspect some unacknowledged differences in the sampling frames, the data were deemed sufficiently similar to carry out transferability analyses. Models were estimated using 1972 survey results; 1972 parameters were then used in conjunction with "true" values of exogenous variables for 1978 to predict modal shares in the later year. In addition, mode choice models were estimated using data from 1978, and various statistical tests performed to compare 1972 models with 1978 models.

The paper has five sections. The first is a general discussion of the notion of model transferability, while the second briefly outlines the approach used in the empirical part of the study. The third section describes the travel environment in Bogota in 1972 and 1978. Empirical results are presented in the fourth section of the paper, which includes a description of basic mode choice estimates and a discussion of the results of statistical and predictive tests of model transferability. Conclusions constitute the fifth and final section of the study.

1. Discussion of Model Transferability

For purposes of clarity, let us briefly define the notion of model transferability. Without loss of generality, we define a basic model as having a dependent or endogenous variable, a set of explanatory or exogenous variables, and a vector of parameters. Mathematically,

$$Y = f\ (X;\ \theta)$$

where: Y denotes the dependent variable;
X is a vector of explanatory variables; and
θ is a parameter vector.

The explanatory variables are known and related to the endogenous variable according to some functional form denoted $f\ (\bullet)$ and a vector of parameter values.

The phrase "model transferability" covers a wide range of situations in travel demand modelling in which some or all model parameters that were estimated with data from a particular population sample are hypothesized to be in some sense identical to those for a different sample. If a model captures all relevant causal elements in a choice process -- that is, the true behavior of the choice maker, and if individual choice-makers act according to a unifying principle of behavior vis-a-vis elements of the choice process, then a model developed to predict behavior of one group of individuals can be transferred to another group, or used to predict another group's behavior based on this second group's relevant characteristics. Mathematically,

$$Y_1 = f\ (X_1;\ \theta_1) \quad \text{and} \quad Y_2 = f\ (X_2;\ \theta_2)$$

where:

Y1 and Y2 denote dependent or choice variables for samples 1 and 2, respectively;
X_1 and X_2 are vectors of explanatory variables related to the choice process;
θ_1 and θ_2 are parameter vectors for samples 1 and 2 respectively.

If a model developed for sample 2 is fully transferable to sample 1, then

$$\theta_1 = \theta_2 = \theta \quad \text{and} \quad f(\bullet) = g(\bullet)$$

The parameter vector can be partitioned into subsets, for example, θ_{1k} and θ_{1m}, θ_{2k} and θ_{2m}, and a model is partially transferable if

$$\theta_{1k} = \theta_{2k} = \theta_k \quad \text{for all } k;$$

over the parameter set θ_k, and not transferable over the parameter set θ_m if

$$\theta_{1m} \neq \theta_{2m} = \theta_m \quad \text{for all } m;$$

Types of Transferability

Empirical tests of model transferability fall into three general categories: (i) analysis within an urban area using a single data set collected at a particular point in time which tests for transferability between market segments (differentiated by one or more explanatory variables) or between randomly selected sub-samples; (ii) analysis within a single urban area using two data sets collected at different points in time which generally tests for overall transferability of model parameters over time and transferability between market segments over time; and (iii) analysis between urban areas using data sets from each area, sometimes collected at different points in time, which tests for spatial transferability of parameters and spatial transferability of market segment parameters.

Tests of within city, single data set transferability are an essential element of the normal process of model calibration and validation. In this sense, speaking of market segmentation tests as an analysis of model transferability is something of a misnomer; rather, market segmentation is used to analyze robustness of key parameters across various subgroups within a single population. Failure of a model to "transfer" between market segments in general indicates a direction for improving the specification. In performing segmentation tests, the fact that the model describes the choice behavior of the population of a single city at a particular point in time serves as a unifying principle for the analysis.

Tests of temporal stability are more problematic and admit of richer interpretation than market segmentation tests. For most urban dwellers, life is a dynamic process of change and adaptation. Technological change impacts the physical and economic environment, while information and communication broaden an individual's concept of opportunities. Changes in the economic environment may cause significant changes in transport supply and operations, and the magnitude and distribution of per capita income may shift dramatically over time. Even if a model truely captures the causal structure of travel behavior at a particular time, the behavior itself may not be static, while the model usually is. Additionally, if important exogeneous effects (for example, income levels) change dramatically, accurate forecasts would require that the model "work" for a segment of the population that was not included in estimating base parameter values simply because the group did not exist in the base year. Clearly, in order for the model to forecast accurately in such case, parameter values for the new population segment must be the same as those for pre-existing segments -- a possible but questionable circumstance.

It may well be that the essential question to be asked in assessing temporal model stability concerns the extent of change in the fabric of urban life; a test of temporal transferability can fail because (i) the model adequately reflected behavior in the base year, but the behavior in itself changed in the interceding years; or (ii) the model adequately reflected the behavior of individuals acting under base year conditions, but a (unanticipated) change in conditions caused a change in actions/responses. We cannot measure the "behavioral" content of the model; by assumption, travellers behave in the way that best suits them (e.g., maximizes an underlying utility function). We can, however, measure changes in the urban environment and traveller behavior by measuring changes in important explanatory variables. By implication, a static demand specification is less likely to apply across time in a rapidly changing environment than in a more stable environment. The greater the rate of change in the travel environment, the more case we must exercise in using typical travel demand models for analyzing transport policy impacts. Recall that temporal transferability is more than just a desirable characteristic of a model -- it is essential for making accurate forecasts. Because of this, it is important to develop an understanding of the validity of conventional travel demand models over time. In doing so, we must remember that transferability is not defined by structural characteristics of the model alone (i.e., functional form, included variables), but also by many external factors (i.e., changes in the transport environment or living conditions of the populace).

The final category, so-called spatial transferability (which often implicitly includes a temporal element), has all the problems of temporal applications plus a wide range of others. Cities are different; they have different histories, different cultures, different traditions. Residents in various cities develop very different patterns of travel that reflect dissimilarities in the urban environment (such as crime, inclement weather and the like) that are difficult to account for in conventional inter-urban travel models. We cannot, for example, assess the impact of climate on mode choice in a particular area because climate is an attribute of the region; it varies only between regions. Further, although a rose is indeed a rose is a rose regardless of the suburban garden that nourishes it, the Washington DC Metro system is a far cry from Mexico's Metro or the underground system in Buenos Aires, although each would be labelled "rail rapid transit" in a typical mode choice model. In addition, the structure of local government and transport-related fiscal concerns can vary greatly between cities, and accordingly may have a significant (and different) impact on the supply side of the market.

The argument for continuing work in spatial transferability rests on pragmatic concerns regarding estimation costs rather than necessarily improving the accuracy of model forecasts (as with temporal transferability). The cost of data collection and estimation can be high; many of these costs could be avoided if models can be either partially or fully transferred between locales without significant loss of predictive power. Recent studies (see, for example, Koppelman, 1984, or Gunn, Ben-Akiva, and Bradley, 1984) explore techniques for "updating" coefficients estimated using data from one locale to be used in predicting traveller behavior in another locale. These techniques generally require less data than is necessary for full model (re)estimation.

To briefly summarize the discussion -- transferability tests based on market segmentation are useful for improving model specification and the overall accuracy of the model in fitting aggregate shares in the estimation sample. Further, thoughtful and carefully constructed market segmentation tests can help to develop a deeper understanding of urban travel behavior. In contrast, spatial transferability is likely beyond the present capability of transport modelling techniques; while some variant of partial transferability with appropriate parameter updating may be possible between cities with roughly similar physical and economic environments, a similar public transport system, spatial structure of employment and general level of amenities (to name but a few crucial dimensions), fully successful spatial transferability given the present state-of-the-art of demand modelling is problematic. However, the rationale for transfering a model between different locales is primarily pragmatic, driven by cost and data considerations, so spatial transfer is often very desirable despite reservations. In contrast, temporal transferability seems to be of paramount importance for accurate policy forecasts. Empirical tests of temporal transferability are seldom performed. Those that have been done generally conclude with a simple "transferable/not transferable" statement, and do not question why model parameters are not temporally stable nor assesses the impact of a lack of stability on the models as policy tools. The empirical work presented in this paper attempts to rectify this and concommitantly make some small contribution to improving the state-of-the-art of applied transport demand modelling.

2. Overview of Transferability Tests

Tests for model transferability fall in one of two general categories; (i) statistical tests of parameter equality; and (ii) tests of goodness of fit (prediction-success) or accuracy of prediction. A recent paper by Koppleman and Wilmot (1981) provides a complete overview of a wide range of statistical tests and some assessment of tests of prediction accuracy. They contrast the two types of tests:

> "Tests of model parameters are designed to evaluate, either subjectively or statistically, the degree to which the estimated models are equivalent. Tests of mode predictions are designed to evaluate, either subjectively or statistically, the accuracy with predictions of a transferred mode describe travel behavior in an application context relative to predictions which would be obtained from a model developed in the new context." (p.3)

A model may fail a test of parameter transferability, yet provide highly accurate predictions in the "new context." Certain coefficients have a greater impact on the predictive quality of a model than others, and statistically significant differences in less important coefficients can have less impact on accuracy of aggregate share predictions than less significant differences in more important parameter values. Alternatively, a model in an old context which is statistically undistinguishable from a similar model in a new context may nonetheless provide very inaccurate forecasts of aggregate shares in the new context.

Ben-Akiva (1979) differentiated between tests of pragmatic transferability and statistical transferability. Pragmatic transferability is somewhat similar to Koppelman's "tests of model predictions", although Ben-Akiva developed a broader conceptualization of "sufficiently similar" results that draws on both statistical and pramatic concerns and reflects

> "an attempt to strike an "optimal" balance between the analysis resources and the accuracy of predictions, within the constraints of time, money, data and professional and computational capabilities." (p.7-8)

Past transferability analyses tended to stress statistical tests of transferability, and neglect more pragmatic concerns. While a primarily statistical approach is well suited to analysis of market segment transferability because the parent population is hypothesized to be the same across the market segments and data is collected and coded in a consistent fashion, spatial or temporal tests of transferability likely require a more flexible, inductive approach.

Empirical work presented in the fourth section uses both statistical tests of parameter stability and pragmatic tests. The tests of parameter stability are directed towards a slightly different end than previous tests, which focused on the question of whether traveller behavior (represented by model parameters) is the same in two periods, or, whether some aspects of behavior is the same. We feel that we cannot know whether behavior is the same or different; we can however determine the extent of change in important travel-related characteristics and form hypotheses about the cause of these changes. We hope our model is robust with respect to these changes as well as robust with respect to differences in coefficients caused by changes in urban travel patterns exogenous to the model. By identifying possible sources of statistical instability (e.g., coefficients which change) or predictive inaccuracies (e.g., differences in coefficients or groups of coefficients which cause predicted choice probabilities to diverge from actual shares), we can identify directions for future work to improve the accuracy of model forecasts.

3. Travel Characteristics of Bogota, 1972 and 1978

The data used in the study comes from two basic sources. The first, a survey undertaken in 1972 under the direction of Consultecnicos, Ltd. of Bogota, encompassed three major areas: (i) household demographics; (ii) transport; and (iii) economic conditions. Approximately 4,700 households were interviewed, containing 26,500 persons, some 15,300 of whom made trips (a one percent sample of the population). Travel information was requested in only 22 of the 63 traffic zones in the city. The data is considered reasonably good, although there is some evidence that upper income households were oversampled, and non-work related travel under-reported. The second data source is the World Bank - DANE City Study survey, conducted jointly by the World Bank and the Colombian National Statistics Office (DANE) in December, 1978. It covered some 3,000 households in Bogota, and included sections on (i) household and dwelling characteristics; (ii) demographic characteristics of all household members; (iii) worker characteristics; (iv) the unemployed; and (v) vehicle ownership and journey to work characteristics. Questions pertaining to non-related travel were not included in the survey.

Bogota is the largest city in Colombia, and the national capital. It is a dynamic and rapidly growing city, presently undergoing extensive development and change across a number of urban sectors, including transportation. The selected socioeconomic indicators presented below (Table 1) exemplify the dramatic changes which occurred in the city even over so short a period as the six year interval between 1972 and 1978. For example, population grew by 5.7 percent per annum, while average incomes grew by some 31 percent overall, or nearly 6 percent per annum. Labor force participation rates increased slightly, and average household size fell. Both employment and residential density gradients decreased, particularly in the central city; the percentage of total employment located in the central city fell from 40.7 to 33.2 percent, although actual employment densities (jobs/ha.) increased from 163.3 to 210.9 jobs per hectare. This suggests the employment outside the central city increased rather substantially over the six year analysis period. In addition, the number of mid-to-upper income residents living at the urban periphery increased significantly, indicating an increasing trend towards suburban living patterns.[2/]

On the transport side, the average number of automobiles per household increased from .167 to .217 between 1972 and 1978, which, given the high level of population growth, indicates a signficiant increase in the number of private vehicles operating in the city. Average work trip travel times increased overall (35.4 minutes per one-way leg in 1972 versus 42.3 minutes per one-way leg in 1978), which is not surprising in light of the previously mentioned decentralization of employment and residential locations. In addition, average speeds appeared to decrease throughout the city, likely due to congestion effects, which also serves to increase average travel times. Average travel times on public transit increased more than travel times in private vehicles.

There are a wide range of transport services available to Bogota's travellers. On the private side, these include automobiles, bicycles, motocycles, and of course, travel on foot. Some 12 percent of all work trips were made in automobiles (car driver or car passenger) in both 1972 and 1978, while 8 percent of workers walked or cycled to work in 1972 in contrast to 13 percent in 1978.[3/] Public modes include taxis and collective taxis (which

---

2/ The conventional view in third world cities is that the wealthy live in central locations and the poor live at the periphery in urban areas of the developing world. While in 1972, Bogota's wealthier residents tended to live in more central locations, there was a significant increase in the share of mid-to-upper income households located on the periphery between 1972 and 1978. Not surprisingly, many of these households owned an automobile.

3/ We suspect that upper income households were oversampled in the 1972 survey, which would serve to inflate automobile shares and underestimate walk shares. Thus, we are hesitant to claim any real increase in walk travel between 1972 and 1978, and find the lack of

carried 2 percent of work trips in 1972 and 1.5 percent in 1978), full-sized 68-passenger buses, similar to U.S. school buses (carrying 69 percent of all work trips in 1972 and 52 percent in 1978), and 24-passenger busetas and microbuses, which carried the remaining 9 (1972) and 18 percent (1978) of riders, respectively. The smaller buses offer faster and more comfortable services than full-sized buses, and are accordingly priced somewhat higher.

Urban passenger services are primarily suppied by the private operators in Colombia; at present, there are some 30 to 35 private companies operating buses, busetas, and microbuses, and one (small) public company. Service levels are extremely high and route coverage generally good throughout the city. For example, in 1972, over 85 percent of the households surveyed reported living within 100 meters of a bus line. Fares, route assignments, and minimum service levels are administered by municiple authorities.

Mode choice was highly correlated with income levels in both 1972 and 1978. In fact, income has in general been found to be a very important variable in describing urban travel behavior in developing areas (see, for example, Swait, et. al., 1983, Kozel, 1981, Thobani, 1984). For example, automobile use was prevelent only in the upper 20 percent of the income distribution, while slightly poorer travellers used taxis or busetas, and the poorest either walked or rode standard buses. The most striking change between 1972 and 1978 pertains to buseta patronage, which increased substantially at the lower end of the income distribution. This likely reflects a (successful) effort on the part of buseta operators to expand their market; in 1972, buseta services were only offered in wealthier areas of the city -- by 1978, most of the city was served, and the number of smaller capacity buses operating in Bogota had increased from 1550 to 3400 vehicles

4. Empirical Results

Basic Mode Choice Models

Table 2 shows multinomial logit work trip mode choice estimates for Bogota in 1972 and 1978. (Data used for model estimation are choice-based samples of full survey data sets.) Travel alternatives include (i) car driver/car passenger, (ii) taxi/collectivo, (iii) standard bus, (iv) buseta/microbus, and (v) walk. The car driver and car passenger categories were combined because intra-household ridesharing was rare in Bogota (nearly all travellers who reported auto trips lived in automobile owning households), and the question of who drove and who was a passenger was considered comparatively unimportant in the face of model simplification. Initial estimates in which the drive and car passenger alternatives were kept seperate evidenced a high degree of parameter similarity between the two auto alternatives, and no significant loss of explanatory power when the alternatives were combined.

---

increase in automobile travel frankly suspect.

Level of service variables include total travel time and cost divided by the wage rate. Socio-economic variables include household income, a dummy variable for heads of household, and a measure of car competition (number of vehicles divided by number of workers). A single dummy variable controlling for the quality of microbus/buseta services is also included in the models. It accounts for variability in microbus service levels and uncertainty regarding actual route structure in 1972. Service levels had stabilized by 1978, so the variable does not affect aggregate shares in the later year.

Level of service variables are considered the crucial policy variables in planning studies; improvements to the transportation system are reflected in these variables. They represent both the monetary and time price of travel, and should accordingly have negative coefficients. The total travel time parameter is very similar between the two models, while in contrast, the cost coefficients (defined as direct travel costs factored by the hourly wage rate) are notably dissimilar. This difference is likely caused by data and structural constraints; the vast majority of travellers used public modes (bus and buseta) or walked to work, which all had flat fares in 1972. Transfers (which necessitate an additional fare) were rare. Lack of cost variation in the estimation sample made it difficult to estimate a robust cost coefficient (see Kozel, 1981). Without cost variation, the variable simply represents the reciprocal of income.

In comparison to analyses typical of more industrialized countries, mesures of household income had a strong and significant impact on work-related travel patterns in Bogota. Further, the magnitude of estimated income parameters remains highly significant and stable over time, in spite of a 31 percent increase in real income.

Income may act as a proxy for the degree of choice potential or opportunity set open to urban residents (Swait et al, 1983). Unlike many rural areas in developing countries, cities function as active market systems, with money serving as the primary medium of exchange. The poorest people may well walk to their place of employment because they have no other choice; motorized transport is simply beyond their means. As income constraints become less binding, some trips -- particularly for work -- are made by bus. Members of the "middle" class are comparably choice-rich; while still very poor, they ride the more comfortable busetas in lieu of buses, and may even occasionally make discretionary trips on public transit. Only the very wealthiest have a range of travel choices that resemble those common to more industrialized countries -- they often drive or use taxis, and travel for non-discretionary as well as discretionary purposes.

Changes in model shares that appear income-determined are not always so, however. For example, the substantial increase in buseta ridership between 1972 and 1978) was not primarily due to a shift in the mode-specific income elasticity; rather, increased ridership indicates a swelling of the middle and lower-middle classes (as evidenced by real growth in GNP) and a relaxation of service supply constraints in lower income areas of the city. In all likelihood, lower income travellers would have utilized busetas more intensively had the service been available to them in 1972.

Parameters specific to the walk alternative reflect the earlier mentioned decentralization of residential location patterns of upper income households between 1972 and 1978. In 1972, workers from wealthy, automobile-owning households favored both walking and busetas (in addition to their automobile) over regular buses. Workers from these households tended to live in central areas of the city, generally along a radial sector to the north of the CBD. By 1978, automobile owning households evidenced a decided tendency not to travel to work on foot, quite possibly because they lived on average further from their place of employment and the disutility of walking was not fully captured in the time coefficient.

Pooled Likelihood Ratio Tests

Likelihood ratio tests were performed on a pooled 1972/78 sample to test whether selective parameter constraints across the time interval caused significant loss (in a statistical sense) in explanatory power. For purposes of comparison, we first estimated a pooled model with all parameters constrained, including the alternative specific constants. 4/ Not surprisingly, the hypothesis of overall statistical equality of coefficients was resoundingly rejected -- the resulting chi-square statistic was 152 with 17 degrees of freedom. If parameter constraints are relaxed for the constants alone, the new likelihood ratio statistic is 30 with 12 degrees of freedom, again indicating a significant loss of explanatory power due to parameter constraints. Results of these tests indicate that it is necessary to examine the models further to identify sources of dissimilarities.

From the previous section, it is clear that one of the glaring differences between the 1972 and 1978 equations is in the cost coefficients -- based on 1972 survey results, the cost/wage effect was small and statistically insignificant. Using the 1978 data set, however, the coefficient was not only of a more credible agnitude, but was highly significant. We suspect that much of difference is caused by data problems and possible scale effects in LOS variables between the two years; in 1972 a significant majority of travellers in the estimation sample rode flat fare modes (buses and busetas), introducing little cost variation in the data, and quite possibly affecting the estimate of the cost coefficient. (see Kozel, 1981, for a more detailed discussion) We estimated a pooled model with both intercept and cost/wage constraints relaxed. The results were somewhat disappointing; the value of the test statistic dropped only to 26 (d.f.= 11), still indicating significant differences between the two equations.

It was thus necessary to examine other variables in the equation in order to understand changes in travel behavior between 1972 and 1978. Parameter differences were evident for some of the

---

4/ Choice-based biases were corrected by adding a vector of correction values to the equation and constraining its coefficient to be equal to one.

socio-demographic variables -- viz, car competition and the dummy variables for heads of household. In particular, relative effects for the walk and buseta modes were very different in the two equations. A likelihood ratio test to evaluate the statistical significance of the perceived differences (unconstrained parameters included the alternative-specific constants, cost/wage, the car competition variable specific to the walk and buseta alternatives, and the household head dummy variable specific to the walk alternative) yielded a chi-square value of 14 (d.f.= 8). On this basis, we cannot reject the hypothesis of parameter equality across the remaining set of constrained coefficient values.

A final test entailing further relaxation of socio-demographic parameter constraints (namely, head of household and car competition specific to the drive/shared ride alternative) was performed. Test results (chi-square value of 8 with 6 degrees of freedom) indicate a fairly high degree of stability over remaining parameters (primarily income and travel time).

In summary, the combined results of the likelihood ratio tests suggest that travel time effects and income coefficients were by and large stable over time in Bogota, while (possibly) cost and many socio-demographic effects were not. Unfortunately, it is not possible to know how much changes in individual traveller's behavior contributed to differences, and how much was caused by dissimilarities in the sampling frame or data collection between the two years -- recall that the underlying hypothesis in the statistical analyses is that both samples are drawn at random from the same parent population. If the samples were drawn differently even if the parent population is the same (and we have some reason to suspect that they were), the likelihood of rejecting the hypothesis of parameter equality will be artificially inflated. Accordingly, the results of our analysis must be viewed with some caution.

It is reassuring, however, to have found relative stability of time and income effects in the face of possibly substantial differences in sample constitution, data collection techniques, and real and significant changes in the underling distribution of important explanatory variables. This suggests that our models, while by no means perfect, may be capturing some crucial aspects of mode choice behavior. The question which must next be raised concerns the impact of parameter differences on the accuracy of model forecasts. If ultimate levels of predicted model shares are relatively insensitive to statistical differences, the differences are not pragmatically important. Alternatively, statistically small differences in crucial explanatory variables may have a large impact on the predictive accuracy of the model. The next section explores the quality of model predictions under the various parameter constraint scenarios just described.

Model Predictions

We observe actual shares and measures of independent variables in 1978. The naive test of full temporal transferability entails using all 1972 parameters and the actual values of independent variables in 1978 to forecast 1978 aggregate shares.

This is the best a policy analyst could do in 1972 assuming perfect information on changes in exogenous variables -- inaccuracies in forecasts would then necessarily be caused by lack of parameter stability rather than errors in predictions of the explanatory variables. It was clear from statistical tests, however, that a number of other variables (cost/wage, model intercepts, socio-demographics) entered the demand equations with statistically different parameter values in 1978. This suggests that model simulations which use only 1972 parameters may not provide very accurate forecasts of 1978 aggregate modal shares, and that judicious replacement of some 1972 coefficients with "true" (1978) values may improve forecasting accuracy.

Table 3 shows actual and predicted modal shares (predictions were made using sample enumeration techniques) for various configurations of 1972 and 1978 coefficients and true (1978) values of the explanatory variables. Model 1 uses all 1972 coefficients, including adjusted intercepts. Interestingly, predicted shares are not so divergent from actual shares, although the model does not discriminate well between buseta and bus users (bus is over-predicted, while buseta is noticeably underpredicted), and walkers are under-predicted. Simply replacing 1972 intercept values with those from 1978 (Model 2) changes the aggregate forecasts rather dramatically; transit shares edge closer to actual values, while the auto drive/passenger and walk alternatives both tend to diverge further from actual values. The results of this simulation presage a problem that affects most of the model simulations -- the drive/passenger alternative tends to be consistently underpredicted by some 30 to 40 percent, while the walk alternative is usually over-predicted by roughly the same percentage. An improvement in the forecast accuracy of one alternative (either walk or drive/shared ride) is generally accompanied by a worsening in the forecast accuracy of the other. There are three possible reasons for this; (i) exogenous shifts in residential and work locations change travel patterns in a way not fully captured by the model; further, many of the poorest households have variable employment locations and mixed commuting patterns that are not adequately reflected in the model; (ii) differences in sampling frame, particularly as concerns income coverage (see Mohan, 1981); and (iii) walk travel in developing areas is in general not well understood and often seriously under-reported.

Models 3 through 5 in Table 3 involve successive replacement of 1972 coefficients with 1978 values, and generally result in progressive improvements in the accuracy of model forecasts. In addition to alternative-specific intercepts, Model 3 replaces the 1972 cost/wage coefficient with its 1978 value. Models 4 and 5 replace some (Model 4) and all (Model 5) of the soci-demographic parameters with 1978 values. Transit predictions generally converge on true values with successive replacements, while auto and walk predictions improve slightly but not in any clear systematic fashion.

Table 4 presents the simulation results of an experiment with coefficient values. In a typical disaggregate mode choice model, time and cost bear an important relationship to each other -- taken together, they constitute an overall price effect which includes both the monetary cost of a trip and the implict value of the time spent travelling. Because the 1972 and 1978 cost/wage parameters were distinctly different in value from one another while time parameters were not, replacing only one and not the other could introduce problems. Accordingly, Table 4 shows predictions from Models 4 and 5 from Table 4 replacing both price coefficients with 1972 values. The results are encouraging; the new models appear to provide more accurate forecasts than the unrevised (cost/wage at 1978 levels, motorized and non-motorized travel times at 1972 levels) versions. Even the drive/shared ride and walk alternatives are more accurately predicted under the revised specification. Thus, it appears that models which use all LOS and income variables from 1972 provide more accurate predictions of aggregate shares in 1978 than do models which use some (cost/wage) LOS coefficients based on 1978 estimates and some from 1972. This suggests a scaling difference along price dimensions rather than underlying behavioral changes in the population.

Table 5 shows aggregate share predictions of income segments for a selected set of the 1972/78 parameter combinations presented in Tables 3 and 4. These results reveal an important finding -- in general, travel choices of workers from households at the lower end of the income distribution are better predicted using primarily 1972 parameters than are households at the upper end of the income distribution. By inference, the travel characteristics of this group appear to have changed very little between 1972 and 1978. This result is perhaps not surprising when one considers the negative correlation between poverty and choice; poor households appear to be isolated in a particular environment that changes minimally with the passing of time. The travel behavior of wealthier households apprears more dynamic -- Model 7, the most adultrated simulation (in the sense of replacing 1972 parameter estimates with 1978 values) clearly provides better predictions of aggregate model shares of wealthy households than those less adultrated. It makes intuitive sense that base (1972) parameters would in some way work best for those elements in the populace who have limited choices and live in a fairly static environment, and be less useful in predicting the behavior of travellers in a choice-rich, rapidly changing environment. As corraborating evidence, note that while we had some difficulty distinguishing between walkers and bus riders at the lower end of the income distribution, the primary mode choices of poor households, forecasting difficulties were pervasive across nearly all modes for workers from households at the upper end of the income distribution.

5. Conclusions

Mode choice models for Bogota were found to be temporally transferable along certain economic dimensions (e.g., travel time and income) and generally not transferable along either direct cost or socio-demographic dimensions (e.g., household size and composition, auto availabilty). The latter effects are hypothesized to reflect taste changes or shifts in market demand schedules. Further, travellers from low income households exhibited more stable behavior than those from wealthier households, likely due to the severe economic and social constraints generally typical of the poor in developing areas (or, alternatively, the positive association between wealth and richness of the travel environment). In addition, both supply and demand factors were important in evaluating model transferability.

It is difficult to draw strong policy implications from the empirical results. In a general sense, the models failed to predict well over what must be considered a reletively short period with reference to typical transport planning applications. In particular, automobile and walk trips were difficult to forecast accurately. Bus and buseta ridership, which did change subtantially over the analysis period, was much better predicted, possibly because the modes are primarily travel choices of mid-to-lower income households, whose behavior shows more indication of temporal stability than that of wealthier households, but also likely because they are utilized by a significant majority of the general population. It is difficult to determine the extent to which data problems contibuted to the apparent lack of model transferability, and how much can be attributed to actual behavioral changes. Our sense is, that while measurement problems and differences in sampling frame had some impact, the evidence suggests signficant changes in the travel behavior of certain segments in the population that are not well captured in the base year models. Thus, caution should be exercised in interpreting demand model forecasts, even over a reletively short time period, in a rapidly changing environment such as Bogota.

References

Atherton, Terry J. and Moshe E. Ben-Akiva. (1976) "Transferability and updating of disaggregate travel demand models", Transportation Research Record: 610.

Gunn, Hugh F., Moshe E. Ben-Akiva, and Mark A. Bradley. (1984) "Tests of the scaling approach to transferring disaggregate travel demand models", Paper presented at the Transportation Research Board annual meetings, January, 1985.

Hensher, David A. and Lester W. Johnson. (1977) "A two-period analysis of commuter model choice: The predictive capability of individual choice models", Logistics and Transportation Review: 13(4), pp. 361-375.

Koppelman, Frank S., and Chester G. Wilmot. (1981) "Transferability analysis of disaggregate choice models", The Transportation Center, Northwestern University, Evanston, Ill.

Kozel, Valerie J. (1981) "Travel demand models for developing countries: The case of Bogota, Colombia", Urban and Regional Report No. 81-26, World Bank, Washington, D.C.

McCarthy, P. S. (1981) "Further evidence on the temporal stability of disaggregate travel demand models", mimeo, Department of Economics, Purdue University.

McCoomb, Lloyd A. (1983) "Analysis of the transferability of disaggregate demand models among ten Canadian cities", mimeo, Transportation Development Centre, Transport Canada, Montreal, Quebec.

Mohan, Rakesh. (1981) "The people of Bogota, who they are, what they earn, and where they live", World Bank Staff Working Paper No. 390, World Bank, Washington, D.C.

Pachon, Alvaro, and Valerie J. Kozel. (forthcoming 1985) Transportation in the Developing Metropolis: A Case Study of Bogota, Colombia, A World Bank Research Publication, Johns Hopkins Press.

Silman, L.A. (1981) "The time stability of a modal-split model for Tel-Aviv", Environment and Planning A: 13, pp. 751-762.

Swait, Joffre, Valerie J. Kozel, and Moshe E. Ben-Akiva. (1983) "Travel behavior in Brazilian cities", Paper presented at the World Transport Conference in Hamburg, West Germany, April 21-25, 1983.

Talvitie, Antti, and Daniel Kirshner. (1978) "Specification, transferability and the effect of data outliers in modeling the choice of mode in urban travel", Transportation: 7, pp.311-331.

Thobani, Mateen. (1984) "A nested logit model of travel mode to work and auto ownership", Journal of Urban Economics: 15, pp.287-301.

Watson, Peter L. and Richard B. Westin. (1975) "Transferability of disaggregate mode choice models", Regional Science and Urban Economics: 5, pp. 227-249.

TABLE 1

Comparisons of Socioeconomic Indicators
Bogota, 1972 and 1978

	1972	1978
Total Population	2,630,000	3,800,000
Average Annual Household Income (Col$/year)	$3,055	$3,998
Persons Per Household	5.4	5.3
Workers Per Household	1.56	1.61
Percentage Total Employment Located in the Central City	40.7	33.2
Employment Density (jobs/ha.) in the Central City	163.3	210.9
Automobiles per 1000 Households	167	217
Average Work Trip Travel Times (minutes/one-way trip)		
- Automobile	23.5	27.6
- Bus	40.6	54.8
- Buseta/Microbus	33.5	47.8
All Modes	35.4	42.3

Notes:
1. In 1978, nominal household income was Col.$13,094 per annum.

Source: Pachon and Kozel, 1985.

TABLE 2

Work Trip Mode Choice Models
Bogota, 1972 and 1978

(Available alternatives include drive/ shared ride, taxi/collective, bus, buseta and walk)

Independent Variables	Estimated Coefficients [1/] Bogota, 1972	Bogota, 1978
Travel time (mins) for all motorized modes	-.0257 (1.3)	-.0675 (2.5)
Travel time (mins) - Walk	-.0688 (6.0)	-.1083 (9.4)
Cost (col.$)/wage rate (col$/min)[2/]	-.3013 (1.0)	-1.083 (4.0)
Gross family income (Col$/month)		
-Drive/passenger	.00013 (3.2)	.00017 (6.6)
-Taxi/collective taxi	.00018 (4.2)	.00018 (5.8)
-Bus	- 0 -	- 0 -
-Buseta/microbus	.00010 (2.7)	.00011 (6.1)
-Walk ($\leq$ 30 min. trip)	-.00003 (0.4)	-.00002 (0.6)
Head of Household (1=head, r otherwise)		
-Drive/passenger	1.616 (5.2)	2.192 (7.8)
-Taxi/collective taxi	- 0 -	- 0 -
-Bus	- 0 -	- 0 -
-Buseta	- 0 -	- 0 -
-Walk	0.263 (0.6)	0.672 (2.5)
Car competition variable (# vehicles/# workers)		
-Drive/passenger	3.450 (4.0)	1.107 (2.2)
-Taxi/collective taxi	- 0 -	- 0 -
-Bus	- 0 -	- 0 -
-Buseta	2.333 (3.4)	-0.117 (0.3)
-Walk	2.702 (2.4)	-1.707 (2.1)
Buseta/microbus accessibility dummy variables		
-Poorly served	-0.465 (1.4)	- 0 -
Alt-specific constants		
-Drive/passenger	-1.696 (3.0)	-2.410 (4.5)
-Taxi/collective taxi	-2.167 (5.7)	-3.749 (6.9)
-Bus	- 0 -	- 0 -
-Buseta	-1.302 (6.2)	-0.761 (6.2)
-Walk	0.754 (0.9)	2.287 (4.9)

TABLE 2 (cont.)

	Bogota, 1972	Bogota, 1978
Model Statistics		
Log likelihood		
At zero	-894	-1538.
At convergence	-524.	-1014.
Likelihood ratio index	.414	.340
Number of cases	732	1244

Sample constitution: 1972 3/

Number of travelers in sample who choose:

(1) Drive/passenger	223 (305 have altrn in choice set)
(2) Taxi/collective taxi	38 (338 have altrn in choice set)
(3) Bus	321 (733 have altrn in choice set)
(4) Buseta/microbus	115 (520 have altrn in choice set)
(5) Walk	35 (590 have altrn in choice set)

Sample Constitution: 1978

Number of travellers in sample who choose:

(1) Drive/passenger	232 (417 have altrn in choice set)
(2) Taxi/collective taxi	21 (614 have altrn in choice set)
(3) Bus	467 (1239 have altrn in choice set)
(4) Buseta/microbus	378 (1239 have altrn in choice set)
(5) Walk	146 (909 have altrn in choice set)

Notes:

1. Asymptotic t-statistics are in parentheses.
2. Measured in 1972 Col. $.
3. All individuals in households owning automobiles have drive/passenger in their choice set. Households living in inner areas of the city (Rings 1-3) have taxi/collectivo. All travellers are assumed to have access to standard buses. Households living in regions served by busetas (according to route maps) have the buseta/microbus alternative -- all households in 1978, while travellers living within 12 km. their work location are assumed able to travel there on foot.

TABLE 3

Simulation of Model Forecasts; Bogota, 1978

Mode	Percentage Mode Split					
	1978-Actual Mode Shares	1. All Parameters Constrained	2. All Parameters Constrained Except Intercepts	3. All Parameters Constrained Except Intercepts and Costs	4. All Parameters Constrained Except Intecepts, Cost, Car Comp. (walk, buseta) and Head Dummy (walk)	5. All Parameters Constrained Except Intercepts, Cost, Car Comp. (all) and Head Dummies (all)
Drive/Passenger	12.5	13.1	10.4	9.2	7.8	9.2
Taxi/collective	1.6	3.1	1.4	1.1	1.6	1.5
Buseta/microbus	18.4	13.0	19.7	19.3	17.3	16.6
Bus	54.1	61.4	49.8	50.8	56.4	54.4
Walk	13.4	9.4	18.7	19.7	16.9	18.4

Note: Percentages may not total to 100.0 due to rounding errors.

TABLE 4

Simulation of Model Forecasts; Bogota, 1978

Mode	1978-Actual Mode Shares	Forecasting Simulations: Percentage Mode Split	
		All Parameters Constrained Except Intercepts, Car Comp (walk, buseta) and Head Dummy (walk)	All Parameters Constrained Except Intercepts, Car Comp. (all) and Head Dummies (all)
Drive/shared ride	12.5	8.9	10.3
Taxi/collective	1.6	1.9	1.8
Buseta/microbus	18.4	17.9	17.3
Bus	54.1	55.1	53.2
Walk	13.4	16.1	17.5

TABLE 5

Simulation of Model Forecasts-Overall and by Income Level; Bogota, 1978

	1978 Actual Mode Shares	Percentage Mode Split		
		Model 1.	Model 2.	Model 5.
Low Income Households (less than 1972 Col.$3000/mo.)				
Drive/shared ride	1.6	2.4	1.7	1.1
Taxi/collective	0.7	1.4	0.6	0.3
Buseta/microbus	13.9	10.1	15.7	13.6
Bus	67.0	76.1	62.7	63.5
Walk	16.7	10.0	19.3	21.5
Middle Income Households (1972 Col.$3000/mo to 8100/mo)				
Drive/shared ride	10.2	11.4	8.6	7.3
Taxi/collective	1.0	2.1	0.9	0.7
Buseta/microbus	20.7	13.2	19.9	16.6
Bus	55.3	63.4	50.9	56.6
Walk	12.9	9.9	19.8	18.8
Upper Income Households (More than Col.$8100/mo)				
Drive/shared ride	36.5	35.9	29.8	27.3
Taxi/collective	4.2	8.0	3.9	5.0
Buseta/microbus	22.8	17.8	26.8	22.2
Bus	28.1	31.0	24.0	33.6
Walk	8.4	7.2	15.5	11.9

5
Stated preference methods

Behavioural Research for Transport Policy
pp. 269-288. © 1986 VNU Science Press.

Sensitivity to Level of Service: Evidence from Stated Preference Work.

by J J Bates, John Bates Services

INTRODUCTION

The response of travellers to differing levels of service is of concern to policy-makers in two different ways. In the first place, there is the need to predict likely changes in demand, and in the second place, it is often important to be able to evaluate investment options which differ in the level of service offered. With evaluation, we are not merely interested in the change in demand, but also in the benefit (or disbenefit) perceived by existing users of a transport system.

In the UK and certain other countries, these two areas have been largely integrated by means of the concept of generalized cost (eg McIntosh & Quarmby 1970). In most work carried out in the late sixties and early seventies, generalized cost was usually a simple linear function of travel time and travel cost, in which the travel time was converted to cost units by multiplying by a 'value of time'. Although there are many aspects of time valuation which are controversial both in theory and in practice, the basic concept is sufficiently well known that the discussion in this paper will largely be conducted in these terms.

With the subsequent development of discrete choice theory and random utility models, it has been possible to extend the notion of generalized cost considerably, and nowadays it would be more conventional to interpret generalized cost as 'utility'. Earlier problems about the units in which these concepts should be measured have been resolved (see Gunn (1983)), though care is still required, particularly for forecasting. However, the main concern of this paper will be to examine the implicit tradeoff between components of utility inasfar as they affect travel choices, with special attention being paid to values of time.

2. RELATIVE WEIGHTS FROM REVEALED PREFERENCE STUDIES

The lessons applicable to studies which have produced values of time are relevant to the general issues of Revealed Preference (RP) models. Traditionally, values of time have been derived from choice data, where both the actual choice made and the components of the various alternative journeys are known for every respondent in the sample. This means identifying all the options (or at least a subset of them)open to each respondent, postulating a set of variables which might be expected to 'account' for the actual choice made, and measuring these for each option, for each respondent.

This in itself is a tall order, and although it is possible to satisfy the requirements in certain circumstances, it is usually expensive in terms of survey time and cost. Further, although in principle there are many situations in which travel choices could be observed, practical conditions limit the field severely. 10 requirements formulated by Harrison & Quarmby (1969) (and subsequently given more rigorous status within the context of random utility theory) need to be met in order to have some expectation of success. Although these conditions were drawn up in the particular context of estimating values of time, most of them have wider applicability. They are

(i) There must exist real choices
(ii) Choices, where they exist, must be fully perceived, and there must be grounds for believing that travellers know something about the alternatives.
(iii) The orientation of the analysis is to explain behaviour, and attempts must be made to extract the effects of other relevant variables.
(iv) Different types of time should be distinguished.
(v) The form of model implicit in the analysis technique must be consistent with the behavioural hypothesis underlying it.
(vi) There must be perceptible differences between alternatives.
(vii) Each variable which it is believed may be important in an explanatory equation should exhibit sufficient variance in the data, and the variables expressing time and cost should not be collinear.
(viii) The analysis technique must show sufficiently high level of explanation of behaviour.
(ix) The population analysed must be assumed homogeneous in the characteristics not measured
(x) The sample analysed must show non-trivial proportions making the different choices.

The result is that the great majority of studies which have been used to estimate values of time have been commuting mode-choice studies, supplemented by a small number of route-choice studies. In fact, revealed preference cannot be used at all in a number of travel contexts of interest, because no discernible choices exist. This applies to a large number of social and recreational trips made by car, where, even though other modes are strictly feasible, they are considered so inferior (or not considered at all) that there is effectively no choice.

Where the revealed preference approach has been successful, it has permitted some discrimination between various aspects of the journey. Following a number of studies, it has become conventional to separate travel time into elements such as walking, waiting, transfer etc., and a general consensus exists that 'out-of-vehicle' time components tend to be more highly weighted (ie, cause proportionately more disutility) than in-vehicle time. (For a useful summary of results in this area, see Bureau of Transport Economics (1982), Table 8.4). Nevertheless, in spite of a convention whereby out-of-vehicle time tends to be valued at twice in-vehicle time, the results of individual studies are rarely accurate enough to discriminate among the specific components of out-of-vehicle time.

The segmentation of travel time components is typically associated with public transport modes, but it is possible to think of it in the context of car travel as well, distinguishing for instance time spent on the motorway, time spent in free-moving but non-motorway traffic, and time spent in congested traffic. Similarly, it is possible to segment costs, particularly between 'out-of-pocket' costs such as tolls and parking meters, as opposed to occasional costs such as that of filling up the tank with fuel. However, very little work has been done in this area, largely because of the data problems, it would appear.

It must also be admitted that in spite of the considerable attention which has been paid to the problem of modal choice, and the methodological advances that have been made, especially in the area of random utility theory, the amount of useful explanation actually provided by such models is small. Typically, a large proportion of the 'correctly predicted' choices in any such model will be due to the modal constant, in spite of significant coefficients being obtained on certain level-of-service variables. To what extent this is due to data problems, and to what extent it reflects the fact that a high proportion of travellers do not perceive any sensible choice is a moot point.

3. REQUIREMENTS FOR UNDERSTANDING REACTIONS TO LEVEL OF SERVICE CHANGES

It would appear to some extent justified to accuse transport research professionals of having unduly concentrated on the phenomenon of modal choice, because of its relative modelling feasibility, while at the same time achieving relatively little in terms of useful results for policy. For, since the presence or absence of a car is a crucial determinant of the choice of mode between public and private transport, and the impact of transport authorities on the acquisition of cars is relatively slight (although it is possible to influence car acquisition, as argued in Bates & Roberts (1981)), more might be achieved in transport planning by setting out directly to provide a better service within the modes chosen (thus taking a relatively passive attitude towards demand). It is not inconceivable that such an approach would have repercussions on mode choice, in the longer term.

The declining demand for public transport in the face of the advance of the car has tended to put public transport undertakings on the defensive, which has led to a debasement of the service provided. To some extent this is inevitable because of resource constraints, but there seems to be considerable scope for a readjustment of priorities within these constraints. A good example is in the relationship between frequency (and thus waiting time) and access to the system (walking time etc.). It may be possible to concentrate routes on particular parts of the network, thus increasing access times, but at the same time running a more regular and frequent service, thereby reducing waiting time. How would passengers respond to such a change?

Another case concerns the tradeoff between journey time and service reliability. It may be possible to improve reliability (in terms of, say, reducing the standard deviation of the travel time) although average journey times would be increased. Would such a change be perceived as a benefit or a disbenefit?

Examples of this kind are not confined to public transport. Motorists might be prepared to accept a longer travel time (making use of a bypass for instance) in order to avoid driving in congested circumstances. There are also many tradeoffs to be made on the cost side, with parking charges, increased petrol tax, tolls etc.

The issues raised as examples here all relate to variations <u>within</u> the chosen mode, and although some guidance can be given on certain of these issues as a result of research into modal choice, for most of the points raised here existing modal choice models are insufficiently detailed, not because of conceptual inadequacy, but because of the cost and difficulty in obtaining a sufficient data base on which to estimate reliable coefficients. If we are serious in wishing to investigate within-mode options, we require alternative techniques, of which Stated Preference (SP) is an increasingly popular possibility.

4. EXPERIENCE WITH STATED PREFERENCE

As has been pointed out in Louviere et al (1981), the advantages and shortcomings of revealed preference and stated preference are largely complementary. Given that similar model forms can be estimated for both approaches, the crucial difference resides in the nature of the dependent and independent variables (in this context, the dependent variable is the expression of choice, while the independent variables are the attributes of the alternatives): with SP the status of the choice is in some doubt as to its reliability, while with revealed preference (RP) there is virtual certainty that the preferred mode was actually chosen; with RP on the other hand the researcher is largely at the mercy of the configuration of the independent variables, both in terms of their reliability and the extent of internal correlations, while in the case of SP they are clearly specified, and can be chosen according to certain rules to reduce or eliminate correlations.

There exists considerable prejudice against the idea of making use of data which is in some way introspective by treating it as straightforwardly reliable. This prejudice is probably strongest among economists, for whom it has been almost axiomatic that revealed preference data is the only data relating to demand that is of any value. There is, at any rate, evidence to show that certain kinds of introspective data relating directly to demand (eg "How often would you buy a product which had characteristics x,y,z?") are unreliable. To have any prospect of being useful, questions must be asked in such a way that the answers are not obvious, the purpose is not too apparent, but at the same time the task presented to the respondent is interesting and within his capability. In relation to this last criterion, it is important to pay attention to response bias, since it is possible that only certain kinds of people are willing and able to perform the tasks in an SP experiment.

The design of the 'experiment' on which the alternatives are based is the most crucial part of the SP study. It is necessary for the alternatives presented to meet three important criteria.

1. They must be commensurate with reality from the perspective of the experience of the respondent.

2. They must offer sufficient choice and range in attributes for the respondent to be able to rank them.

3. They must be sufficiently sensitive to differences in individual tastes not to cause all individuals to make the same (obvious) choices.

At present, no very rigorous method has been proposed for meeting these criteria, and in practice various assumptions are made about likely variations among people's tastes, and by means of a trial and error process suitable values are obtained.

In addition, it is helpful to include among the alternatives certain combinations such that, on any rational choice process, some alternatives should be preferred to others. This can be used as a check to see that respondents are taking the tasks seriously. It is often possible to devise other, more or less rigorous, tests of 'rationality'. This does, of course, beg the question of whether individuals do in fact make rational decisions!

A major concern about SP which does not yet appear to have been satisfactorily resolved relates to how respondents react to the actual values of the attributes presented. An original suspicion that respondents might merely take account of whether an attribute had a high or low value, without being concerned about the exact value, seems largely to have been allayed, as will be discussed in the case studies in the next section. But there remain a number of dangers. One is of presenting different values of an attribute that for practical purposes are indistinguishable (because of 'thresholds of perception' etc.) Thus a difference between a journey time of 30 minutes and one of 32 minutes may be of no practical importance to the respondent, while he may nonetheless reply to an SP question in a 'rational' way. Another point is that there is evidence from social research about the existence of thresholds on certain journey times. To the extent that straightforward linear expressions are used in analysis, the results could be seriously misleading.

In fact, SP is well able to allow for non-linearities. There is however a cost attached to this, since it requires multiple-valued attributes, to some extent complicates the experimental design, and, more seriously, increases the number of options that need to be presented. It is nevertheless the author's firm belief that when dealing with continuous variables like cost and time, binary-valued attributes are often inadequate.

5. TWO CASE STUDIES

In this paper we will refer to two central studies. One, reported in Bates & Roberts (1983a,b) was in connection with a study of Long-Distance Travel for the Dutch Ministry of Transport; since the earlier studies, the work has been updated, and the results will be briefly presented here. The other, reported in part in Bates (1984), was in connection with a study on the Value of

Time for the UK Department of Transport; currently further work is being undertaken in this connection. Readers are referred to these references for greater detail about the context of the studies and how the surveys were carried out.

Space precludes a fuller discussion of a number of analytical issues, but in particular it should be noted that the 'repeated measurements' relating to each respondent have been treated as independent observations. It is not known to what extent this constitutes a problem, but it is at least likely to reduce artificially the size of the standard errors associated with estimated coefficients.

In addition, in both studies, mode choice questions have been asked according to a five-point scale. For each pair of journeys, respondents had to decide which mode they would use, and how strong their preference was, according to the scale:

1 Definitely prefer mode 1
2 On balance, prefer mode 1
3 Indifferent
4 On balance, prefer mode 2
5 Definitely prefer mode 2

This is a limited scale by the standards of advocates of 'functional measurement analysis', where scales of 1 to 20 or 1 to 100 are typically used (see for example Louviere et al (1981)). However, it is not certainto what extent the apparent increase in information resulting from such scales is vitiated by unreliability associated with the more difficult task imposed on the respondent.

Finally, where SP data has been of the ranking type, the results have been analysed by the ordered logit model (see eg Beggs, Cardell & Hausman (1981)). Again, there are some unresolved issues connected with this approach.

A. The North Kent survey

This survey, which was carried out on behalf of the UK Department of Transport in the context of the Value of Time Study, was designed to allow a mode choice model to be calibrated for long-distance commuters where there was a genuine choice between coach and rail travel to Central London, the coach being slower but cheaper than the train.

Two types of SP question were included. Firstly, a mode choice question based on five attributes: main in-vehicle time (coach or train), other in-vehicle time, waiting time, walking time, and total cost. These attributes were varied in a systematic way, and in addition the sample was divided at random into two sets who were given different valued alternatives, though the experimental design was kept constant. The second question was of the 'within-mode' type, and related to the attributes in-vehicle time, waiting time, walking time, interchange time (for rail users) and variation in overall journey time (for coach commuters).

The preliminary analysis (reported in Bates (1984)) indicated that for the whole sample, values of both waiting and walking time were substantially lower than in-vehicle time on the main mode - a result which is at variance with the general body of research findings in this area. In comparison with the RP results (shown in Figure 1), the walking times agreed closely, but the RP value for waiting time was more in agreement with conventional experience (at about two and a half times the value of main mode in-vehicle time).

Further investigation of these results is in hand, but at the time of writing no further information is available. There is a suspicion, however, that the values offered for walking and waiting times in the experimental design were too insignificant in relation to the typically long in-vehicle times in the sample (between one and two hours). This may have led to respondents placing very little weight on the out-of-vehicle time components. It must also be borne in mind that the form of the mode choice question was not necessarily the best for obtaining values of walking and waiting time.

However, evidence of thresholds did emerge. When respondents were presented with the same pattern of differences between the modes, but the absolute values of cost, time etc. were increased, there was clear evidence of an increased tendency to choose the faster mode, rail. The impression given was that when journey times were moving to unacceptable limits, people were prepared to pay more money, thereby demonstrating non-constant values of time.

In addition to the mode choice question, the within-mode questions were analysed. In this case, the relative weighting of both walking and waiting time to in-vehicle time were greater than 1.0, but not by a very large amount. On the other hand, for rail travellers, there was an indication that interchange time was perceived as a rather greater disutility. It is hoped that some further disaggregation will reveal clearer distinctions, particularly when the actual circumstances of the respondents' current journeys are taken into account.

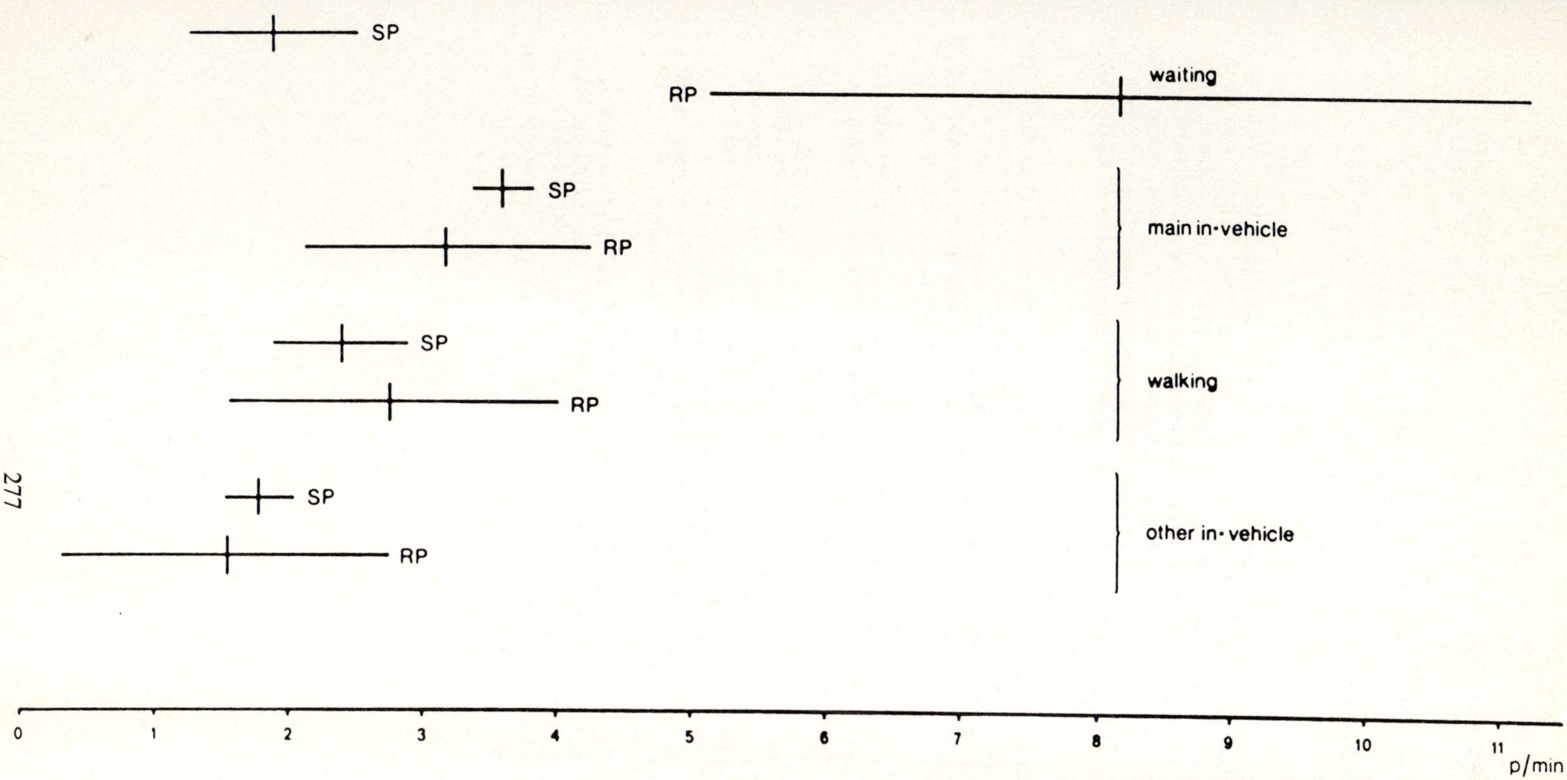

Figure 1. Estimated values of time, with 95% confidence intervals, revealed preference and stated preference compared

B. Long Distance Travel Survey

An initial sample was selected in September 1982, and the results of the analysis are reported in Bates & Roberts (1983a). Since then, the survey has been repeated and a new sample drawn. However, the values of the attributes have been altered, thus offering the opportunity of investigating the extent to which the results of the analysis are dependent on the values used in the experiment.

For historical reasons, we refer to the September 1982 survey as the 'Phase IIA Survey', and to the later survey as 'Phase IIB'. In addition, the Phase IIB sample was subdivided at random into two segments who were given differently valued sets of attributes. It is thus possible to compare the three resulting samples, Phase IIA, Phase IIB (i) and Phase IIB (ii). The sample sizes were 108, 76 and 73 respectively.

The questionnaire was designed to relate to the journey to work, with particular emphasis on the choice between car and train. Apart from obtaining a limited amount of socio-economic data, the core of the questionnaire was three questions, each relating to hypothetical choices.

Question 1 offered alternative journeys (characterized by time and cost) by the two modes car and train. There were ten cards offered in this question.

Question 2 related to car journeys only: the design was based on the four variables in-vehicle time, presence or absence of congestion, egress time (from parking) and cost. Eight cards were offered to the respondent, containing different values of these four variables, and the task was to rank them in order of preference.

Question 3 was an equivalent question for rail journeys. In this case there were sixteen cards, and the design was based on six variables - access time, service interval, in-vehicle time, the necessity to interchange, egress time and the cost.

Analysis of Question 1

Each of the four factors was entered at two levels - "high" and "low". The comparative values are set out in the Table below.

TABLE 1 : QUESTION 1. LOW AND HIGH VALUES FOR VARIABLES

Variable	Phase IIA Low	Phase IIA High	Phase IIB(i) Low	Phase IIB(i) High	Phase IIB(ii) Low	Phase IIB(ii) High
Car Cost	12	22	15	19	13	21
Train Cost	8	16	10	20	8	22
Car Time	40	70	40	70	40	65
Train Time	45	75	50	65	55	60

NB: Costs are expressed in guilders, and relate to the round trip. Times are expressed in minutes and relate to one-way journeys.

In the analysis of Phase IIB data, only a generic specification was attempted; in other words, it was assumed that time and cost would be valued the same, irrespective of what mode they applied to. The model thus reduces to a linear function of the differences between the values in the design for time and cost, and casting it in the standard utility formulation, we have

$$DU_k = a + b_c DC_k + b_t DT_k ,$$

where, by convention, we will measure all differences as "train - car". With this convention, the b coefficients should be negative, and a positive value of a would indicate a preferential bias in favour of train. In the equation above, k refers to the replication in the experiment.

In Table 2, we give the results of the model-fitting for the two versions (i) and (ii) of Phase IIB and for Phase IIA.

TABLE 2 : ESTIMATED COEFFICIENTS ON GENERIC SPECIFICATION

Data set	ASC (train) [a]	Time $[b_t]$	Cost $[b_c]$	Value of time * (fl/hr)
IIA train	0.4788	-.04130	-.1420	8.73 £
IIA car	-0.4215	-.04271	-.1313	9.76 £
IIB(i) train	0.6171	-.03818	-.2002	5.72
IIB(i) car	-1.025	-.05226	-.1940	8.07
IIB(ii) train	0.3647	-.04070	-.1757	6.96
IIB(ii) car	-0.8029	-.05101	-.1578	9.69

NOTES
"car" and "train" refer to the subsamples using these modes.

* The value of time is calculated by dividing the time coefficient by the cost coefficient, and multiplying by 30; this takes of account of the fact that the cost relates to round-trips.

£ Note that while the IIB results were obtained by linear regression, these results were obtained from logit analysis; it has been shown (Bates & Roberts 1983) that the difference is small.

For reasons of space, the standard errors have been omitted. However, as a guide, the t-ratios returned by the program had the following order of magnitude. For the Phase IIA surveys the values were:for the constant, 10.0; for the cost and time coefficients, around 21.0. For the IIB surveys the values are slightly lower, reflecting the reduced sample sizes. However all these values can be expected to be overestimates, because of the repeated measurements problem referred to earlier.

Taking these results together, there are some variations in the coefficients from one sample to another, and in the resulting values of time, but overall there is considerable consistency. The value of time for train users is always somewhat lower than that for car users, though the values for train users show more variation. The pattern results from the fact that the car coefficient on time is generally larger than that for train, while the cost coefficient is smaller.

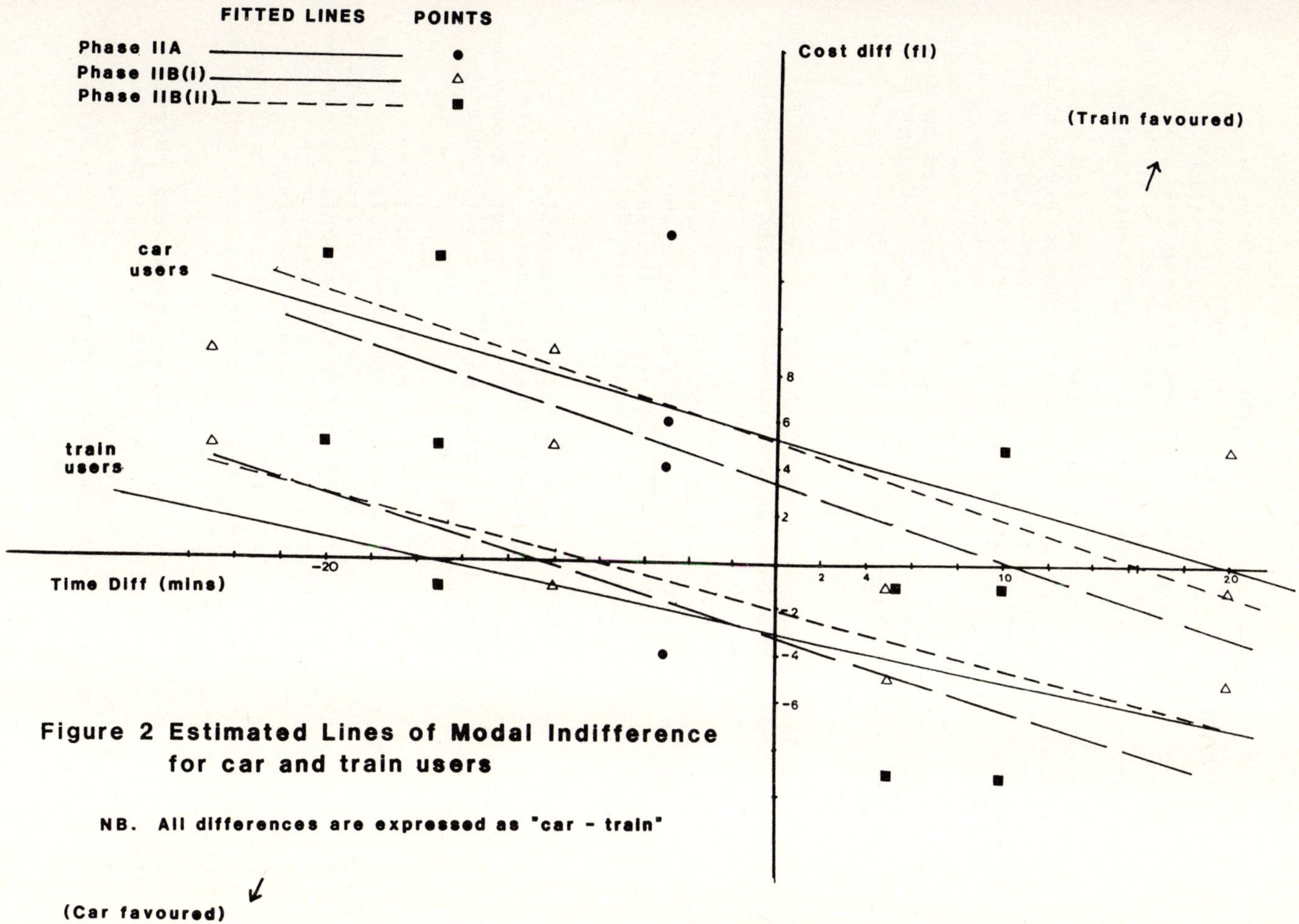

Figure 2 Estimated Lines of Modal Indifference for car and train users

NB. All differences are expressed as "car - train"

The ASCs for car users are always negative, indicating a bias against train, other things being equal, while those for train users are positive.

In Figure 2, the lines DU = 0 (the line of modal indifference) are plotted in the (DT,DC) plane for all the estimated values, together with the actual points presented in the experiment. It can be seen that the results of the three surveys are very close, and the main distinction is that between car users and train users.

A major concern following the Phase IIA work, and one which resulted in the splitting of the Phase IIB sample into subsamples (i) and (ii), was that respondents might not be taking sufficient notice of the actual values offered to them on the cards, but might have produced the same responses simply by noting whether values were "high" or "low". On such a hypothesis, the results would be much more strongly conditioned by the experimental design per se than by the values chosen for the attributes.

The two opposing hypotheses, then, can be stated as follows:

a) The responses will be the same irrespective of the card values, provided that the experimental design (in terms of high and low) is held fixed. In this case the marginal utilities for any given replication would remain the same, independent of the attribute values offered.

b) The card values will be the sole determinants of the response. In this case, the estimated coefficients should be the same, regardless of the attribute values chosen.

Although it cannot be said with certainty that the coefficients are the same across the three data sets, and thus there may be some suspicion that the ultimate results depend on the values offered in the experiment, it can be safely concluded that hypothesis a) is rejected. A comparison of the relative scores given for each replication indicates clearly that the relative preference for train or car is affected by the card values, and in an intuitively reasonable way.

Thus the results are generally encouraging, and suggest that train users making long distance journeys have an average value of time of around 7.5 fl/hr, while corresponding car users have a value of around 9 fl/hr.

Analysis of Question 2

Table 3 gives the high and low values for each attribute, in the three data sets.

TABLE 3 : QUESTION 2. LOW AND HIGH VALUES FOR VARIABLES

Variable	Phase IIA		Phase IIB(i)		Phase IIB(ii)	
	Low	High	Low	High	Low	High
In-veh time	35	65	35	65	45	60
Congestion	0	1	0	1	0	1
Egress time	2	10	4	8	2	10
Cost	12	22	14	20	12	22

NB Costs are expressed in guilders, and relate to the round trip; times are in minutes, for one-way trips. Congestion is a (0,1) variable.

It will be seen that the principal differences are that the cost difference has been narrowed in Phase IIB(i) and the in-vehicle time difference hs been narrowed in Phase IIB(ii).

The estimated models, using a straightforward linear utility function, are set out in Table 4.

TABLE 4 : ANALYSIS OF QUESTION 2. COEFFICIENTS AND VALUES OF TIME

Dataset	In-veh. time	Cong-estn	Egress time	Cost	Value of in-veh time	Value of egr.time
IIA train	-.10857	-2.3661	-.2128	-.3067	10.62	20.81
IIA car	-.08264	-2.4675	-.2026	-.2180	11.38	27.89
IIB(i)trn	-.11049	-2.0636	-.2622	-.3523	9.41	22.31
IIB(i)car	-.11728	-2.3203	-.2820	-.3742	9.40	22.61
IIB(ii)tr	-.19438	-3.0579	-.2045	-.3257	17.91	18.88
IIB(ii)ca	-.16411	-2.4703	-.2385	-.3114	15.81	23.03

The coefficients show in general considerable regularity across the data sets, but there is one worrying exception: the in-vehicle time coefficients for Phase IIB(ii) are much higher than for the other data sets. There is a suspicion that this is related to the reduced difference in the attribute values.

Referring back to the discussion about Question 1, and the two alternative hypotheses about how the questions are answered, there are corresponding versions of the two alternatives which can apply to Question 2. On the first hypothesis, the marginal disutility attached to moving from the low to high values of any given attribute would be constant, irrespective of the card values offered; on the second, the coefficients would be constant. We have seen that in general the coefficients are constant, but that this is not so for in-vehicle time. However, a further investigation shows that the increase in the in-vehicle time coefficients for Phase IIB(ii) does not quite compensate for the reduction in the difference, in the values offered.

An interpretation which would be consistent with the evidence (and which is also supported by our investigation of the North Kent data) is that there are important non-linearities in the acceptability of various journey times. One way of postulating this effect is that there is a considerable disutility of increasing in-vehicle time by 15 minutes, and that given this, increasing by 30 minutes is not appreciably worse. An alternative, and probably more credible, approach is to accept that there is an upper threshold on acceptability as far as in-vehicle time is concerned, and this might well be of the order of 60 minutes. On such an interpretation, comparing the values in Table 3 for IIB(i) and (ii), a shift from 35 minutes to 45 minutes might be perceived as a moderate disutility, but in either case, moving to a value of 60 or 65 minutes would be extremely inconvenient.

It should be noted that in all cases, egress time is valued more highly than in-vehicle time, and with the exception of the IIB(ii) results, where we have already suggested that in-vehicle time is probably overvalued, egress time is valued at about double in-vehicle time, a result supported by the conventional wisdom about out-of-vehicle time.

Analysis of Question 3

The values for the attributes are set out in Table 5

TABLE 5 : QUESTION 3. LOW AND HIGH VALUES FOR VARIABLES

Variable	Phase IIA Low	Phase IIA High	Phase IIB(i) Low	Phase IIB(i) High	Phase IIB(ii) Low	Phase IIB(ii) High
Access time	5	15	8	15	6	12
Serv. Intvl	10	30	10	30	5	35
In-veh time	30	60	30	60	40	55
Interchange	0	1	0	1	0	1
Egress time	2	10	4	8	2	10
Cost	8	16	12	18	9	21

NB Costs are expressed in guilders, and relate to the round trip; times are in minutes, for one-way trips.

The estimated coefficients, again using a straightforward linear utility function, are set out in Table 6; the related values of time are given in Table 7.

TABLE 6 : ESTIMATED COEFFICIENTS FOR QUESTION 3

Dataset	Access time	Service Intervl	In-veh time	Inter-change	Egress time	Cost
IIA tr	-.1361	-.02571	-.05839	-.1460	-.09087	-.2724
IIA ca	-.1090	-.02404	-.05846	-.1025	-.10095	-.1652
IIB(i)tr	-.1028	-.04412	-.06818	-1.0278	-.12867	-.2534
IIB(i)ca	-.1468	-.03424	-.06988	-.5497	-.11455	-.2323
IIB(ii)tr	-.1465	-.02921	-.09227	-1.1541	-.09534	-.2901
IIB(ii)ca	-.1004	-.04414	-.08013	-.7933	-.10546	-.2650

TABLE 7 : IMPLIED VALUES OF TIME FOR QUESTION 3 (fl/hr)

Dataset	Value of in-veh time	Value of accs.time	Value of egrs.time
IIA train	6.43	14.99	10.01
IIA car	10.62	19.79	18.33
IIB(i) train	8.07	12.19	15.27
IIB(i) car	9.02	18.98	14.85
IIB(ii) train	9.54	15.10	9.86
IIB(ii) car	9.07	11.32	11.89

Looking first at the values of time, there are a number of somewhat unexpected results, but the general pattern is clear. With only one exception, access and egress time values are always considerably higher than in-vehicle time, and car in-vehicle time values are in most cases higher than those for train users. On the other hand, there is no clear pattern about the relative weighting of access and egress time.

Turning now to the coefficients, there is one clear difference between the Phase IIA and IIB results, and that relates to the interchange variable. In Phase IIA this was given an extremely low weight, and in terms of the implied marginal utility attached to each factor, it was by far the least significant factor in affecting the rankings. In the Phase IIB surveys, a pattern arises whereby for train users, the weight placed on the need to make an interchange is increased by a factor of about 8, while for car users it is considered less important, but still much more than was found in Phase IIA. It seems likely that this difference is due to the nature of the actual experience which respondents have of interchange in the various corridors where interviews took place.

There is again some suggestion that the coefficients on in-vehicle time in the IIB(ii) results have risen as the difference between the low and high values is narrowed, corresponding to what was found with Question 2. In this case, however, there has been a corresponding change in the cost coefficients, so that the values of time have not increased.

The 'service interval' variable has simply been included in the analysis by using the values in the design (relating to the length of time between services) as if some direct value could be placed on them; this is unlikely to be a very satisfactory assumption, and in particular we have refrained from calculating "values of time" for this variable.

6. CONCLUSIONS

The analysis reported in this paper has concentrated less on the actual results than on some general questions about respondents' ability to carry out the tasks put to them in an SP questionnaire. The general conclusion is that respondents (at least the proportion who co-operate!) are both willing and able to carry out the kinds of task described here.

The analysis carried out to date by the author is fairly rudimentary, and much more could be done in terms of segmentation and improved model specification. In addition, there is a need for reconciliation of the various results. More important, however, is the general conclusion that there appear to be non-linearities and thresholds in people's behaviour, and that these can only be satisfactorily analysed by including multiple-valued attributes.

This in its turn will complicate the experimental design and the process of selecting attribute values, which serves to emphasize the following conclusion. SP offers great advantages in allowing us to analyse areas of choice which are difficult to identify in revealed preference contexts. Nonetheless, it is not a simple 'off the peg' tool. The design of the experiment, the choice of the attributes, and the selection of values are significant tasks, requiring a great deal of preparation before taking to the field. In particular, piloting becomes a more lengthy process, and in planning an SP survey, it is recommended that adequate time be left between the pilot and the main survey.

While the technique remains controversial, the present analysis is certainly sufficiently encouraging to recommend further studies. It is particularly hoped that more attention will be paid in the future to alternative journeys within a given mode, and the tradeoff between the component attributes of that mode, as opposed to the previous emphasis given to modal choice.

REFERENCES

McIntosh PT & Quarmby DA (1970), Generalized Costs and the Estimation of Movement Costs and Benefits in Transport Planning, UK Department of Transport MAU Note 179.

Gunn HF (1983), Generalized Cost or Generalized Time, Journal of Transport Economics and Policy Vol xvii no. 1

Harrison AJ & Quarmby DA (1969), The Value of Time in Transport Planning - A Review, European Conference of Ministers of Transport, Round Table no. 6, OECD Paris.

Bureau of Transport Economics (1982), The Value of Travel Time Savings in Public Sector Evaluation, Australian Government Publishing Service, Canberra

Bates JJ & Roberts M (1981), Indicators for Future Car Ownership and Use, Transport Policy and Decison Making, Vol 1, pp 275-90

Louviere JJ, Henley DH, Woodworth G, Meyer RJ, Levin IP, Stoner JW, Curry D and Anderson DA (1981), Laboratory-Simulation versus Revealed Preference Methods for Estimating Travel Demand Models, Transportation Research Record 794. pp 42-51

Bates JJ & Roberts M (1983a), Recent Experience with Models fitted to Stated Preference Data, PTRC Summer Annual Meeting, Brighton, paper N8.

Bates JJ & Roberts M (1983b), Data Collection for Choice Models based on Hypothetical Transport Alternatives, Second International Conference on New Survey Methods in Transport, Hungerford Hill, Australia

Bates JJ (1984), Values of Time from Stated Preference Data, PTRC Summer Annual Meeting, Brighton, paper H2

Beggs S, Cardell NS & Hausman J (1981), Assessing the Potential Demand for Electric Cars, Journal of Econometrics (16), pp 1-9

Behavioural Research for Transport Policy
pp. 289–305.

Functional Measurement of Route Choice Behavior: A Study of Commuting Cyclists in the Netherlands

Mark A. Bradley
Cambridge Systematics Europe, The Hague
and
Piet H.L. Bovy
Institute for Town Planning Research, Delft

ABSTRACT

An application of scenario-based, or "stated preference" survey and analysis techniques is described in the context of cyclists' route choice. Route choice modelling with observed choice data is hampered by the cost of processing network data, and by the difficulty of assessing the alternative routes and the perceived attributes of the routes considered by individual travellers. An alternative approach is to obtain stated evaluations of well-defined hypothetical routes. Such data was collected from commuting cyclists in the city of Delft in the Netherlands, and analyzed using functional measurement to measure the relative importance placed on route attributes such as time, traffic level and surface quality. Though the techniques used are well founded in the marketing and psychology literature, the route choice context raises issues which are particularly important for their application in transport policy analysis. The stated preference framework is discussed, followed by a description of the route choice survey and results, concluding with a discussion of the applications of such an approach in policy analysis.

1. Introduction

An understanding of the relative influence of time and cost versus qualitative factors in route choice is valuable in several types of transport policy analysis. Alternative plans for new roads or cycling facilities may involve changes in travel times which must be weighed against "costs" or benefits in terms of other choice factors. These tradeoffs are important in predicting the usage of a planned facility and assessing the benefits to those users. Such information may also be useful for large, network-based studies, to define the paths which best represent travel mode/destination alternatives, and to assign the predicted flows for those alternatives to the network.

Qualitative factors are particularly important to bicyclists. Bicycle trips are generally shorter than auto trips, have a wider variety of routes with comparable travel times, and are more exposed to the external environment. For cycling, however, the general lack of network data, both qualitative and quantitative, is more acute than for auto travel. The alternative to collecting such data is to offer hypothetical travel options in survey form, each option defined in terms of the attributes assumed to be most important. This general method is what we term the "stated preference" approach, as opposed to "revealed preferences" inferred from choices among real options.

The scenario-based approach, used in this study to assess cyclists' tradeoffs among important route attributes, is termed "functional measurement". The approach was developed in the field of mathematical psychology (Anderson 1981), and has been developed in applications to many choice contexts, some within the transport modelling field (Lerman and Louviere 1979). Thus, our purpose is not to argue or extend the theoretical validity of the techniques, although the behavioral assumptions are made clear. Rather, the objective of this paper is to provide empirical evidence of the utility and efficiency of the stated preference approach in route choice analysis and application.

2. The Study Approach

In the past, route choice analysis has followed two main approaches. In the "motivational/attitudinal" approach, travellers are asked to state their reasons or motivations in selecting routes in a network. The results are often in the form of qualitative evaluations of the adequacy and importance of individual route attributes, and of overall beliefs about alternative routes (Carpenter 1977, Huchingson et.al. 1977). Such studies provide useful input to other approaches by identifying choice factors, helping to define market segments, providing operational definitions for qualitative factors, and giving insight into the relevant choice sets and choice constraints facing individuals. While it is possible to relate intended or actual choices to perceptions and attitudes (Wachs 1967), it is difficult to apply such results in assessing policies which can be characterized only as changes in observable route attributes.

With the "revealed preference" approach, observable route characteristics are related directly to observed route choices, often using an individual random-utility maximization framework, such as that of multinomial logit estimation (Hamerslag 1981, Ben Akiva,et.al. 1984). This approach requires knowledge of the set of alternative travel routes considered by each person, as well as objective data for all salient attributes for each route in the set. Collection of such data is generally difficult and costly, and does not ensure that the route characteristics used in modelling consistently represent the subjective measurements of those attributes made by travellers. This problem is especially relevant to route choice, where many route characteristics are monitored continuously, and may vary a great deal over the course of travel. Choice set constraints, such as the need to travel to secondary destinations, are also difficult to infer from revealed preference data.

A variation on the revealed preference approach requires travellers to state the choice alternatives considered for a familiar choice context, and to provide their perceived values along a given set of attributes for each alternative. With this information, it is possible to estimate the relative importance and interactions of perceived route attributes in behavior.

The stated preference approach is analagous to this form of revealed preference analysis, with the difference that the set of choice alternatives and their attributes are given in a hypothetical context, and the response is thus a rating, ranking, or stated choice among the alternatives rather than a verifiable choice. Although the responses are not subject to the perception processes and choice constraints of actual choice contexts, Louviere and others (1981) found that the two approaches, carried out on the same sample, resulted in very similar tradeoffs among important mode-choice factors, with the stated preference data giving estimates of greater statistical significance. In a larger mode choice study, which included bicycle commuting, models from scenario-based data validated well against the sample's actual choices among their self-described mode alternatives (Wis.DOT 1981).

The relative advantage of the stated preference approach, in most cases, is the controlled nature of the choice scenarios. This feature allows greater freedom in defining choice contexts, alternatives, and attributes; as well as direct comparison of the responses across individuals. The ability to obtain multiple responses from each individual also reduces sample size requirements. With these advantages, comes the liability that the success of the approach depends largely on the consistency of the hypothetical alternatives and the corresponding sets of attributes with their perception in actual choice situations.

The route choice context presents particular problems in the translation to a uniform set of attributes. A route can be thought of as a set of links which are delimited by intersections along the route. Each link may have different characteristics, and the set of attributes important in evaluating links may differ from that of intersections. Qualitative attributes may even vary along a single link.

If each link and intersection is assumed to contribute linearly to the overall utility of a route, then partially overlapping alternatives can be compared solely along those portions that differ. This assumption was used in the study to avoid making the experiment too complex. Routes were compared along a single set of attributes, each of which was assumed to be homogenous along each alternative. People were asked to assess various routes as if they were single links. Conceptually, the relative preference between two routes can then be modelled as a function of the important differences in non-common links. This conceptual solution, however, does not preclude peoples' difficulties in perceiving routes in this way. Various presentational approaches can be used to confront this problem, as described below.

First, however, a situational context and set of route attributes and levels must be defined. For this study, it was thought best to choose a situation that would be familiar to the respondents: the home to work trip for regular bicycle commuters. This trip purpose generates a large proportion of urban bicycle travel in the Netherlands, much of it along regularly chosen routes.

The set of route attributes was chosen after assessing the results of previous research (Upcott 1976, Axhausen 1984). Based on such work and previous research in Delft; travel time, surface quality, traffic level, and cycle facility type were selected. Other potentially important factors, such as weather and land use along the route, were defined in the background information for the choice context, or stated to be the same along all routes.

Three levels were chosen for each attribute to allow the estimation of non-linear (quadratic) effects. For the quantitative variable, time, a base value of 12 minutes was chosen, roughly the median of the survey respondents' reported one-way travel times. The high and low levels were defined as 15 and 9 minutes, a range equal to half the base level, and encompassing the majority of self-reported times. This range was considered large enough to be perceivable in actual choice situations, but not so great as to overshadow the influence of changes in the qualitative attributes.

The levels for facility type and surface quality were defined as recognizable types of cycle network construction, as shown in the verbal descriptions included in the survey (Figure 1). Though the definitions were made as mutually independent as possible, there is bound to be some correlation in the perception of these two attributes (i.e. separate bike paths are most likely to have an adequate surface).

The traffic level factor was the most difficult to define, as can be seen from Figure 1, a translation from the Dutch. To impart a clearer perception of this and the other qualitative factors, one half of the respondents were given photographs to portray each level of traffic, surface quality, and facility type. One half were also asked to categorize their own most-common route in terms of the given verbal and (in some cases) pictorial descriptions.

To create route choice scenarios, the factors and levels were arranged orthogonally to allow the independent effect of each attribute to be estimated. To estimate all main (quadratic) effects, and all interaction terms, evaluations of all 81 possible route configurations would be required. To limit the difficulty of the experiment, a one-ninth fractional factorial design (9 routes) was included in each survey. This design allows the estimation of the main-effects for each individual, with all interactions terms confounded with other terms. As

FIGURE 1

***** DESCRIPTIONS OF FACTORS AND LEVELS GIVEN *****

FACTOR	LEVEL	GIVEN VERBAL DESCRIPTION
Facility Type	Physically Separated	This portion of the roadway is meant only for bicycles and mopeds, and is totally separated from other traffic by curbing or plantings. Pedestrians are also provided with a separate walkway.
	Reserved On-Street	This is a full lane of the roadway, reserved for bicyclists and marked with a white stripe on the surface. Now and then there are autos parked on this lane.
	Non-Existent	There is no separate space for bicyclists on the street. They must ride in the same lanes as other traffic.
Surface Quality	Smooth	The surface over the whole route is good asphalt with no large cracks.
	Moderate	In this case, about half of the route is asphalt and the other half brick. There are occasional bumps and cracks.
	Rough	This route has a brick surface from beginning to end with bumps and cracks which one must try to avoid.
Traffic Level	Light	A car or bicycle comes along this route from time to time. There are few crossing pedestrians. Cyclists can easily ride next to each other. You never have to stop for others.
	Moderate	Quite a few cars use this route, but this is not a hinderance to cyclists. You can still ride along with other cyclists when it does not present a problem.
	Heavy	Many cyclists and autos ride along this route. It is very busy here. It is often difficult to enter or cross. You must also wait for other traffic and crossing pedestrians. It is not possible to ride next to other cyclists.
Travel Time		This refers to the total time that you are travelling, including all of the delays encountered on the route. The average travel times used in this survey are:
	Short	- 9 minutes,
	Medium	-12 minutes, or
	Long	-15 minutes.

individual-level estimates were desired for segmentation rather than for strict tests of functional form, the simple design was deemed adequate.

Three distinct sets of routes were designed to provide greater variation in the alternatives. Together, the three main-effects designs form a one-third fractional design (Kocur et.al. 1982, Master Plan 8), allowing aggregate estimation of the two-way interactions between the qualitative variables (or of facility type with the other three variables). All other two-way and higher interaction terms are confounded with the other terms. The three main-effects designs were distributed evenly across the sample, and across other survey design variations.

All routes were presented as pairwise choices against a common base route. Preferences were measured on a seven-point verbal scale, with the extremes representing a clear choice, and the midpoint indicating no preference. The base route was chosen to have the "medium" level of each factor, so that preferences would be distributed evenly as possible around the midpoint. Recognizing that people may interpret the scale in different ways, each individual's responses were normalized, according to their internal mean and variance, before aggregating responses.

The functional measurement technique is based on the assumption that the responses are a linear function of the relative utility of the choice alternatives, allowing direct estimation of all effects provided for in the experimental design. At the individual level, there are thus nine orthogonal equations to determine the effects of the eight possible deviations from the base route; a task which can be done algebraically.

Models based on evaluations across individuals can be estimated using ordinary-least squares regression of the route scores against dummy-variables representing the route attributes. The dummy variables can be set to yield piecewise-linear estimates of the factor coefficients (Timmermans 1984), or quadratic terms and interactions (Louviere and Woodworth 1983). The aggregate form of the utility equation, in terms of the arrays of route attributes, $\underline{X}$, and their relative weights, $\underline{W}$, is:

$$U_{ij} = f(\underline{W}.\underline{X}_j) + e_{ij}$$, for individual i and alternative j.

Most standard estimation techniques, such as OLS regression with scaled data, or multinomial logit with discrete choices or ranked data (Chapman and Staelin 1982), rest on the assumption that the error terms, (e), are independent and identically distributed across all alternatives (j's) and individuals (i's). A more general specification of the error term might be:

$$e_{ij} = a_i + f(\underline{v}.\underline{X}_j)$$, for individual i and alternative j.

The first error component (a) reflects the correlation in the responses resulting from repeated measures within individuals. Repeated measures designs can be analyzed using certain analysis of variance procedures (e.g. MANOVA), or complex iterative regression procedures. The second component reflects situations where the variance of the error term is a function of the attributes of the alternatives. Such variations may arise if certain attribute levels are more difficult to perceive or evaluate than others, an effect which may be inherent in the choice context or due to survey presentation. Another possible cause is "taste variation", if the perception and influence of factors not included in the model are correlated more with certain attributes than with others. Again, such correlations may be influenced by the presentation of the attributes in the survey context. To address this possibility, an iterative generalized least-squares approach is necessary to specify an error variance-covariance structure for scaled data, or a probit approach for discrete choice data.

As the problems outlined above may apply to the data collection approach we have described, it is likely that the statistical siginicance of the results will be overstated. As there has been no comparitive evidence which shows that the independent error assumption will bias the relative importance of the attributes, however, we have chosen to use analysis techniques which demonstrate the practicality of the stated preference approach. It would be a useful exercise to create data sets with various error correlation structures using simulation, and then to test the relative success of different response methods and analysis techniques in reproducing the underlying utility function. In this paper, however, we can only bring out instances where the where the survey results may be affected by the issues raised above.

3. The Survey Results

The survey experiment was administered to 134 cyclists in group sessions at the Technical University at Delft. For ease of administration, only university staff were contacted. Consistent with the scenario context, the sample was restricted to those who used the bicycle for commuting at least two days per week. Only those people living in or very near Delft were surveyed, keeping actual travel times within the general range used in the scenarios. Within these constraints, the sample provided an adequate cross-section for characteristics such as age, sex, car and transit availability, and income (proxied by occupation).

Following the scenario comparisons, each respondent was asked to map their own route, and to assess the importance of the four attributes in their daily route choice, the clarity of these attributes in the survey presentation, and the overall difficulty of relating the scenarios to their own choice situations. The responses to these questions indicate that the choice factors were generally considered important and perceived clearly, and that the scenarios provided a sufficiently realistic context for stating preferences. While these evaluations are not completely objective, they can be used to investigate differences in individuals' responses to the experimental portion of the survey.

After screening for missing data and clearly illogical responses, 128 of the 134 completed sets of route evalauations were used to calculate factor importance weights for each individual. The marginal utilities for the qualitative factors were then normalized relative to the travel time utilities, and compared across various segmentations of the sample. Selected results are shown in Table 1.

The overall difficulty of evaluating the hypothetical routes appears to have had little effect on the relative weightings, as none reported more than moderate difficulty. The next two breakdowns indicate that as the qualitative factors were perceived more clearly, they generally were found more important relative to travel time. Based on this assumption, the inclusion of photographs or the categorization of an actual route, in conjunction with the verbal descriptions, appears to have helped considerably in clarifying the factors.

The results also indicate, however, that using both additional presentation approaches in combination was generally no more effective than using either one separately; even less effective in the case of traffic level.Perhaps the two tasks brought out inconsistencies between the homogeneous routes suggested by the pictures and the variable nature of factors associated with actual routes. The identification of certain route attributes with actual routes familiar to some subjects is an important issue, because this could also associate non-included route factors, possibly causing "taste variation" bias, as discussed previously. In this regard, the use of photographs seems a better way to clarify the attributes than relating them to real alternatives.

For policy analysis, the most interesting variations in the responses are those which can be associated with the travel patterns and personal charactersitics of the individuals. Segmentation with respect to the self-assessed importance of the various attributes brings out the largest variations, as one would hope. The average weight on each of the three qualitative factors was two to four times as high for those who stated that the factors were important in their daily route choice, relative to those who indicated the opposite. Correspondingly, those who found the time variable important placed less importance on the qualitative factors (see Table 1). In additon, a breakdown conditional on the actual route type of the respondent (for those who were asked), shows that those who chose the "worst" level of the qualitative factors in real situations tended to place the least importance on them in the scenarios. The trend is not evident for actual travel times, which are determined mainly by non-choice factors.

As for other travel characteristics, the number of days per week by bicycle has no clear effect, however, the small proportion of people in the '3 days or fewer' group seem to place the largest relative value on qualitative factors. This group might correspond to the notion of cyclists as 'choice' users, sometimes selecting the bicycle over faster modes to gain benefits along other dimensions. In some countries, virtually all commuting cyclists are assumed to fit this characterization. In this sample, however, one fifth of the subject reported having no other feasible mode available, and only slightly more than half reported that traveling by car (perhaps in addition to other alternatives) was possible. The 'captive' cyclists appear to place markedly greater importance on cycling time than the car-available group, as one would expect.

TABLE 1
RELATIVE FACTOR UTILITIES FOR SAMPLE SEGMENTS

(All values normalized with respect to time, in minutes)

Factor:	FACILITY TYPE	SURFACE QUALITY	TRAFFIC LEVEL	ALL QUAL. VARIABLES
Tot. Sample (N=128)	4.5	6.6	5.0	16.1
SEGMENTED BY:				
Overall Difficulty				
- None (N=31)	5.4	7.1	4.4	16.9
- Little (N=70)	4.4	5.9	4.3	14.6
- Moderate (N=26)	3.4	8.3	7.1	18.8
Factor Presentation				
- Photos (N=35)	7.1	8.5	6.4	22.0
- Actual Rte (N=37)	3.2	6.4	5.9	15.5
- Both (N=44)	3.5	6.8	3.5	13.8
- Niether (N=22)	2.8	4.0	3.5	10.3
Clarity of Factor *				
- High (N= 123)	4.6	(115) 6.7	(100) 5.4	(123) 16.2
- Moderate (N= 2)	0.0	(10) 6.0	(20) 3.5	(0) 0.0
- Low (N= 1)	0.0	(1) 0.0	(6) 1.6	(3) 5.1
Import. of Factor *				
- High (N= 71)	6.5	(117) 7.1	(86) 5.9	(95) 9.5
- Low (N= 57)	1.9	(11) 2.1	(42) 3.1	(33) 35.3
Level-Actual Route*				
- "Best" (N= 45)	4.7	(22) 5.3	(18) 3.9	(28) 13.3
- Medium (N= 14)	1.9	(43) 7.5	(40) 4.7	(21) 15.6
- "Worst" (N= 9)	1.6	(4) 3.0	(11) 1.7	(21) 13.9
Days per week by bike				
- 3 or fewer (N=21)	5.5	7.0	6.0	18.5
- 4 (N=28)	3.8	4.9	4.7	13.4
- 5 or more (N=78)	4.5	7.1	4.7	16.3
Other Modes Available				
- None (N=25)	1.5	4.8	2.8	9.1
- Walk only (N= 8)	4.4	6.6	6.1	17.2
- Bus/walk (N=13)	6.9	5.5	3.9	16.3
- Car/other (N=79)	5.2	7.5	5.9	18.6
Income group				
- High (N=67)	2.6	5.1	4.0	11.7
- Low (N=49)	7.3	9.2	7.0	23.5
Age group				
- Under 30 (N=18)	2.8	3.9	2.7	9.4
- 30 to 40 (N=38)	3.0	5.5	3.5	12.0
- 40 to 50 (N=30)	7.4	8.8	6.9	23.1
- Over 50 (N=42)	4.5	7.3	6.0	17.8
Sex				
- Male (N=100)	4.6	6.7	5.3	16.6
- Female (N= 25)	4.4	6.6	4.2	15.2

*- The number of observations refer to each factor separately. The last column is segmented by data for the time factor.

Income has often been found to be the most important variable in explaining variations in individuals' marginal utility of time. Here, the university occupations reported by the subjects have been divided into two income classes, with the lower income group generally placing less relative importance on travel time than the higher income class. Further cross-segmentations indicated that age and car availability had little contribution to this effect. Age, alone, however, does show an effect. The higher age groups appear more sensitive to the qualitative variables, which might indicate an increased consciousness of comfort and safety. The last segmentation, by sex, shows no clear trend, nor did additional cross-segmentations of the sample.

Despite variations across indiviuals, a fully-aggregated model yields reasonable and accurate estimates of the attribute effects. A simple plot of the evaluation of each type of route, averaged across the entire sample, gives an overall picture of the importance placed on the route attributes, and of possible non-linearities and interactions. The plots in Figure 2 show the average scores by surface condition, traffic level, and travel time, conditional on each type of facility. The linear effects of travel time and surface condition are greatest, and roughly equal. The plot also shows a consistent non-linearity in traffic level and, possibly, negative interactions for both surface quality and traffic level with facility type. It is probable that these are not true interactions in the behavioral sense, but due to an interaction in the perception of these attributes from the survey presentation.

The regression results in Table 2 give the estimates of these relative effects, normalized to minutes of travel time. In theory, the significance of the aggregate terms should be adjusted according to the error in the interaction of each term with each individual, and will tend to be overstated unless the repeated measures nature of the data is taken into account. In practice, a conservative approximation of this adjustment is often to multiply the standard errors from regression by the square root of the number of observations per subject (Louviere and Woodworth 1983). Thus, all t-statistics presented below have been reduced by a factor of three, and should be considered as a cautious test of the 'true' significance of the estimates.

The first model in Table 2 was estimated in a piecewise linear fashion, treating each level of the qualitative variables separately. Interactions were estimated for all cases where two qualitative variables either 'improved' or 'worsened'. The main effects have relative contributions as would be expected from Figure 2. All are well-estimated except for the effect of 'light traffic', which has an influence only in combination with 'high surface quality'. This non-linearity in the effect of traffic level may indicate a threshhold phenomenon which should be considered in policy analysis.

FIGURE 2

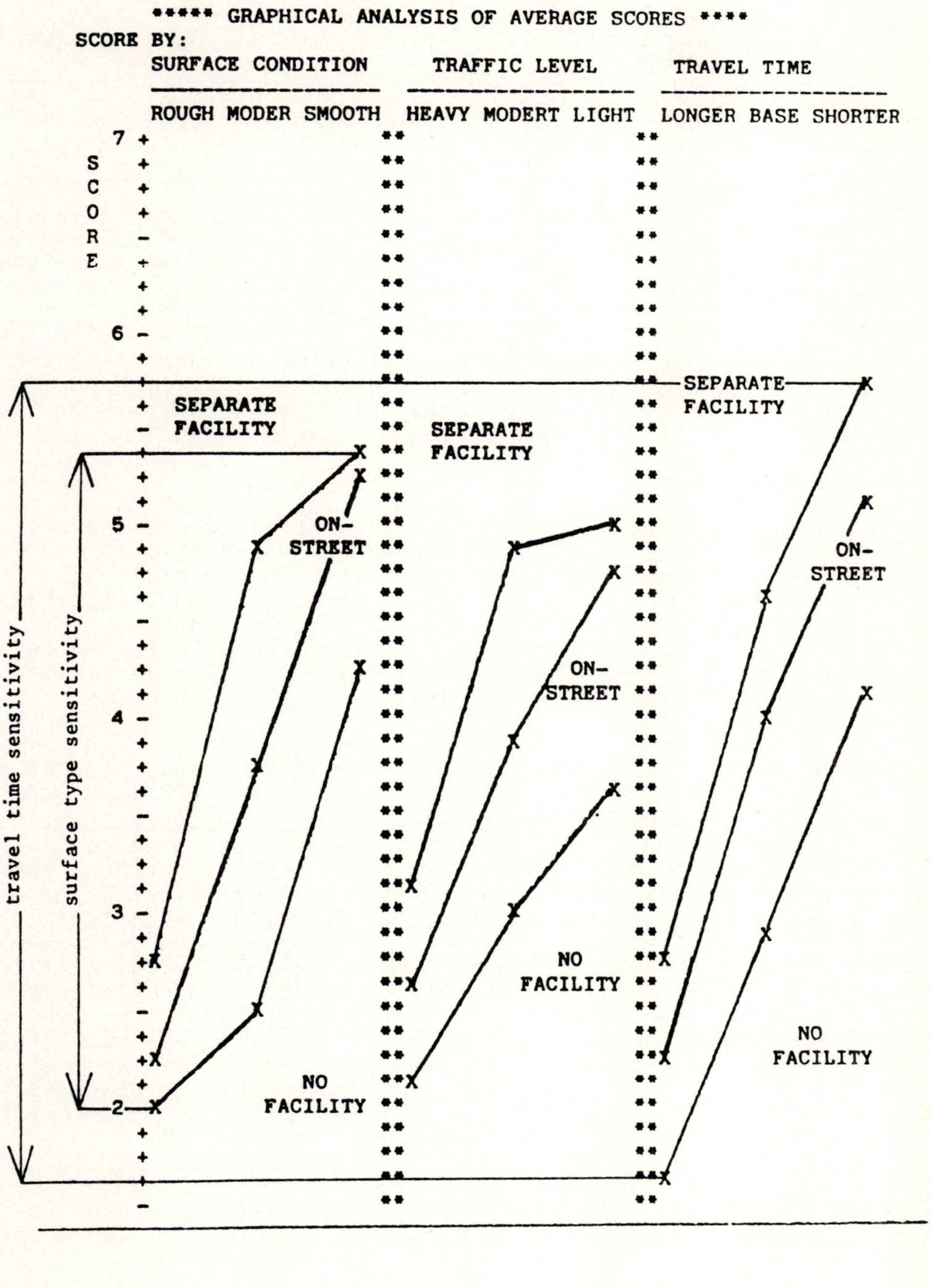
***** GRAPHICAL ANALYSIS OF AVERAGE SCORES ****
SCORE BY:
SURFACE CONDITION
TRAFFIC LEVEL
TRAVEL TIME
ROUGH MODER SMOOTH
HEAVY MODERT LIGHT
LONGER BASE SHORTER
SCORE
7
6
5
4
3
2
SEPARATE FACILITY
ON-STREET
NO FACILITY
travel time sensitivity
surface type sensitivity

TABLE 2 - AGGREGATE REGRESSION RESULTS

All estimates normalized to utility in minutes of travel time.
T-statistics in () adjusted for repeated measurements (see text)

Model:	Piecewise-Linear		Quadratic	Linear
Observations:	1152		1152	1152
R-squared :	0.62		0.61	0.59
Travel Time	6.00 (5.8)		6.00 (8.5)	6.00 (8.6)
Surface Quality				
- High	2.32 (2.2)	Linear	6.63 (5.5)	5.91 (8.2)
- Low	-4.34 (4.1)	Quadr.	-0.22 (0.6)	
Surface/Traffic				
- High/Light	3.05 (1.5)	Linear	2.04 (1.0)	
- Low/Heavy	1.05 (0.5)			
Traffic Level				
- Light	0.22 (0.3)	Linear	3.69 (3.2)	3.85 (5.5)
- Heavy	-3.52 (3.3)	Quadr.	-0.61 (1.5)	
Traffic/Facility				
- Light/Path	-0.54 (0.5)	Linear	-1.63 (0.9)	
- Heavy/None	1.13 (0.9)			
Facility Type				
- Separate Path	2.08 (2.2)	Linear	5.18 (4.3)	3.21 (4.5)
- None (street)	-3.13 (3.2)	Quadr.	-0.30 (0.8)	
Facility/Surface				
- Path/High	-1.93 (1.5)	Linear	-4.23 (2.1)	
- None/Low	2.30 (1.8)			
Constant	1.02		0.03	-0.07

While the piecewise linear specification provides the most information about responses to this particular range of levels, it is difficult to relate the estimates to more general cases. The second model approximates the level-specific case using quadratic functions and two-way interaction terms for the three qualitative factors. The loss of explanatory power, in terms of R-squared is not significant, and all main-effects are well-estimated. As suggested by Figure 2, the quadratic term for traffic level and the interaction between facility type and surface quality are the most significant of the other effects. The positive interaction between surface quality and traffic level is marginally significant (an upper limit t-statistic of 3.0), and, in this case, it is unlikely that such an interaction would arise solely from the survey descriptions.

The final model, based only on linear terms for the four factors, is able to explain nearly as much of the variance in the responses as the quadratic specification. Together, the marginal

TABLE 3 - SEGMENTED REGRESSION RESULTS

All estimates normalized to utility in minutes of travel time. T-statistics in () adjusted for repeated measurements (see text).

Segment:	Captive Users	High Income Under 40	High Income Over 40	Low Income Under 40	Low Income Over 40
Observations:	225	207	261	153	306
R-squared :	.69	.70	.54	.59	.54
Travel Time	6.00 (5.3)	6.00 (5.4)	6.00 (3.6)	6.00 (2.4)	6.00 (3.4)
Surface Quality	5.13 (4.4)	4.45 (3.0)	5.72 (3.4)	9.48 (3.6)	6.81 (3.7)
Traffic Level	3.10 (2.7)	2.93 (2.6)	4.39 (2.6)	4.64 (2.0)	4.76 (2.7)
Facility Type	2.47 (2.2)	2.15 (2.0)	2.80 (1.8)	4.43 (1.8)	4.95 (2.7)
Constant	-0.15	-0.07	-0.04	-0.01	-0.04

utilities between the extreme levels of the qualitative terms add to 13 minutes, with almost half of the contribution from surface quality.

This linear model was used in segmentation analysis to facilitate comparisons across groups of individuals. Based on the comparisons of Table 1, five groups were selected. First, the bicycle 'captives' were separated from the rest of the sample. Then, the remaining observations were classified according to the income group and the age of the respondents. The estimated models in Table 3 show clear differences across the segments. Sensitivity to the qualitative factors, relative to time, decreases with captivity and income, and increases with age, as expected. One unexpected result is the relative importance of surface quality in the lower income, low age group. A possible reason is that the people in this group are most likely to use light-frame, multi-speed bikes, which are more sensitive to surface conditions, an effect also noted by Axhausen and Smith (1984).

The true validity of these models in explaining behavior can be judged only with respect to independent data on observed choices. There were, however, several steps taken to increase our confidence in the results. In addition to the scaled responses analyzed here, responses were also collected in the form of percentage probabilities of pairwise choices, and in the form of rankings across entire sets of nine routes. Each subject performed all three tasks on different sets of routes.

The percentage scores were first transformed to represent utilities, using a simple logit transformation, and then normalized according to the mean and variance within individuals. The resulting values were used to estimate the model specifications reported above, with very similar results for the significant effects.

The ranked data were transformed to represent discrete choices, and multinomial logit estimation was performed, assuming independence across alternatives. Although this estimation approach does not take into account all aspects of repeated measures data, the main-effects estimates were very similar to those from the scaled data. The resulting, segmented models were then applied to predict the rankings for a small number of subjects not included in the estimation, giving a rank-correlation similar to that achieved for the estimation sample (see Bradley and Bovy 1984 for details). A closer approximation to the discrete choice situation, not tested here, can be achieved by arranging the alternatives into balanced choice sets, with the presence of each alternative in the various sets made orthogonal using a factorial design. Discrete choices from each set can then be aggregated and analyzed using weighted least-squares regression (Louviere and Woodworth 1983).

As another test of internal validity, several respondents were asked to return after a period of one month, and to complete the same experiments again. Over eighty-five percent of the interval responses were identical to those given originally.

External validity can be judged somewhat through comparison to the results of similar studies. In a concurrent stated preference analysis in Wisconsin, Axhausen and Smith (1984) identified very similar tradeoffs among the same four attributes (using distance instead of time). The authors also found slope, land use, and cycling experience to be influential. These factors, however, exhibit much less variation in the Netherlands, and were assumed not to affect the route choice of most cyclists. In another stated preference experiment, also in Wisconsin, but this time for car vs. bicycle mode choice (Wis.DOT 1981) time, surface, traffic, and facility were shown to have roughly the same relative influence as reported above.

4. Policy Applications

Standing alone, the stated preference approach appears to give stable estimates of the tradeoffs among specific choice factors in a specific context, at a cost usually much lower than that of alternative methods. The information on tradeoffs within different market segments is useful in ranking alternative policies, and in identifying advantaged and disadvantaged groups of users affected by those policies. There are, in addition, several ways in which analysis of the type described above could be extended to assess specific travel routes and policies.

For more detailed policy analyses, the models must be calibrated to predict behavioral responses in actual choice situations. This procedure involves using data on actual choice attributes reported by the respondents to linearly scale the utility equation. The resulting models can be applied in general policy analysis, using a technique such as pivot-point logit analysis (Kocur, et.al. 1982).

The models could also be calibrated by coding the routes in a limited network according to the modelled attributes, and scaling the utilities to predict aggregate flows on the routes. The data requirements for this procedure are considerably less than for disaggregate revealed preference modelling. In this study, we required subjects to map their actual routes, and to categorize these routes according to the design variables. Adequate information of this type could be used to code the network of most commonly used routes for subsequent calibration and policy analysis. If the subjects were asked to map and define a set of alternative routes as well, the model calibration could be done on this data, and more data would be available to code the routes for policy applications. The authors are investigating this approach in studying drivers' route choice.

Policies and projects which improve a given route in a network affect current users of the route, any users diverted away from other routes, and any trips along the route which are generated by the increased accessibility to travel destinations. Using a tradeoff analysis experiment, Morisugi and others (1981) were able to apply models of the type presented above to forecast the number of users diverted to a qualitatively improved highway, and to estimate the monetary-equivalent benefits to both existing and diverted users.

As the models above apply directly only to route choice (trip assignment), potential trips along the route generated by changes in mode choice, destination choice, and trip frequency behavior must be estimated through other means. Such behavior is often analyzed using a system of linked models, such as nested logit (Sobel 1980). Simplified, least-generalized-cost route choice models are commonly used both to define the representative paths for the mode/destination alternatives, and to assign the flows predicted from the model system. If important non-time and cost variables are neglected at the path generations stage, the model results may be biased at each subsequent level of analysis (Jansen and Bovy 1983). If these qualitative variables were coded for a section of the network under investigation, then information on tradeoffs between time and non-time route choice factors could be valuable in the estimation and application of revealed preference model systems. This is just one example in which information from the stated preference and revealed preference approaches can be used together in application.

5. Conclusions

The survey and analysis described above, performed in a very short period and at relatively low cost, has provided sensible and stable estimates of tradeoffs among selected bicycle route characteristics. Of the qualitative factors studied, surface condition appears to be the most important relative to travel time, particularly in combination with high traffic levels. The sensitivity to all of the qualitative factors, relative to time, appears to vary with the age, income, and availability of alternative travel modes. Most of these results correspond to those obtained in other study contexts.

Results such as these may be useful for simple policy assessments, and, with the use of additional data and validation, for more detailed analysis and forecasting. The results also indicate, however, that the presentation of the choice context and attributes can have a significant effect on individuals' tradeoffs. Research into the methods of defining and presenting qualitative route factors appears to be an important priority in extending these techniques to a wider range of route choice contexts and policy applications.

Acknowledgements

This research was performed partly under contract to the Dienst Verkeerskunde of the Rijkswaterstaat (Dutch Ministry of Transport and Public Works), with special thanks to Mr. Gert van der Meer. Thanks are also due to Mr. Dick den Adel of the Delft Institute for Town Planning Reseach, who managed all of the survey production and administration, and to Hugh Gunn and Ben Jansen for their constructive criticisms.

References

Anderson, N.H. 1981. Foundations of Information Integration Theory. New York. Academic Press.

Axhausen, K.W. 1984. Route Choice Behavior of Bicyclists. M.Sc. Thesis. Univ. of Wisconsin, Dept. of Civil & Env. Eng.

Axhausen, K.W. and R.L. Smith. 1984. "Bicycle link evaluation: A stated preference approach". Draft Report. Univ. of Wisconsin, Dept. of Civil and Environmental Eng.

Ben Akiva, M., et.al. 1984. "Modelling inter-urban route choice behavior". Proc. Ninth Int. Symposium on Traffic and Transportation Theory. Utrecht. VNU-Science Press.

Bradley, M.A. and P.H.L. Bovy. 1984. "A stated preference analysis of bicycle route choice". Proc. Twelfth Annual Meeting of the PTRC. Sussex.

Carpenter, S.M. 1977. "Drivers' route choice project: pilot study". Paper 085/WP42. Transport Studies Unit, Oxford Univ.

Chapman, R.G. and R. Staelin. 1982. "Exploiting rank ordered choice set data within the stochastic utility model". _Journal of Marketing Research_. v.19. p. 288.

Hamerslag, R. 1981. "Investigation into factors affecting the route choice in Rijnstreek-West with the aid of a disaggregate logit model". _Transportation_. v.10, p.373.

Huchingson, R.D., et.al. 1977. "Survey of motorist route selection criteria". _Highway Research Record_. 643, p.45.

Jansen, G.R.M. and P.H.L. Bovy. 1983. "Travel time measures and shortest route assignment: theoretical and empirical studies". Proc. Eleventh Annual Meeting of the PTRC. Sussex.

Kocur, G., et.al. 1982. "Guide to Forecasting Travel Demand with Direct Utility Assessment". Washington. UMTA. U.S. DOT.

Lerman, S.R. and J.J Louviere. 1979. "Using functional measurement to identify the form of utility functions in travel demand models". _Transportation Research Record_ 673.

Louviere, J.J., et.al. 1981. "Laboratory simulation versus revealed preference methods for estimating travel demand models". _Transportation Research Record_ 794.

Louviere, J.J. and G. Woodworth. 1983. "Design and analysis of simulated consumer choice or allocation experiments: An approach based on aggregate data". _Journal of Marketing Research_. v. 20. p.350.

Morisugi, H., et.al. 1981. "Meaurement of road user benefits by means of a multi-attribute utility function". _Papers of the Regional Science Ass._ v.46. p.31.

Sobel, K.L. 1980. "Travel demand forecasting with the nested multinomial logit model". 59th Annual Meeting of the Transportation Research Board. Washington.

Timmermans, H. 1984. "Decompositional multiattribute preference models in spatial choice analysis: a review of some recent developments". _Progress in Human Geography_.

Upcott, W.F. 1976. _Cyclists' Route Choice: A Behavioral Analysis of the Journey to School_. M.Sc. Thesis. Cranfield Institute of Technology.

Wachs, M. 1967. "Relationships between drivers' attitudes toward alternative routes and driver and route characteristics". _Highway Research Record_. 197. p.70.

Wisconsin Dept. of Transportation. 1981. "Development of Wisconsin Work Trip Models for Forecasting Modal Choice". Madison. Division of Planning and Budget.

Behavioural Research for Transport Policy
pp. 307–326. © 1986 VNU Science Press.

THE USE OF ATTITUDE MODELS AND STATED PREFERENCE MODELS IN PRACTICAL TRANSPORT ANALYSIS

Eric Kroes
Netherlands Institute of Transport
Robert Sheldon
Steer, Davies & Gleave Ltd.

1. Introduction

Practical transport planning needs methods which are able to predict the behavioural response of travellers that is induced by physical changes in the transport environment. For that purpose, econometric models of travel choices have been developed, describing the travellers' choice process between those alternatives which they perceive to be available to them. Following developments in social psychology and market research, choice models based upon subjective responses of individuals relating to travel options have recently been introduced in transport analysis. From this general and diverse family of models, two main streams of approaches have emerged, which are usually described as attitude models and stated preference models. Although these approaches have many features in common, they also differ considerably in aspects which are important in the context of transport planning. The objective of this paper is to discuss the distinction between attitude models and stated preference models, and to suggest fields of application in transport planning where either method can be most successfully applied.
The paper starts with a short description of both methods. Then the virtues and weaknesses of both methods will be discussed from theoretical and practical viewpoints, followed by recommendations concerning their practical use. The paper concludes by suggesting some further areas for application of the stated preference method.

2. Attitude models and stated preference models

This section aims to clarify the distinction between the basic attitude models and stated preference models, without dealing with either method in detail. For a more thorough description of the basic attitude models we refer to eg. Held (1981), while the use of stated preference models in a transport context has been described in eg. Hensher (1982) and Kroes and Sheldon (1983).

When we consider fig. 1, where a general framework for the travel choice process after Golob (1980) is given, two main groups of variables can be distinguished: objective variables, observable characteristics and behaviour, and subjective psychological variables, incorporating elements of perception, affect and conation.

Fig.1 Psychological framework for the individual's mode choice (after Golob 1980)

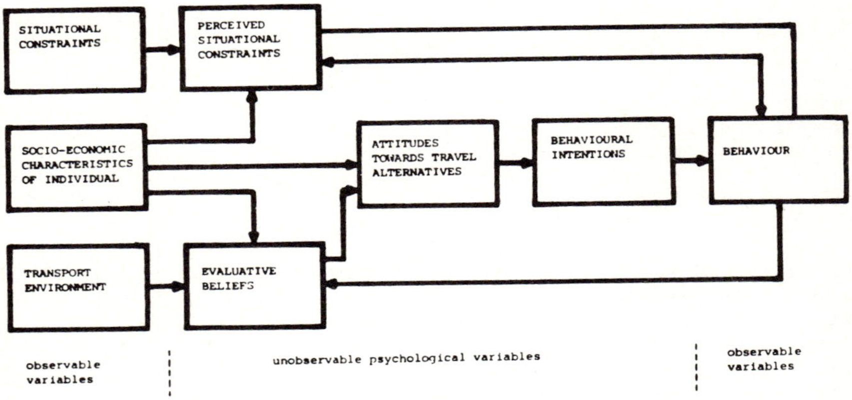

Traditional econometric choice models, based upon utility theory, are usually calibrated using observable variables only, to avoid difficulties in forecasting. Attitude models and stated preference models on the contrary, use either psychological variables only, or a combination of observable and psychological variables.

The basic attitudinal models, of which the Fishbein and Ajzen model (Fishbein and Ajzen 1975) is best known, are derived from attitude theory, and produce attitudes or behavioural intentions as the model output. These models deal with the evaluative part of the choices process, as it is indicated in fig. 2. The input to the model consists of a set of evaluative beliefs of respondents relating to existing choice alternatives, with which the respondents are familiar (see Appendix 1 for an example of an attitude survey form).

Fig. 2 The basic attitude model structure

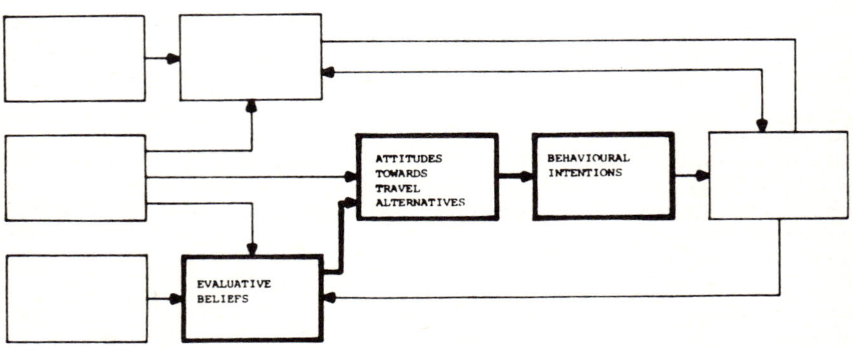

These evaluative beliefs are usually measured on 5-point or 7-point scales and relate to all salient attributes which can be verbally defined, including eg. qualitative attributes. The number of attributes to be included in the model is not fixed, and can be extended to some twenty or more. The model uses a linear additive combination rule to combine the separate evaluative beliefs into attitude scores, one score for each alternative. In this combination rule a set of weights is employed (one for each attribute), consisting of self-assessed strength-of-belief-scores, usually measured on 5-points or 7-points scales.

The next stage in the modelling procedure involves the application of a maximizing choice rule, where the respondent is assumed to choose the alternative with the highest attitude score*). Note that the model output is obtained by a compositional procedure, where behavioural intentions are composed of individual evaluative beliefs.

The basic stated preference models are derived from decision theory, and rely on the concept of utility. These models produce preferences (attitudes) for the choice alternatives and choice-probabilities (behavioural intentions) and deal with the same evaluative part of the choice process as the attitude model (see fig. 3). Input to the model is a set of descriptions of (objective) conditions, which the respondents are asked to evaluate (see Appendix 2 for an example). This set of descriptions consists of

Fig. 3 The basic stated preference model

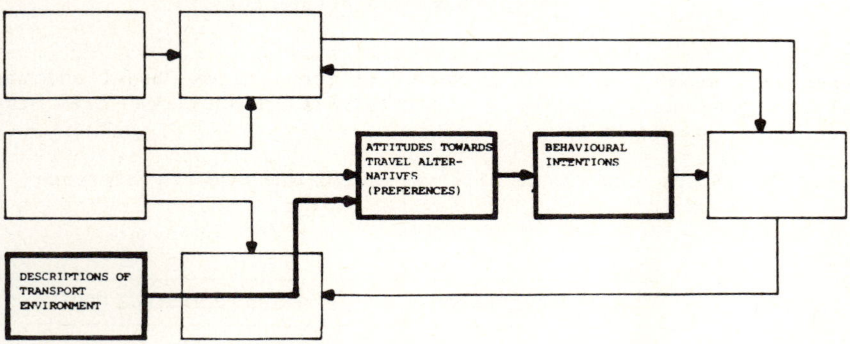

systematically varied alternatives, controlled in an experimental design, which may extend beyond real life conditions (in practice existing conditions with which the respondents are familiar should be included in the design, in order to create a realistic context for the evaluation task). The respondents are asked to rank or rate the alternatives offered to them, or to indicate their most likely choice. The model is then calibrated using statistical procedures,

*) In the original Fishbein and Ajzen model, the choice rule has an extention, where behavioural intention includes social and personal norms and the motivation to comply with these norms.

which relate the variance in the preference scores or choice likelihoods to the variance in the described conditions, and thus involves a decompositional procedure. The attributes to be included in the model are limited in number, in order to keep the description of the alternatives understandable for the respondents. Both quantitative and qualitative attributes can be included. Stated preference models typically use a linear additive combination rule, and a maximizing choice rule. The weights in the combination rule are empirically derived in the statistical decomposition procedure.

Having briefly described only the basic attitude models and stated preference models, we should emphasize that in practical transport research in the last fifteen years many hybrid models have been developed (for an extensive overview see Golob and Golob 1982). However, in order to clarify the distinction between these groups of models, we restrict our theoretical discussion in this paper to the basic models only.
This does not imply that the discussions and conclusions given in the next sections are only valid for these basic models. The conclusions relating to the attitude model are, in our opinion, valid for virtually all models incorporating self-assessed subjective evaluations. The conclusions relating to the stated preference model can be considered useful for the majority of these models.

3. Attitude models versus stated preference models

When we compare the basic elements of the attitude model and the stated preference model, it is obvious that there are many analogies (see table 1).

Table 1 Basic elements of attitude models and stated preference models

ELEMENT	ATTITUDE MODELS	STATED PREFERENCE MODELS
model input	subjective evaluations	objective characteristics
weights	self-assessed	empirically derived
combination rule	usually linear additive	usually linear additive
choice rule	maximum attitude score	maximum utility
model output	behavioural intentions	preference/choice

Both models usually use linear additive combination rules and maximizing choice rules. Moreover both models produce subjective outputs which are closely related to behaviour. However, the attitude model relies on subjective measures as the input, while the stated preference model uses descriptions of objective measures. Nevertheless Van Raay (1977) concludes that the attitude model (as proposed by Fishbein) is isomorphic with the multi-attribute utility model (such as the stated preference model). Note that both models can incorporate situational constraints and socio-economic differences between individual travellers by means of segmentation techniques.

Although the basic theoretical elements of both models are very similar, there are considerable differences in practical aspects which are relevant for the application of these models in practical transport planning. Since these are of prime interest in this paper, we deal with these issues in some more detail (see table 2).

Table 2 Practical aspects of attitude models and stated preference models

ASPECT	ATTITUDE MODELS	STATED PREFERENCE MODELS
types of attributes	✓	✓
number of attributes	✓	*)
subjective evaluation of characteristics	✓	*)
link to objective characteristics		✓
redundancy	*)	✓
missing attributes		✓
causal structure		✓

*) see text

1. Types of attributes to be included in the model.

Both the attitude model and the stated preference model can deal with quantitative attributes such as travel time and cost as well as with more qualitative attributes such as comfort, convenience. This is an important advantage over traditional econometric choice models which can only deal with variables which can easily be measured using engineering type measures. This is one of the reasons why an increased interest in attitude models arose in the mid 1970's (see eg. Stopher, Spear and Sucher 1974, Neveu, Koppelman and Stopher 1978).

2. Number of attributes to be included in the model.

The attitude model can accomodate a virtually unlimited number of attributes, because of its compositional nature (all relevant parameters are obtained through direct measurement). The patience of the respondents and their ability to give useful answers are the only real limitations here. In stated preference models the number of attributes is limited to a maximum of some five or six, in order to keep the evaluation task for the respondents managable (Kroes and Sheldon 1983). However, this maximum number can be raised to some fifteen or sixteen attributes by using linked approaches (eg. Steer, Davies and Gleave 1984).

3. Subjective evaluations of objective characteristics

The attitude models use evaluative beliefs as the main input, thus providing explicit information relating to cognitive and affective elements relating to travel characteristics. In stated preference models information concerning evaluative beliefs is usually not explicitly obtained, although evaluative elements are implicitly represented in the estimated importance weights. In fact the way in which the different options are usually described imposes strong assumptions about the traveller's perception, by supplying the respondents with complete and exact information regarding the options. The validity of this assumption is doubtful, as it has been demonstrated that there are perceptional differences between users and non-users of travel alternatives (eg. Golob, Horowitz and Wachs 1979). There are, however, ways to deal with this problem in data collection for stated preference models, however, for instance by using gaming approaches (Payne and Steer 1981).

4. Link to objective characteristics

The evaluative beliefs which are used in attitude models do relate to objective characteristics of the transport environment, but this relationship is not explicitly modelled. Attempts do to this have shown some succes (eg. Louvière and Meyer 1982), and recent evidence from Kawakami and Hirobata (1984) is very interesting. However further research is needed before reliable and transferable models predicting subjective evaluations from objective information will be available. For practical transport planning this means that, although attitude models may be succesful in the ex-post prediction of choice behaviour for a given sample of respondents, they will generally not be transferable to make forecasts of choice behaviour under changed conditions. This is because the required input for the model, the evaluative beliefs relating to changed conditions, cannot as yet be estimated with a reasonable degree of accuracy.
In stated preference models the link between objective characteristics and subjective evaluations is implicitly taken into account, thus making the models suitable for prediction of response to changed conditions. In fact the considered changes in the transport environment may be part of the experimental design itself, thus allowing for controlled departures from existing conditions.

5. Redundancy

Attitude models employ those attributes considered to be salient to the choice process under consideration. It is implicitly assumed that the attributes as interpreted by the respondents are not redundant (mutually independent). In practice redundancy (interdependencies) may accur, causing biasses in the model predictions. However, factor-analytic approaches may be used to solve this problem.

The experimental design which is used for stated preference models will allow the complete avoidance of interdependencies, or the investigation of selected interactions.

6. Missing attributes

Attitude models are likely to produce systematically biassed predictions when relevant attributes have been omitted. In stated preference models this problem is avoided, because the effects of missing attributes will be absorbed in the estimated alternative specific constants in the utility function.

7. Causal structure

Both attitude models and stated preference models implicitly assume a unidirectional causal link, with causality leading from evaluative beliefs to behaviour. In practice this assumption is highly questionable for the attitude model, because several theories suggest feedback mechanisms:

- learning theory (Fishbein 1967)
- cognitive dissonance theory (Festinger 1957)
- selfperception theory (Bem and McCornell 1970)
- habit theory (Goodwin 1977)

Levin (1981) describes the attitude-behaviour relationship in basic attitude models as circumstantial rather than causal. Practical evidence from a variety of sources supports the hypothesis of bi-directional causality in the attitude-behaviour relationship (eg. Golob, Horowitz and Wachs 1979, Dobson et al (1978), Tischer and Phillips 1978, Kroes 1981).

Some of this evidence will be discussed in the next section. What is important here is the consequence of the existence of such feedback phenomena. Golob, Horowitz and Wachs (1979) conclude that those attributes displaying strong cognitive dissonance are likely to be potentially the most cost-effective candidates for planning or marketing campaigns aiming to induce changes in travel behaviour. On the other hand the very existence of feedback phenomena makes the single equation attitude model inappropriate to investigate the effects of changes in the transport system on travel choices, unless feedbacks are explicitly accounted for.
In stated preference models the problem of the direction of causality is avoided, because the experimental approach employed in the datacollection ensures that the stated preferences relating to the different options are given in response to the presentation of these options (Levin 1981).

The seventh point mentioned, concerning the causal structure of the attitude models in particular, is less obvious than the other six points mentioned, and deserves some more attention. Therefore the following section will present some results of practical research investigating the causal structure of the attitude-behaviour relationship, together with a discussion of a recently suggested composite approach which might be used to solve the causal problems and the problem of the uncertain link between subjective evaluations and objective characteristics simultaneously.

4. Causality in the attitude-behaviour relationship - practical evidence of the problem and a possible solution

There are several methods to investigate causality in the relationship between correlated variables in non-experimental research:

1. Preference bias tests (Bronner and De Hoog 1978)
2. Cognitive dissonance tests (Golob, Horowitz and Wachs 1979)
3. Path analysis (Blalock 1971)
4. Simultaneous equations analysis (Heise 1975)
5. Cross-lagged panel analysis (Rozelle and Campbell 1969)

Preference bias tests and cognitive dissonance tests are very similar in concept, and consist of statistical tests investigating the differences in mean subjective evaluations of transport alternatives between the users and non-users of those different transport alternatives. When for example train-users rate the train significantly higher on a number of attributes than car-users, while rating the car significantly lower, the preference bias or cognitive dissonances hypothesis is accepted.
Path analysis involves a sequential correlational analysis using OLS regression. A structure of linear relationships between attitudes and behaviour is hypothesized, including feedback links. The strength of the correlations which are found, is then used to accept or reject the hypothesized causal structure.
Simultaneous equations analysis is similar in concept to path analysis, but here the different hypothesized relationships are analysed simultaneously rather than sequentially. In order to do this using a linear model the more complex two-stage least-squares (2SLS) estimation method is required. In the first stage the variables which are assumed to be mutually dependent (the endogenous variables) are estimated as a function of all exogenous variables. Then in the second stage the estimated values for the endogenous variables are used in the estimation of the relationship between the different endogenous variables. The significance of the coefficients in the second stage estimation is used to accept or reject the hypothesized bi-directional causal relationship (see eg. Tardiff 1977).
Cross-lagged panel analysis is related to path analysis, but relies on panel data rather than cross-sectional data (which the other methods use). By analysing correlation between variables which have been measured at different points in time (the so-called cross-lagged correlations) causality hypotheses can be accepted or rejected. Causality is supported when there is a sgnificant correlation <u>and</u> time dependency between the variables of interest (Tischer and Phillips 1978).

Practical evidence concerning the existence of bi-directional causality in the attitude-behaviour relationship has been produced by several analysts, using all five methods mentioned (eg. Golob, Horowitz and Wachs 1979, Dobson et al 1978, Tischer and Phillips 1978). We will present here a summary of the results of two studies in the Netherlands, using simultaneous equations and cross-lagged panel analysis.

In the Park-and-Ride study, mode choice of commuters living in a corridor north of Amsterdam and working in Amsterdam was investigated (see Baanders, Van der Hoorn and Jetten 1981). A two-stage sample was used, with the second stage consisting of home-interviews amongst 400 train- and car users having freedom of choice between rail and car. In the home intervieuws, subjective evaluations (self-assessed importance and satisfaction scores on a 5-point scale) were obtained relating to 15 attributes of the journey by train and car. These data were used in an analysis aiming to examine the attitude-mode choice relationship. Initial tests demonstrated the presence of multicollinearity between the satisfaction scores, and consequently a factor analysis was carried out to reduce these interrelationships by combining the strongly correlated attributes into a smaller number of latent factors. Four factors were extracted for each mode (see table 3). For each respondent the factor scores (latent evaluative beliefs) were calculated, and used in the simultaneous equations analysis together with information on mode choice, transport environment and socio-characteristics.

Table 3 Latent factors in the Park-and-Ride study

FACTOR	ALTERNATIVE	SALIENT ATTRIBUTES
1. quality	train	stress working/reading safety privacy
2. waiting	train	waiting for train waiting for tram/bus reliability
3. independence	train	privacy flexibility
4. residual factor	train	working/reading
5. quality	car	safety stress working/reading reliability cost of travel
6. parking	car	looking for parking space cost of parking
7. indepencence	car	flexibility privacy
8. congestion	car	queueing at tunnels congestion in Amsterdam

The simultaneous equations analysis involved the estimation of a set of linear relationships using 2SLS. The analyses proceeded in a stepwise backward manner, eliminating non-significant relationships. The resulting most satisfactory model is given in a flowchart in fig. 4. The arrows in the chart represent hypothesized causal relationships, and the t-values of the estimated coefficients are placed on the arrows (t-values are given rather than coefficients because the acceptation or rejection of the hypothesis concerning the existence of the specified causal link is of more importance here than the magnitude of the coefficient).

Fig. 4 The most satisfactory simultaneous equations model

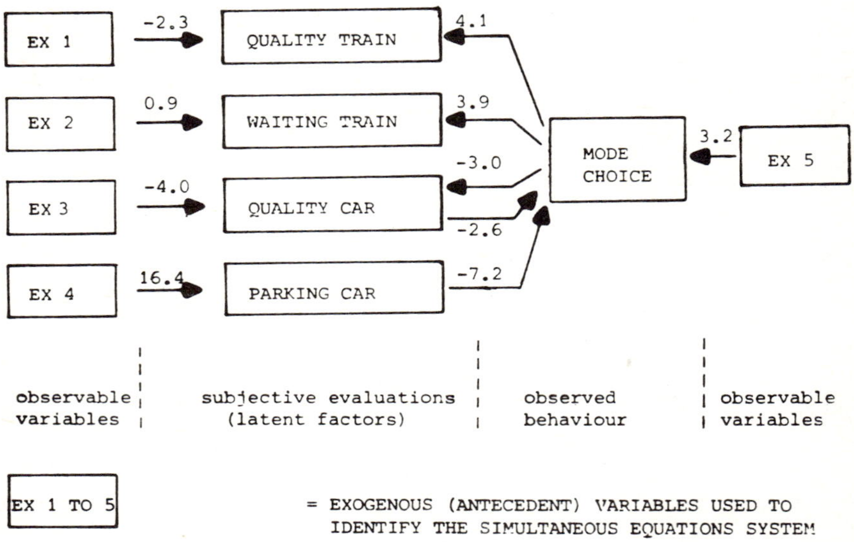

Four out of the original eight factors showed a significant relationship to mode choice, and the other four factors were excluded from the model. From fig. 4 it can be concluded, that the hypothesis concerning bi-directional causalty in the relationship between subjective evaluations and mode choice behaviour is accepted. The evaluations relating to the car appear to influence mode choice significantly but the evaluations relating to the train appear to be determined by, rather than to be determinants of mode choice. For the subjective evaluations relating to the quality of the car a bi-directional causal relationship is found. For a more thorough description and interpretation of these results we refer to Kroes (1981).

The Lek Bridge before-and-after study aimed to establish the effects of the opening of a new bridge in a heavily congested part of the Dutch motorway network (see Van der Hoorn, Kroes and Meijer 1983). The design of this study was very similar to the Park-and-Ride study, and again home-interviews were conducted amongst train- and car-users having freedom of choice. In these interviews several types of subjective evaluations were obtained, including directly measured attitudes towards the journey by train and car. The home-interviews were held amongst the same group of travellers at two points in time between which a change in the transport infrastructure occurred: September 1981 and October 1982. The dataset obtained has been used for a cross-lagged panel analysis investigating the causal relationship between attitudes and mode choice in time. The results of this correlational analysis are presented in fig. 5. In this chart the correlations between six variables are presented: attitudes towards travel by car: in 1981 and 1982, attitudes towards travel by train: in 1981 and 1982, and mode choice (expressed as a dummy variable train = 1) in 1981 and 1982.

Fig. 5 Results of the cross-lagged panel-analyses (A = attitudes B = mode choice)

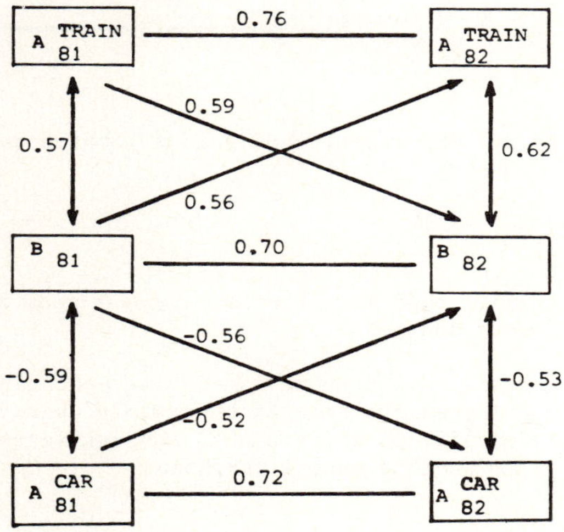

The presented correlation coefficients (Kendall's Tau) are corrected for measurement errors in the attitudinal variables and tested against a "baseline" correlation which can be expected a-priori (Tischer and Phillips 1978). In this analysis the four cross-lagged correlations are of crucial importance. The results suggest, that there are statistically significant lagged causal relationships leading from attitudes (1981) to behaviour (1982). But there are also significant lagged feedback relationships leading from behaviour (1981) to attitudes (1982), while there are no significant differences between the two pairs of cross-lagged correlations. This gives support to the hypothesis concerning bi-directional causality in the attitude-behaviour relationship. For a more detailed description of this study together with details on the statistical tests involved, we refer to NVI (1984).

An interesting practical solution to the problem of bi-directional causality in the attitude-behaviour relationship has been proposed recently by Kawakami and Hirobata (1984). They suggest a composite attitude-econometric model which explicitly incorporates the feedback from behaviour to attitudes. The basic structure of the proposed mode choice model is given in fig. 6.

Fig. 6 Composite attitude-econometric model as proposed by Kawakami and Hirobata (1984)

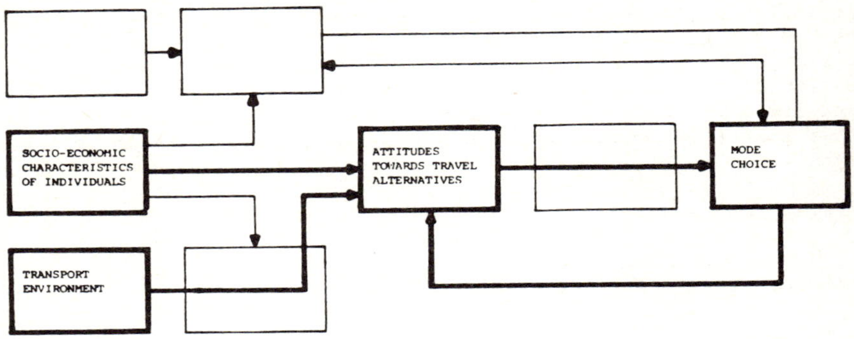

The model is a standard behavioural choice model, based upon random utility maximization theory:

$$P_i = \frac{\exp(V_i)}{\sum \exp(V_j)} \tag{1}$$

Where P_i = the probability that an individual chooses mode i;
V_i = the utility of mode i.

In contrast to the traditional econometric choice model, however V_i is not specified as a direct function of observable characteristics of mode i, but as a function of individuals' subjective evaluations of mode i (together with socio-economic variables):

$$V_i = f(E_i, S) \tag{2}$$

where E_i = subjective evaluations relating to mode i
S = socio-economic characteristics

The individuals' subjective evaluations, in turn, are specified as a function of observable characteristics of the modes (together with socio-economic characteristics):

$$E_i = g_i\ (O_i, S) \tag{3}$$

where O_i = observable characteristics of mode i.

Note that the subjective evaluation function g contains an index i indicating that individuals who use different modes are assumed to have different subjective evaluation functions.
The model is calibrated in a two-stage estimation procedure. The first stage involves the estimation of equation (3), which is interpreted by Kawakami and Hirobata as a psychometric scaling model. In the second stage the predicted subjective evaluations $\hat{E}_i$ are used in the estimation of equation (2), which completes the

calibration of the model. (Note that the model is essentially a simultaneous equations discrete choice model, in theory requiring the use of a full information maximum likelihood estimation method).

5. Recommendations for the use of attitude models and stated preference models in transport planning

It will be clear from the previous sections that we do not particularly favour the use of isolated attitude models for prediction purposes in transport planning. In fact we would argue that such models are inappropriate to forecast longer term behavioural changes in a transport context, due to the problematic causal structure and the uncertain link between subjective evaluations and objective characteristics of the transport environment. However, the development of composite attitude-econometric models of the type discussed in section 4, might prove to be a valuable solution. In addition to that we want to emphasize that the analysis of explicit subjective evaluations (without using these for modelling purposes), can be particularly useful in certain areas of transport planning, such as product planning and quality monitoring.

In the marketing of transport services, such as a public transport system, subjective evaluations of both users and non-users can be used to identify attributes where cognitive dissonance exists, using simple t-tests comparing mean satisfaction scores as discussed in section 4. Smyth (1984) concludes that market campaigns emphasizing the attributes of the private mode judged to be inferior to the public transport alternative would be the most appropriate way of inducing changes in travel behaviour where cognitive dissonance is shown to be present. In addition to that attitude models of the Fishbein and Ajzen type may be used to investigate the effects of normative influences on mode choice. In conditions where such influences appears to be significant marketing campaigns could be developped recognising this role of family influence. According to Smyth (1984) current British Rail advertising strategies are based both upon the concept of cognitive dissonance as well as the influence of family members and other referents on mode choice.

In quality monitoring repeated measurements of user evaluations of a variety of attributes are important data inputs. Especially the analysis of changes in user evaluations will indicate quality aspects of the transport service which require extra attention. For that purpose for instance the Netherlands Railways developed the "Thermograph". This instrument is based upon repeated measurements (twice a year) and analyses of train users' satisfaction and importance ratings, where some 33 different attributes are evaluated. These attributes include a variety of service attributes, such as the audibility of the announcements on the platforms, cleanliness of the waiting rooms, friendliness of staff.

The standard analysis involves the calculation of mean scores and standard deviations for all attributes, as well as t-tests aiming to identify significant changes relative to previous time periods.

For the development of practical transport models, aiming to forecast changes in behaviour due to changes in the transport environment, stated preference models are better suited than attitude models. This is because the causal structure of stated preference models is unambiguous, and there is an explicit link between objective transport characteristics and predicted behaviour. However, in order to create a realistic context for the interviews (which is essential to obtain useful results) the description of the options to be evaluated by the respondents should closely relate to existing behaviour. Also constraints on behaviour due to situational factors and interactions between household members should be identified and taken into account in the experimental design. Provided that these requirements are met useful forecasting tools can be developed using stated preference methods, as has been demonstrated by eg Hensher and Louvière (1983). Work remains to be done on the validation of such models, with the most interesting (and difficult!) test being a comparison between observed and predicted behavioural changes in before-and-after studies. Also the dynamic aspects of these models require further research, particularly the incorporation of short term habit or dissonance reduction effects in the utility function by adding usergroup specific constants as proposed by Bates (1984). Although the basic logit model implicitly incorporates habit-like effects (as pointed out by Gunn 1984) further research is needed to investigate whether this is an adequate representation of behavioural responses to changed transport conditions on the short term and on the long term. Habit-like influences are likely to be less on the long term, due to population renewal, lifecycle changes, etc.

Alternative uses of stated preference models include the estimation of values of time (Bates 1984), the derivation of travel demand elasticities (Kroes and Sheldon 1983) and the quantification of the perceived importance of qualitative factors such as prespecified station improvement programs (Steer, Davies and Gleave 1984) or quantitative factors such as design aspects of cycleways (Bovy and Bradley 1984). Some suggestions for further applications will be given in the next section.

6. Suggestions for some further applications of stated preference models in transport planning

Having discussed the use that can and has been made of attitude models and stated preference models in practical transport planning, we want to conclude this paper by suggesting two areas where stated preference models could potentially be succesfully applied.

The first suggestion involves the integration of a stated preference model with a study, where perceived user benefits resulting from a change in the transport system are to be established and quantified. Although the establishment of user benefits is a complete area of research on its own, we restrict our discussion here to a short description of the approach which we propose. The data collection procedure in the suggested approach involves a series of home interviews, where first the chosen modes and perceived travel times and costs are measured before and after the change in the transport system has occurred. Then the interviewers work out an experimental design around the reported modes of transport, travel times and costs, and the respondents are asked to evaluate the options in the design. These options include the respondents' current and previously chosen modes and the corresponding travel times and costs. Note that the options are translated by the interviewers into values which are relevant to the specific circumstances of each traveller. The respondents' evaluations are then used to calibrate individual choice models, including coefficients for travel time, cost and alternative specific constants in the utility function. The constants account for aspects which are not explicitly incorporated in the design, such as comfort and convenience. By relating the estimated time coefficients and constants to the money cost coefficients, money-equivalents can be calculated for these variables. By relating the differences in reported travel time between before and after situation and the alternative specific constants (if a change of mode has occurred) to the money-equivalents, and by adding these to the differences in reported travel costs, the perceived user benefits can be calculated for each individual and subsequently be aggregated to segments of travellers. This approach is currently tested in a study for the Nederlands Ministry of Transport, where the perceived user benefits resulting from the introduction of a bustaxi system replacing the traditional line bus services in a rural area are to be evaluated.

A second suggestion concerns the estimation of aggregate choice models using a utility function which is "transferred" from a disaggregate stated preference choice model to an aggregate econometric market share model. Kocur et al (1982) have described similar approaches in a validation context. The first step involves the estimation of a stated preference logit choice model using a representative but relatively small sample of respondents. The attributes can include qualitative factors and conditions which are not presently available, as usual with the stated preference approach. Then the estimated utility function of this model is as it were "transferred" to an aggregate logit market share model, which is calibrated using appropriate techniques (WLS, GLS) on aggregate data containing few variables or variables with only little variation. The transfer procedure involves the estimation of scale factors and alternative specific constants as described by Gunn and Pol (1985). In this way a model can be obtained which is sensitive to the policies of interest to the study and is at the same time consistent with aggregate statistics which do not include (or have insufficient variation in) all variables of interest. This approach has been suggested in a study for Manchester Airport, where both authors are involved.

7. Acknowledgements

We want to thank Peter Jones, Tom Golob and Hugh Gunn for their valuable comments on an earlier version of the paper. The research reported in this paper was performed under contract to the Projectbureau Integrale Verkeers- en Vervoerstudies and the Dienst Verkeerskunde of Rijkswaterstaat in the Netherlands.

8. References

1. Baanders A., Hoorn A.I.J.M. van der and Jetten J.T.: "Park-and Ride in North-Holland, a study of the mode choice of commuters in a corridor", paper presented at the PTRC Summer Annual Meeting, 1981.

2. Bates J.J.: "Values of Time from Stated Preference Data", paper presented at the PTRC Summer Annual Meeting, 1984.

3. Bem D.J. and McCornell M.R.: "Testing the Self-Perception Explanation of Dissonance Phenomena. On the Salience of Pre-Manipulation Attitudes", Journal of Personality and Social Psychology, 14, 1970.

4. Blalock H.M.: "Causal Inferences in Non-Experimental Research", The University of North Caroline Press, 1961.

5. Bovy P.H.L. and Bradley M.A.: "A Stated Preference Analysis of Bicyclist Route Choice", paper presented at the PTRC Summer Annual Meeting, 1984.

6. Bronner A.E. and Hoog R. de : "Political Preference: Evaluate and Choose", PhD Thesis, Universiteit van Amsterdam, 1978.

7. Dobson R., Dunbar F., Smith C.J., Reibstein D and Lovelock C.: "Structural Models for the Analysis of Traveller Attitude-Behaviour Relationships", paper presented at the TRB, Washington DC, 1978.

8. Festinger L.: "A Theory of Cognitive Dissonance", Stanford, 1957.

9. Fishbein M.: "Attitude and the Prediction of Behaviour", in Readings in Attitude Theory and Measurement, New York, 1967.

10. Fishbein M and Ajzen I.: "Beliefs, Attitudes, Intention and Behaviour: An Introduction to Theory and Research", Reading, Mass, 1975.

11. Golob T.F., Horowitz A.D. and Wachs M.: "Attitude-Behaviour Relationships in Travel Demand Modelling", in Behavioural Travel Moddelling, eds. Heisher and Stopher, London, 1979.

12. Golob T.F.: "Transit Riders' Attitudes to Bus and Train", unpublished research document, 1980.

13. Golob T.F. and Golob J.: "A Classification of Approaches to Travel-Behaviour Analyses", resource paper for the conference on Travel Behaviour and Values, Easton Maryland, USA, 1982.

14. Goodwin P.B.: "Habit and Hysteresis in Mode Choice", Urban Studies, 14, 1977.

15. Gunn H.F.: "An Analysis of Transfer Price Data", paper presented at the PTRC Summer Annual Meeting, 1984.

16. Gunn H.F. and Pol H.D.P.: "Model Transferability: the Potential for Increasing Cost Effectiveness", paper presented to the 1985 International Conference on Travel Behaviour, Noordwijk, 1985.

17. Heise D.R.: "Causal Analysis", Wiley New York, 1975.

18. Held M.: "Some Thoughts about the Individual's Choice among Alternative Travel Modes and its Determinants", in New Dimensions in Travel-Behaviour Research, eds. Stopher, Meyburg and Brög, Lexington Books, 1981.

19. Hensher D.A.: "Functional Measurement, Individual Preference and Discrete Choice Modelling: Theory and Application", Journal of Economic Psychology, 2, 1982.

20. Hensher D.A. and Louvière J.J.: "Identifying Individual Preferences for International Air Fares", Journal of Transport Economics and Policy, 8, September 1983.

21. Hoorn A.I.J.M. van der, Kroes E.P. and Meijer H.E.R.: "Mode Choice and Mode Captivity in Interlocal Commuting: Two Dutch Studies on Modal Split in Congested Transportation Corridors", paper presented at the World Conference on Transport Research, Hamburg, 1983.

22. Kawakami S. and Hirobata Y.: "Analyses on Changes of Modal Choice Based on New Behavioural Hypotheses", Transportation Policy Decision Making, 2, pp.389-406, 1984.

23. Kocur G., Adler T., Hyman W and Aunet B.: "Guide to Forecast Travel Demand with Direct Utility Assessment", US Department of Transportation, UMTA, 1982.

24. Kroes E.P.: "Attitudes and Mode Choice", Unpublished MSc Thesis, Institute for Transport Studies, Department of Civil Engineering, The University of Leeds, 1981.

25. Kroes E.P. and Sheldon R.J.: "Deriving Travel Elasticities and Other Related Measures Using Stated Preference Techniques", paper presented at the Tenth Transportation Planning Research Colloquium, Zandvoort, 1983.

26. Levin I.P.: "New Applications of Attitude-Measurement and Attitudinal-Modelling Techniques in Transportation Research", in New Dimensions in Travel-Behaviour Research, eds. Stopher, Meyburg and Brög, Lexington Books, 1981.

27. Louvière J.J. and Meijer R.J.: "A Composite Attitude-Behaviour Model of Traveller Decision Making", Transportation Research, Vol. 15B, 5, 1981.

28. Neveu A.J., Koppelman F.S. and Stopher P.R.: "Perceptions of Comfort, Convenience and Reliability for the Work Trip", New York State Department of Transportation, Planning Devision, 1978.

29. Nederlands Vervoerswetenschappelijk Instituut (NVI): "The Lekbridge Vianen before-and-after study", (in Dutch), Rijswijk, 1984.

30. Payne G and Steer J.: "Gauging Passenger Preferences Using a Gaming Approach", Annual Conference of the Market Research Society, Brighton, 1981.

31. Raay, W.F. van : "Consumer Choice Behaviour: An Information Processing Approach", PhD thesis, University of Tilburg, 1977.

32. Rozelle R.M. and Campbell D.T.: "More Plausible Rival Hypotheses in the Cross-Lagged Panel Correlation Technique", Psychological Bulletin, 71, 1969.

33. Smyth A.S.: "Mode Image and Dissonance", paper presented at the PTRC Summer Annual Meeting, 1984.

34. Steer, Davies and Gleave Ltd.: "Research to Evaluate Passenger Priorities for Investment on London's Underground", unpublished report for London Transport, 1984.

35. Stopher P.R., Spear B.D. and Sucher P.D.: "Towards the Development of Measures of Convenience for Travel Modes", Transportation Research Record, no. 528, 1974.

36. Tischer M.L. and Phillips R.V.: "The Relationship between Transportation Perceptions and Behaviour over Time", paper presented at the TRB, 1978.

APPENDIX 1

Example of the type of questionnaire for use with the Fishbein attitude model (from Smyth 1982):

1. Please think about the following items connected with travelling home from work: - then please tick the space which most closely represents YOUR opinion on each

	very good	quite good	slightly good	neither good nor bad	slightly bad	quite bad	very bad
being independent/ free from time-tables	____	____	____	____	____	____	____
a good service	____	____	____	____	____	____	____
a reliable service	____	____	____	____	____	____	____
a 'door-to-door" service	____	____	____	____	____	____	____
a slow trip	____	____	____	____	____	____	____
a cheap trip	____	____	____	____	____	____	____
parking charges	____	____	____	____	____	____	____
having a seat	____	____	____	____	____	____	____
being caught in traffic jams	____	____	____	____	____	____	____

2. Now we would like YOU to consider whether the items below are likely or unlikely when travelling home from work by train. Using the train regularly during next week to travel home from work would mean:

	very likely	quite likely	slightly likely	neither likely nor unlikely	slightly unlikely	quite unlikely	very unlikely
a good service	____	____	____	____	____	____	____
being independent/ free from time-tables	____	____	____	____	____	____	____
a reliable service	____	____	____	____	____	____	____

Using the train regularly during next week to travel home from work would mean:

	very likely	quite likely	slightly likely	neither likely nor unlikely	slightly unlikely	quite unlikely	very unlikely
a slow trip	_____	_____	_____	_____	_____	_____	_____
a cheap trip	_____	_____	_____	_____	_____	_____	_____
a 'door-to-door" service	_____	_____	_____	_____	_____	_____	_____
parking charges	_____	_____	_____	_____	_____	_____	_____
being caught in traffic jams	_____	_____	_____	_____	_____	_____	_____
having a seat	_____	_____	_____	_____	_____	_____	_____

APPENDIX 2

Example of the type of questionnaire for use with a revealed preference choice model (from Kocur et al 1982). Note that interactive interviews of the type proposed by Payne and Steer (1981) are much better suited for this type of research, but cannot easily be reproduced to serve as an illustrative example.

	AUTO FACTORS			WALK FACTORS			PLEASE-ANSWER IN THIS COLUMN: HOW LIKELY ARE YOU TO DRIVE ALONE IN AN AUTO OR WALK? (CIRCLE A NUMBER)				
	Gas Availability	Gas Price	Average Wait Time at Station to Buy Gas	Length of Trip	Amount of Sidewalk on the Way	Season	Always Auto	Probably Auto	In-difference	Probably Walk	Always Walk
SITUATION 1	Ample Supply	$1.30/gallon	5 minutes	½ mile	All the way	Winter	1	2	3	4	5
SITUATION 2	Ration of 10 gallons/week*	$2.60/gallon	5 minutes	½ mile	Part way	Summer	1	2	3	4	5
SITUATION 3	Ration of 10 gallons/week*	$1.30/gallon	20 minutes	½ mile	Part way	Winter	1	2	3	4	5
SITUATION 4	Ample Supply	$2.60/gallon	20 minutes	1 mile	Part way	Winter	1	2	3	4	5
SITUATION 5	Ration of 10 gallons/week*	$1.30/gallon	20 minutes	1 mile	All the way	Summer	1	2	3	4	5
SITUATION 6	Ample Supply	$1.30/gallon	5 minutes	1 mile	Part way	Summer	1	2	3	4	5
SITUATION 7	Ample Supply	$2.60/gallon	20 minutes	½ mile	All the way	Summer	1	2	3	4	5
SITUATION 8	Ration of 10 gallons/week*	$2.60/gallon	5 minutes	1 mile	All the way	Winter	1	2	3	4	5

*If your car gets 15 miles per gallon, you can travel 150 miles per week.

UNDER WHAT SITUATIONS WOULD YOU DRIVE ALONE OR WALK?

Consider a trip short enough so that driving alone in an automobile or walking are realistic choices.

Below are a number of factors describing eight different situations where you are faced with choosing whether to drive alone or walk to make a one half or one mile trip.

Look at each situation across the entire line and please answer in the last column to the right how likely you are to drive alone or walk.

6
Household structure and lifestyle

Behavioural Research for Transport Policy
pp. 329–342.

ENVIRONMENT, CONDITIONS OF LIFE AND TRAFFIC.

Jytte W. Keldborg
Civil engineer
The Danish Ministry of the Environment

Introduction

This paper is a presentation of methods and results from a project, in which the Danish Ministry of the Environment and Haderslev, a municipality of 30.000 inh., cooperated over a period of 2 years. The aim was to emphasize in which ways people in their everyday life are influenced by environmental aspects of the traffic.

In the project the new source of information is detailed interviews with 21 families. The interviews are used as an addition to "classical" planning data and analyses.

The information from interviews and planning data is used to suggest solutions for environmental and traffic problems - solutions which aim to improve the everyday life of the population.

Many of the decisions concerning environment and traffic are made in the municipalities, and it is therefore important to ensure that the background for the decisions includes knowledge of the consequences for people's everyday life.

The final report is relatively short - 55 illustrated pages - and aimed for politicians and planners working with town planning and traffic planning. The working papers total more than 1000 pages, including the interviews, and they are published too, but in a small issue.

WHAT DID WE DO - A DESCRIPTION OF THE PROCESS AND METHOD

The process had 4 clearly separated phases.

Phase 1. What we knew

Phase 1 was a description of which environmental and traffic problems could be found from written sources, mainly planning papers and plans from the municipaltity and the county.

The problems we were looking for concerned

- risk of traffic accidents
- lack of confidence when in traffic
- accessibility
- traffic barriers
- traffic noise
- air pollution
- interference with nature and cultural areas.

From the written sources known connections between environment, conditions of life and traffic were described: Which problems and solutions had been taken into consideration, and where and for whom were the problems known. We made a map of the known problems marking where noise, missing bicycle tracks, throughfare traffic in residential areas and a town center with a mixture of almost all the problems were to be found.

Problems with traffic and environment
- in the municipality of Haderslev

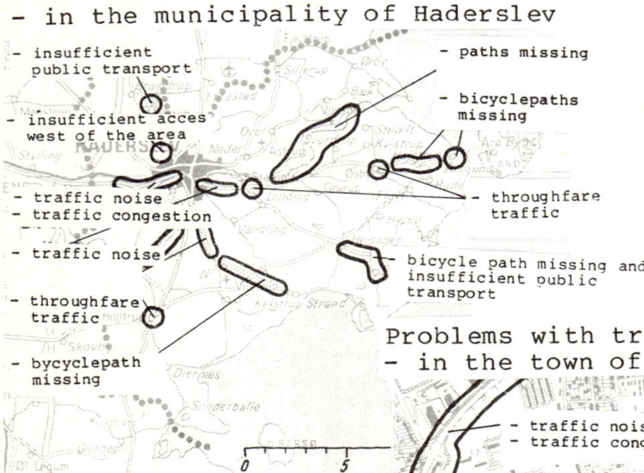

Problems with traffic and environment
- in the town of Haderslev

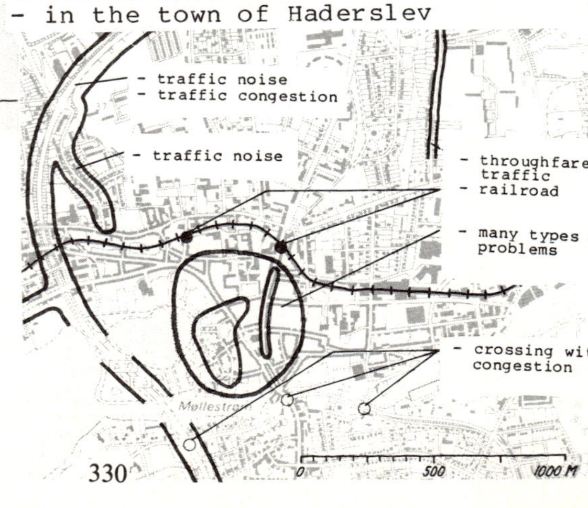

It should be mentioned here that the process of making a municipality plan had started, and included in the written sources was material from a public debate about planning problems.

The question was now if the problems known by the planners and the politicians were the same as those which the families knew from their everyday life.

Are the problems known to the planners? Are there other relevant problems with environment and traffic? And to whom are certain problems important?

Phase 2 - The interviews

The use of intensive interviews as a method of obtaining information for the planning process is new in Denmark.

In this project it was clear from the beginning that the resources were limited. We already had a relatively good knowledge of the problems - what we needed was a number of detailed interviews to analyse which problems people felt were the most important, the causes of the problems, how they fitted into the daily routines, and how problems for one member of a family affects other members.

It was therefore decided that the number of families which could be interviewed would be about 20.

The interviews were meant to illustrate how different types of families lived with the traffic and the environmental problems.

The knowledge of the problems from phase 1 and the data from the ongoing planning process in the municipality lead to two sets of criteria for selecting families for interviews:

Household criteria:

- type of family (number of persons - with or without children etc.)
- need of transportation
- means of transportation (car ownership, bicycle ownership, use of public transportation).

Geographic criteria:

- living in the center of town
- living in suburbs
- living in a village
- living in the country

The combination of all these criteria resulted in a "wishing list" of families, stating the characteristica of 21 unknown families. An example:

In an area of social housing north east of the town center we would like to interview a family with a single parent working in town and with two children - one in kindergarten, one going to school. The family is to have bicycles, and use public transportation, no car.

The families were then picked out, using informal channels. Head teachers, local clubs of different sorts, shopkeepers etc. were contacted and asked to name one or two families with the wanted characteristica. It was an easy way to find the families, and a good introduction to the interviews. (We did consider using registers, but it is fortunately illegal to combine so much informations from different registers, and it would have been difficult to obtain a warm welcome in the families).

The interviews were made with a list of subjects in mind, but the structure of the interviews depended to a great extend on the families themselves. The interviews took place in the homes, with all members of the household together, and they lasted app. 2 hours.

THE CHECK-LIST OF THE INTERVIEWER

Conditions of life and need of transportation

- the size of the household
- age of the members
- work/education
- economy
- health/handicaps
- car(s), bicycles, use of busses
- a little history
- habits of shopping, leisure activities, holidays

Conditions of life and behaviour during the day

Trip chains:

- which trips are made by each member of the household every day
- which trips are made regularly (Sunday trips, weekly trips etc.)

The description of the trips are elaborated with information concerning:

- the choice of route
- problems from the traffic along the route and at the stops (at home, at work etc.)
- environmental effects of one's own transportation on the surroundings
- choices of transportation and destinations
- the influence of the seasons of the year
- possible individual solutions
- possible communal solutions.

Conditions of life and the system of transportation

- air pollution
- traffic noise
- safety/confidence
- interference with nature and cultural areas.

Conditions of life, the environment, transportation and politics

Opinions on traffic and environment or priorities between:
- the environment and traffic
- public transportation and private cars
- needs of transportation and the geographical distribution of towns and villages
- bicycle-paths and roads etc etc.

Questions for reflection

Why are certain solutions suggested? Where is the problem generated?
Are there other solutions?
Why does the family live here?
Were the problems known in advance?
Are there actual choices in the trip-chain of a day?
Is transportation a positive or negative part of everyday life?

This is where the interviewed families live
- in the town
- in the surrounding area

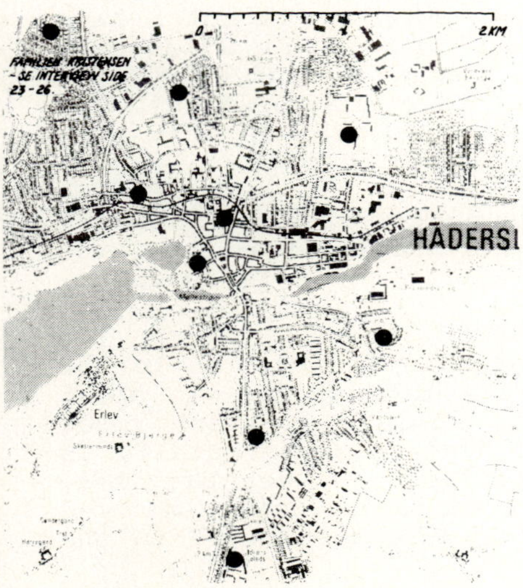

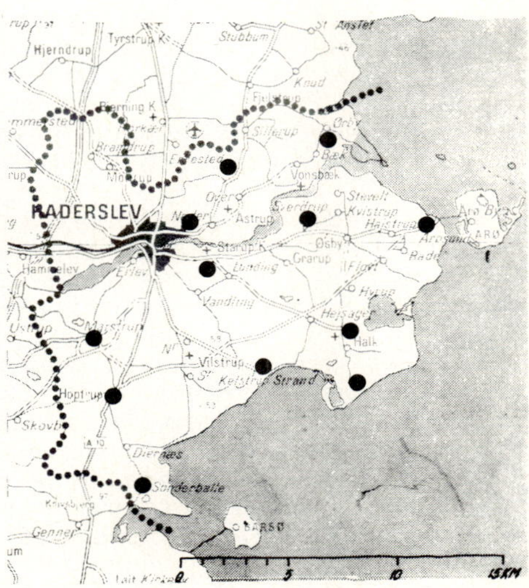

The interviews were taped, and during the interview, maps of trip-chains were made and notes were taken.

Of all 21 families, 20 were willing to be mentioned by name and address, only one wished to be anonymous.

Phase 3 - Opinions on environment, conditions of life and transportation

After the interviews we made, the conversation was edited into a paper of 10-30 pages including maps, photos etc. The interviewer rewrote some of the conversation and mixed it with many qoutations. The edited interview was then mailed to the families who made some corrections and permitted them to be printed with the full name and adress, and in many cases a photo of the family.

Furthermore there was for each interview made a "trailor" of half a page, to make it possible for the reader to find opinions from a type of family, one is especially interested in.

An index of all the problems mentioned was made to make it possible to get a quick overview on the type of problems and to find out in what connection and by whom the problems were mentioned.

The report with the interviews is in all a "phone book" of 400 pages. In an additional small report a short description of the results from the interviews was made, using the same headlines as in the index. Those were the problems for different groups of people, different environmental problems, problems for different modes of transport and problems connected with actual locations in the municipality.

The interviews were read by all the planners, but not by all politicians. (It should be mentioned, that it is not a hard job to read the interviews. They are written in a language pretty far from the usual "planners language", and the quotations are making the text much more alive than usually). The politicians have been helped selecting themselves what they wanted to read by using the trailers and the index.

There was in editing the interviews no censorship as to selecting problems or excluding problems mentioned by the families. Everything they mentioned as a problem or as an information to the planners was printed.

The interviws did bring new and valuable information to the planners and politicians. Much of the information was about local problems, for example about a village in which roads and crossings near the school were dangerous, especially in the morning, because there were conflicts between bicycles, busses, cars and pedestrians. This area is now being improved by the municipality, and the bus stop has been moved to the school entrance.

The result from the interviews as a great number of problems, presented in the frame of the everyday life of the families. They are a mixture of what you from the outside could call small and great problems, but presented in the interviews, their importance was not always valued by the same "yardstick" as the one used by the planners.

A lot of the problems were caused by conflicts between different modes of transport, especially in the town center, where the streets are used not only for transportation but also for small talk, when you meet a friend. In the pavements and in the pedestrian streets, goods from the shops nearby are taking up space, and residents in the dwellings need the streets or a mixture of peaceful and lively surroundings. When as in many other old towns, cars, busses, bicycles and pedestrians are supposed to share a network of very narrow streets, it is inevitable that they feel irritation and danger, and a part of them change their behaviour. (Some of the problems felt by bicyclists in the town are shown on the map page 11).

As examples of the problems mentioned, we could look at some of the statements from families without a car.

The most important type of problem is caused by insufficient planning for bicyclists. There is for instance a short part of the mainroad where no bicyclepath is built - it is on a road administred by the county, and on their maps it is shown that the non-existing little bit is there! And there is no routes through town you could recommend to the bicyclists as "a safe route of quality" - a net of bicycleroutes has never been planned.

The problems mentioned could often seem small, if they were considered isolated, but the summing up of many smaller problems would often result in a lesser use of the bicycles than, the actual distances and activities could be expected from. It seems important to the dayly maintance and the improvements to have a net planned.

The bicyclists living in the country-side are not satisfied either. They are often feeling unsafe on stretches without paths, and this causes in several cases bans on childrens visits to friends, sport etc. Especially in the winter, when the routes are narrowed by the snow, the bicyclists are afraid of using the roads.

The families without a car state, that the important reason why they can do without it, is that the public transport is good and that they use their bicycles the greater part of the year, although it is often felt as dangerous.

The greatest satisfaction with the traffic system was found at a family living in surburban area with no throughfare traffic and with activities for children inside the area. Another family without a car living in a village at the seaside wanted very much a speed limit of max. 40 km/h in the village, and bicyclepaths or safe routes for the children to the other villages nearby, where the children had their school, sports and friends - or more activities in their own village. The problems are not unique, but they are told to the planners and politicans by families and persons who do not take active part in a public debate on planning.

Palle Sørensen - to work by bus, walking and by car.
- an example of a map of the dayly routes

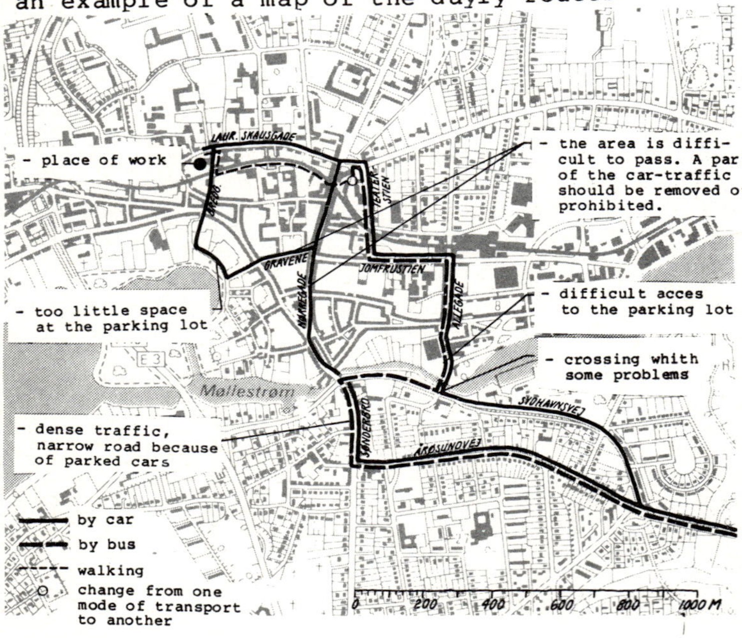

The interviews and the added index and description of the problems are in themselves useful for planners and politicians, because the problems are presented in the daily life of different families. They do give a better understanding of how the systems we plan are connected, of how important it is to look for the source of the environmental and traffic problems, because the best solution could be changing the districts of schools, or a long term solution such as changing the location of kindergartens, football-fields etc.

It was therefore decided that the problems written down in the interviews should be followed by suggestions for solutions.

Phase 4 - From problems to suggesting solutions

The information obtained in the first 3 phases can be used in several ways. We have chosed 12 different themes in order to show how the interviews can be used. We have hereby tried to find the limits and the possibilities in the use of the detailed interviews in the planning process.

The themes cover an extensive range of problems and so do the suggestions for solutions. Among the solutions are clear traffical improvements, cooperation between the traffic planners and other parts of the administration, increased information about the local transportation system, need of further analyses, use of interviews as a basis for local discussions in the planning of a village or a residential area, and finally suggested objectives concerning traffic and environment in the municipality plan.

In an ideal process, the themes could have been chosen by the local politicians, provided that they had read the interviews etc. In this case, however, the civil servants from the municipality and the ministry selected the themes.

The themes were to cover problems:

- of importance to the municipality
- of importance to the ministry of the environment
- mentioned and illustrated in the interviews
- in smaller local places
- in a village or a smaller area
- essential to the municipality as a whole
- known before the interviews
- known from the interviews
- concerning places where it was known that something was wrong, but where the interviews add information and point at other types of problems.

<u>The 12 themes</u> presented were:

1. The daily environmental and transportation problems of a single family.
2. Transportation between the town and the surrounding area.
3. Throughfare traffic in a village.
4. Traffic in an area for weekend cottages and summer holidays.
5. The aggregated environmental and traffic problems in a village.
6. The town center.
7. The use of bicycles in the town.
8. The town and the main road.
9. The circular road close to the center.
10. The railroad line through town.
11. School children.
12. Confidence in traffic.

Each theme presented the problems involved - those known from the existing plans and other written sources and those known from the interviews. Apart from theme 1, quotations from several interviews were used to illustrate the problem.

This was followed by suggestions for a <u>policy</u>, the <u>local possibilities</u> to deal with the problems and <u>solutions</u> - suggestions in which the environment and the conditions of life were given high priority.

Results from two of the themes are presented briefly in the following:

Bicycle routes

Problems for the bicyclist are mostly accidents and lack of confidence, followed by a suppression of transportation needs where no alternative mode of transportation is possible. The distances in a danish municipality are often relatively short. In the town of Haderslev and most of the villages the bicycle is a good alternative to other modes of transportation, if the municipality was provided with a safe net of bicycle routes.

The policy suggested to the politicians was therefore:

- the safety and confidence of bicyclists must be improved considerably
- the making of a net of bicycle routes covering the whole of the municipality must have high priority
- the routes are, if possible, to follow the direct line between destinations
- the routes must be located in nice surroundings, making the trip a positive event
- the net must chiefly consist of stretches where the bicycle traffic to-day is considerable.

The policy suggestion is accompanied by a proposed net of existing and new constructions combined with routes, where the bicycles are given high priority, secured by speed restrictions and prohibition of heavy vehicles.

Suggested net of bicycleroutes in the municipality

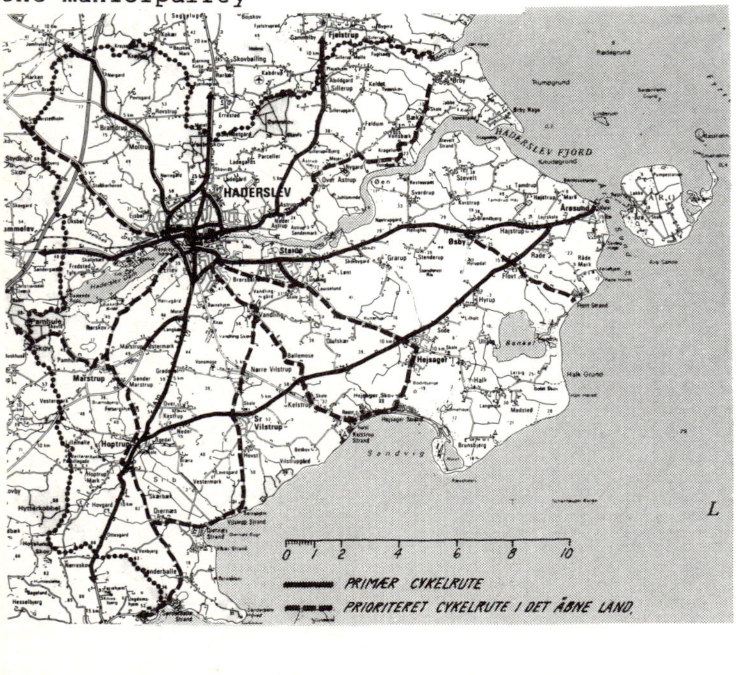

Map of problems for bicylists

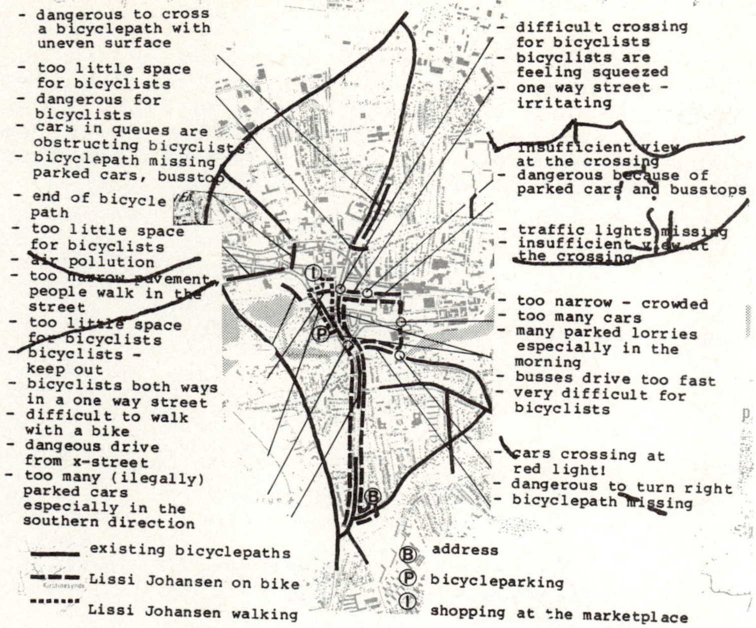

In the town center, a detailed map was made of the problems for bicyclists. This was accompanied by a proposed net of bicycle routes in the town center. To-day only very few of those routes are secure for bicycles.

Bicycleroutes in Haderslev - suggested solution

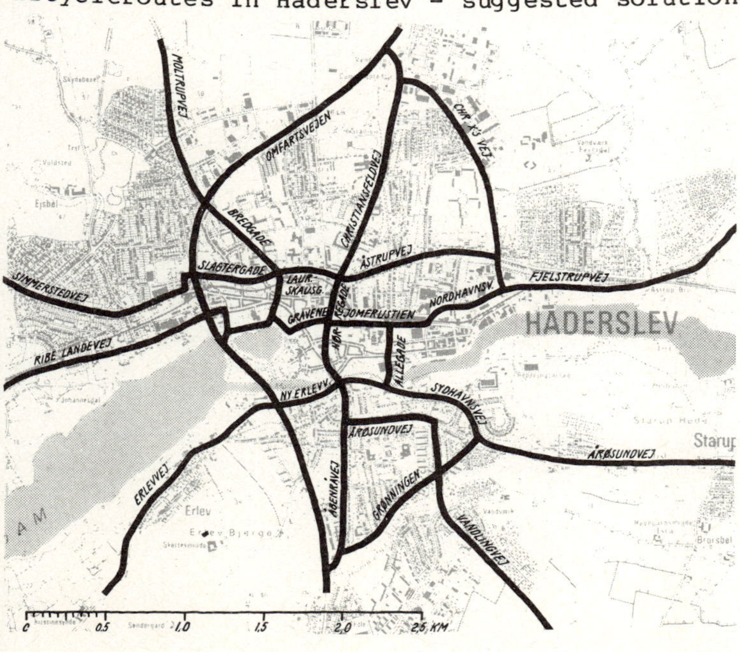

Village with a by-pass - the use of interviews in the planning process

In a village where a by-pass had been built some years ago, there were surprisingly many problems mentioned by the interviewed family. One example was that a tunnel built to secure the school children was considered dangerous and never used by the children, and in several places in the village there were conflicts between throughfare traffic and children.

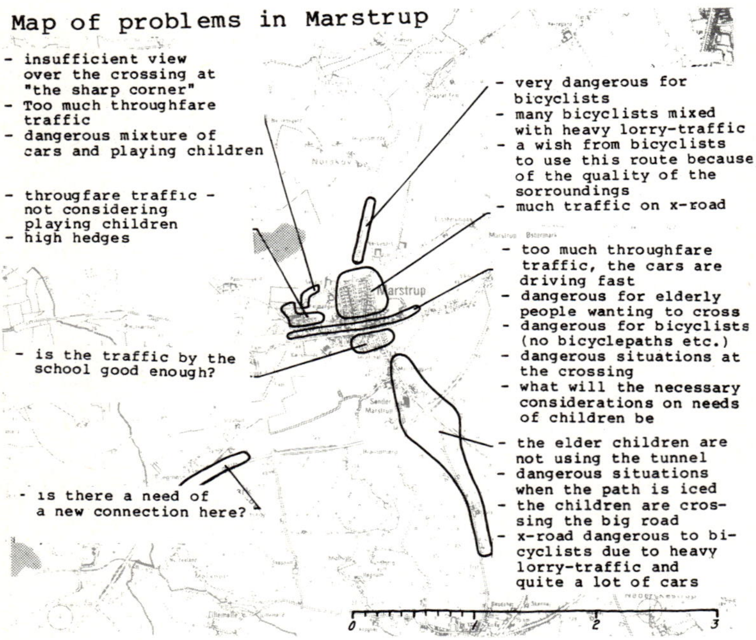

The problems were new to the planners and there seems to be a need for some minor improvements of the existing roads in the village. Bulding a by-pass is not enough.

Using the information from the interviews, solutions were suggested, but of main interest in this theme was a suggestion to use interviews as a basis for local debate. One or two interviews will have a direct appeal to the citizens, they are easy to understand and discuss. They must of course be combined with the "classical" knowledge of the traffic planners.

Principles suggested for Marstrup

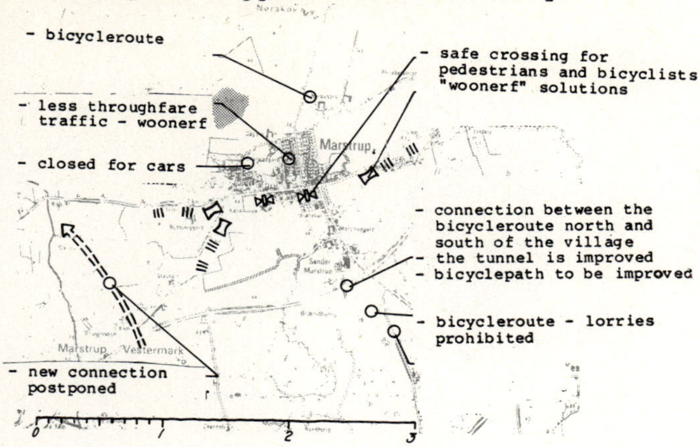

THE PROCESS - CIVIL SERVANTS AND POLITICIANS INVOLVED

The project was made by a working group consisting of civil servants from the Ministry of the Environment, town and traffic planners from the municipality and a consultant firm which is also working at the municipality plan. A steering commitee consisted of 4 managers, 2 from the Ministry and 2 from the municipality.

Furthermore, a group of experts on environment, transportation and conditions of life from universities, counties, other municipalities, ministries etc. was following the process, and discussing the papers at 3 meetings during the 2 years.

The work was only presented to the politicians once during the procss, and there was only limited contact to other parts of the administration of the municipality.

The final report has afterwards been discussed with the politicians, and a great part of the material is now used in the municipality plan and the following plan for roads and bicycle routes.

The final report is distributed to all municipalities in Denmark, and the method of using detailed family interviews is now beeing applied in a few other municipalities.

If the process was to be repeated, the most important change I would suggest would be to involve other parts of the municipality administration at an early stage. And last, but not least - involve the politicians more in selecting themes and discussing problems.

Lissi Johansen on her bike 1983

Det er torvedag, men Lissi tør ikke cykle på Gravene og fortovene er for smalle.

Lissi skal i Brugsen. Hun cykler på Jomfrustien, som er en del af centerringen.

Sydhavnsvej, på vej hjem fra indkøb. Lissi presses af bil, der skal til højre.

Sønderbro. Pladsen er trang og farlig for cyklister med parkerede biler og biler i begge retninger.

Lissi Johansen on her bike 1990

Bilerne er væk fra Gravene. Der er nu god plads til gående og cyklister.

Cykelstien på Jomfrustien var nem at lave. Der var plads nok.

Og på Sydhavnsvej blev der også plads til en cykelsti og 2 kørebaner i hver retning.

Op af bakken er der brug for megen plads, når man cykler. Derfor er det dejligt, at der er kommet cykelsti.

Behavioural Research for Transport Policy
pp. 343–362.

HOUSEHOLD LIFECYCLE: PREDICTOR OF TRAVEL EXPENDITURE

Lidia P. Kostyniuk, Department of Civil Engineering
Michigan State University, East Lansing, Mich. 48824
and
Ryuichi Kitamura, Department of Civil Engineering
University of California, Davis, Calif. 95616

1 Introduction

The household as a social unit is a pooling of economic and human resources and the trips generated by a household result from the out-of-home activities of the household members pursued either individually or jointly. Viewing travel as a linkage between activities and examining the distribution of activities among the members of the household has led to a much greater understanding of travel behavior [1-3] and some practical applications [4]. Despite this, most travel demand estimation methods currently in use do not adequately consider the activity leading to trip making, but rather focus on isolated individual trips (for discussions, see, e.g., references 5 through 10).

The only direct connection between households' needs to participate in activities and conventional travel demand estimation methods is found in trip generation models. Trip generation models are usually categorical in nature and predict the number of trips generated by a given type of household for a particular trip purpose. Common classifiers of households in trip generation models have been: household size, income, and automobile ownership [11]. Residential descriptors, such as the residential density of the zone where the household resides or the type of dwelling unit, have also been used in the past [12-13]. Only in a few cases has a detailed household descriptor such as the stage in the household lifecycle been used in actual planning practice in the United States. A notable exception is the model developed in the Detroit Transportation and Land Use Study (TALUS) [14].

When the number of trips generated by households is examined in cross-sectional analysis, these classifiers usually yield satisfactory goodness-of-fit. Application of the resulting models to forcasting future travel demand, however, is based on the assumption that the basic relationships of household trip making as captured by a particular set of classifiers remain stable over time for as much as 20 years into the future.

The activity and travel patterns of a population evolve in an ever-changing environment. The social, cultural, institutional, and economic environments of industial societies continue to change as evidenced by the reduction of household sizes, increased employment of women, and fluctuations between prosperity and recessions that have occured in the past 20 years. The spatial arrangements of the infrastructure of metropolitan areas have also changed over this time. Extensive freeway networks have been built and there has been a shift of commercial centers into the suburbs. The effects of the ongoing innovations in telecommunications are not fully understood yet. It is in this ever-changing environment that the transportation planner makes predictions about the travel demand of the future. In such efforts it is prudent to base the estimates on the most stable relationships of trip making that can be identified.

This study examines the nature of trip making by focusing on the factors which closely represent the microscopic environment in which daily activity and travel evolve. The analysis uses the lifecycle concept as such a factor. Underlying this is the conjecture that stages in lifecycle are strongly correlated with the needs and desires for out-of-home activities. They are also believed to reflect the constraints imposed on activity participation and time allocation decisions. The variables frequently used to define lifecycle stages, age of the individuals, marital status, and the composition of the household, are also believed to largely determine what the household members require and desire to do. Sex and employment of the individual (or role) are known to strongly affect the allocation of activities within the household [14-16]. Many research results accumulated in recent years point to the significance of the lifecycle and role variables as factors influencing household members' travel and activity patterns [1,17-20].

The variation and stability of travel patterns across lifecycle and gender subgroups are the focus of this paper. Several issues are examined. One is which aspects of travel behavior are correlated with the stage in lifecycle and whether the correlations remained stable over time. Another issue is how adult members of a household adopt to increased childcare and other responsibilities as it moves along the lifecycle. Also discussed in this paper is how these responsibilities are assigned between the male and female members of the household and how the assignment pattern changed over time. Through the examination of thses issues, the study attempts to infer the mechanism underlying the household's activity and travel behavior, to determine temporal variations in travel behavior, and to demonstrate the usefulness of lifecycle variables in analyzing activity and travel behavior.

Empirical evidence is provided through analysis of travel patterns of adult household members obtained from two metropolitan regions in the United States at two points in time. In addition to various travel pattern indicators including number of trips, number of trip chains, trip rate by purpose, and travel time expenditure, intra-household task sharing is examined for maintenance trips. Trends over time in these indicators are observed and the relationships between the changing travel environment and travel behavior are inferred.

The statistical analysis indicates that very few aspects of travel patterns have remained unchanged over the period of time considered in this study. The two regions studied here share some similar directions of change in travel patterns and activity engagement. Detailed analysis of one of the areas shows that the stage in lifecycle influences individuals' travel patterns in a logical and consistent manner over the time period. The study results also suggest that people adopt to increased activity needs by adjusting time allocation to different activities rather than modifying activity engagement altogether.

2 Lifecycle as a Predictor of Mobility

In light of the extensive changes that have taken place in the travel environment, it is important to critically examine if the relations embedded in typical travel forecasting models have remained stable. Among the many investigations into the stability of trip generation relationships and travel time or cost expenditure [e.g., 21-25], only a few claim stability at an aggregate level [e.g., 21]. The fact that most studies are unable to conclude stability in trip generation relations is rather disturbing. This may be because statistical fit to cross-sectional observations, rather than causal relationships based on rigorous understanding of travel behavior, has led the development of most trip generation models. It is therefore appropriate here to briefly review the relationships between travel behavior and those factors frequently selected in previous model development efforts.

Activities can be divided into two categories: those necessary for the existence and functioning of the household and those which are discretionary [26-28]. What constitutes required activities differs depending on the composition of the household and changes over its life [27]. Intra-household assignment of required activities, i.e., the way household members collaborate to perform necessary tasks, is a function of the social and cultural environment of the time as well as of the resources of the household [16]. The amount of time allocated to certain activities depends on the technology available for the tasks and again on the resources of the household to afford this technology or services to perform them.

These factors also affect the attribute of activity most fundamental to travel analysis, i.e., whether the activity is pursued at home or out of home. Where and when activities are carried out is influenced by the spatial and temporal distribution of opportunities, the institutional environment of the time, and the time, monetary, and other resources for travel available to the household. The resources of the household depend on the income, which in most cases is related to the economic environment of the time.

It is evident from the discussion that the conventional household classifiers are related to household activities. The size of the household is a surrogate for the types and durations of the activities it performs. Larger households typically imply more children, which in turn implies more obligations to perform childcare, education, and other activities. It also leads to increased constraints imposed on adult members of the household [1,27]. Income is a measure of the household's ability to aquire technology and services in performing its required activities. Presumably it is also a measure of the household's inclination towards discretionary out-of-home activities. The residential descriptors reflect the spatial distribution of activity locations relative to the residential location while the type of dwelling unit is another indicator of the composition of the household. The household's needs to travel and its ability to satisfy the mobility needs are both reflected in automobile ownership. The greater transportation capability offered by automobiles may in effect reduce various constraints and enable engagement in a wider range of activities.

While these conventional household classifiers are known to be closely correlated with observed trip generation in cross-sectional data,their relations to household activities are neither direct nor entirely logical. For example, mobility may be a function of not only the required activities but also the amount of discretionary time available to the household members. Key factors may then be the number of workers and the presence or absence and age of children. The importance of the former is obvious. The latter may be crucial since it dictates the childcare needs. The conventional classifiers are only loosely correlated with these household attributes and the magnitude of this correlation is perhaps subject to change over time.

The lifecycle descriptor–frequently defined in terms of the marital status of the adult members of the household, and presence or absence of children and their ages–specifies the household composition to finer detail than household size. This variable therefore seems to be more closely related to activities and travel of households. Recent research efforts focused on lifecycle as a descriptor of household characteristics and predictor of travel behavhior [1,20,17-19,29-32]. While some of the studies find this a very useful classfier of travel behavior [1,20,17-19], others do not reach such positive conclusions [31-32]. Clearly the first issue that needs to be critically examined is whether the use of lifecycle descriptors will lead to improvement of the travel forecasting procedure, or it will merely complicate the procedure without any gains. The convincing argument based on the innovative in-depth interviews of households by the Transport Studies Unit at Oxford [1] still needs to be verified with empirical data from different countries, regions, and points in time.

Many additional issues must also be critically examined before the use of lifecycle descriptors in forecasting and policy evaluation. One of the concerns is which classification scheme is to be used, or, how lifecycle stages should be defined. Several studies have addressed this question [17,31], but no concensus appears to exist on which variables are most effective to use (e.g., youngest child's age or the oldest child's age), or how many categories should be used in classifying households.

Another issue often raised about the use of the lifecycle concept is whether future distribution of lifecycle stages can be forecast with reasonable accuracy and if this forecasting can be done in a practical manner. In order to incorporate a lifecycle measure into existing forecasting procedures, one must examine: What is the relationship between the lifecycle stage and other key factors in travel behavoir analysis, such as car ownership? Is it valid to assume a fixed relationship between lifecycle stage and type of dwelling unit? While these questions lie outside the scope of this paper, the empirical results presented below undoubtedly warrant the future effort toward their answers.

3 Research Design

This section offers a brief description of the criteria that were used to develop the controlled sample used in this study. The definitions of the key variables of the analysis are also presented.

Sample Design: The emphasis of this study requires that the sample of analysis allow identification of the effects of household characteristics on the household members' activity and travel patterns–in particular, the impacts of children on adult members of different responsibilities, intra-household division of labor, and the effects therein of gender and employment. Because of this emphasis, the person, rather than the household, is used as the unit of statistical analysis. This emphasis also prompted the decision to, unless otherwise noted, limit the sample to those individuals from households consisting of an adult man and woman who can be considered as married, with or without children in the household (the data files used do not necessarily code marital status of individuals. The two adults of opposite sex are considered as married here when their age difference is less than 15 years). Another criterion used in the sample development is that at least one of the two adult members of household was employed. Only trips of adult individuals (at least 20 years old) are analyzed in the study.

Data Sets and Screening Criteria: The analysis uses data sets from two metropolitan areas: Rochester, New York, and Detroit, Michigan. Two surveys taken at different time points are available from each area and used in this study.

Rochester Data Sets (1963 and 1974): The main body of this study is based on the analysis of home interview survey data from Rochester, taken in 1963 and 1974. Most tabulations presented in this paper are limited to the subset of households as described above, i.e., nuclear households with two adults of which at least one was employed, and exclude single-adult households, households with more than two adults, households of individuals of the same sex, and households with no workers. The 1963 and 1974 samples contain 7,972 and 1,763 individuals, respectively. These individuals' trip records are carefully examined of their consistency and completeness using a set of criteria similar to those used by the authors in previous studies [8,18,20]. The intent of this study, however, is to retain as many observations as possible. Accordingly, the sample size occasionally varies from table to table.

Detroit Data Sets (1965, 1980): The 1965 and 1980 data files from the Detroit metropolitan area were used by the authors in previous analysis of urban time-space paths and temporal stability of travel patterns [8,18,20,33-34]. The Detroit data are used here to facilitate the inter-regional comparison of temporal stability in activity participation and travel expenditure. Detailed description of the data files are given in references 8 and 20.

Defining Key Variables: Lifecycle stages used in this study are defined in terms of the age of the male adult of the household, the presence of children in the household, and if present, by the age of the youngest child. This lead to the five stages for nuclear families as shown in Table 1. The age of the male adult is used only when no children are present in the household, while the lifecycle stage of a household with children is determined solely by the age of the youngest child. This definition originates from the TALUS study of the 60's [21] and is quite similar to the one reported in reference 1. Underlying the classification scheme is the assumption that adult members of households with children of a given age group behave similarly regardless of their age.

The households in the Rochester sample are classified according to the number of workers, either one or two. Typically, the worker of a one-worker household is a male. Another variable of critical importance is the role of an individual, which is defined in terms of the individual's sex and employment [27]. The number of workers and role are not used together as classification variables because of the strong association between employment and sex among members of one-worker households.

Car ownership is not considered in the statistical analysis of this study. There are several reasons for this. The first is a pragmatic one that sample size requirements prohibited any higher dimensionality of the analysis. Second, as a consequence of the maturing of motorization in 1974, there are only four households in this sample of two-adult households with workers to which no automobiles were available. Third, car-ownership effects may be diminishing in 1974 among the two-adult households as documented elsewhere [35]. And finally, car ownership may be viewed as one aspect of travel behavior rather than a factor that determines the behavior. The reader, however, is cautioned to keep in mind this omission of car ownership and also the selective sample used in the analysis when interpreting and generalizing the findings of this study.

4 Changes Over Time In Socio-Demographic Factors

Some of the changes that took place between the two time points covered by the Rochester and Detroit data sets are reported in Table 2. The distribution of lifecycle stages shown in Table 2-a indicates taht the fraction of households with the youngest child younger than 5 years of age (Stage 3) has decreased while that of couples without children has increased (Stage 2 and Stage 6). The result is perhaps a reflection of the continuing trend in the United States toward smaller households. The percentage of young households without children (Stage 2) is much larger among two-worker households than among one-worker households, both in 1963 and 1974. This is clearly a result of choice made by households as to the female adult members' labor-force participation. The percentage of households with preschool children (Stage 3) has decreased substantially in 1974 among the one-worker households, while an increase from 11.7% to 15.1% is observed among two-worker households.

The second half of Table 2-a shows a dramatic increase of two-worker households among those with preschool children (Stage 3). More women with preschool children, who have traditionally been responsible for childcare, are in the labor force in 1974. The changes in labor force participation by sex are compared in Table 2-b between the Rochester and Detroit data sets. Both areas show the same trend of increasing labor-force participation by women. Note that the 1980 Detroit statistics reflect the depressed economy of the area at that time (additional tabulation for the Detroit region can be found in reference 20).

The percentage of the population licensed to drive and car owership increased between the two time points. More individuals, especially women, are licensed to drive, and virtually all households in this sample own at least one car in 1974 (Tables 2-c and 2-d). A possible consequence of the increased mobility is the increased fraction of women from one-worker households who made at least one trip on the survey day as shown in Table 2-e.

5 Changes Over Time In Travel Pattern Indicators

Corresponding to these and many other changes in socioeconomic, demographic, transportation, land use, and societal factors are many changes in travel behavior of urban residents between the two time points. Table 3 summarizes these changes in terms of typical indicators of travel patters for both Detroit and Rochester data sets (the Detroit data files do not contain records of individuals who did not make trips and the statistics shown in the table are averages per tripmaker, not per person, for both areas. The Rochester data are retabulated here to include all household lifecycles and to make the two tabulations comparable. Only adult individuals are included in the tabulation).

Overall, both data sets show decreases in the trip pattern indicators in the second time period. This is particularly the case in Detroit where a decrease of 15.1% is observed for the total number of trips. Similar decreases can be observed for the number of sojourns (i.e., number of stops made to pursue activities) and number of trip chains (i.e., a home-to-home sequence of trips made by an individual). The mean number of sojourns per trip chain also shows some decrease in the second time period. The total travel time expenditure decreased in the Detroit area while a slight increase is observed for the Rochester area. The severly depressed economic state of the Detroit area is undoubtedly reflected in the decreases in the travel pattern indicators of 1980. The economy may influence travel behavior more substantially than it has been believed to (also see reference 36).

The average number of trips by trip purpose, or the mean frequency of out-of-home activity participation by activity type, also shows some variations between the two time points (the second part of Table 3). Changes in the frequency of work activity are different between the two areas; the decrease in Detroit is presumably a reflection of economy, while the increase in Rochester is perhaps due to the increased women's labor force participation. Personal business activity shows relatively small changes, while school activity shows a remarkable increase in both areas and suggests an increased population of adult students. The increased frequency of eat-meal activity found also in both areas may be a reflection of changes in lifestyle. Part of the increase in Rochester may be caused by the increase in employment, which is often accompanied by lunch trips as evidenced in many analyses of trip chaining [33,37]. The increase found in Detroit where employment decreased in 1980 and work-trip rate is consequently decreased by 11.4%, however, implies that this cannot be attributed solely to workers' lunch trips. The increase of course is parallel to the rapid development of fast-food outlets which can be now ubiquitously found along commercial corridors in the United States.

The remaining three activity types, shopping, social-recreation, and serving passengers, all show decreased trip rates in the second time period in both areas. The large decrease in shopping in the Detroit area may again be due to the depressed economy. The slight decrease in Rochester may be attributable to household appliances and consumer products which had become available in the second period and reduced the needs for frequent convenience shopping. The increase in the number of working women is also another possible contributing factor. The large decrease in social-recreation may be due to substitution of in-home activities for out-of-home activities [20]. The larger decrease in Detroit may again be a reflection of the depressed economy. Finally, the significantly decreased rate of serve-passenger trips can be attributed to the increased population of drivers and multi-car households.

This section has portrayed temporal changes in socioeconomic and travel characteristics. It is clear that many aspects of travel behavior have changed over time and that many of these changes are related to also many changes in factors that influence travel bahavior. This examination of travel patterns from two metropolitan areas at two different time points casts a serious doubt on the notion that travel behavior remains stable over time. The discussions above suggest definite relationships between changes in travel behavior and changes in the travel environment.

6 Effects of Lifecycle, Sex, and Number of Workers

This section focuses on the travel behavior of adult individuals classified by lifecycle stage, sex, and number of workers in household. The Rochester data sets are used in the analysis. Individuals' travel patterns are characterized in terms of the number of trips classified by activity type, number of trip chains, and time allocation to activities and travel.

Total Number of Trips and Trip Chains: The total number of trips per person increased slightly from the overall average of 3.50 in 1963 to 3.55 in 1974 (not significant at $a = 0.1$). Figure 1, which summarizes the mean total number of trips by the number of workers (1 worker versus 2 workers), lifecycle stage, sex and year, shows that men in two-adult households make more trips than women ($a < 0.01$, both 1963 and 1974); an observation made in previous comparisons of travel patterns between men and women [38-39].

This gender difference also exists even when male and female adults of a household are both employed. The 1963 average daily trip rate of males from two-worker households is 4.23 and that of females is 3.50 ($a < 0.01$). The difference, however, seems to be narrowing; the corresponding 1974 values are 4.02 and 3.69, respectively ($a < 0.1$). The only, but important, exception is the two-worker households in 1974 that had children (Stages 3 to 5), in which the personal trip rates are almost identical between males and females. It is interesting to note that a previous analysis of the 1965 Detroit data did not indicate gender differences in this respect among single-person households [40].

Males of one-worker households with children (Stages 3 to 5) made fewer trips in 1974. On the other hand, younger females (Stages 2 through 4) show quite notable increases in the number of trips. The difference between the sexes has decreased drastically for lifecycle stage 3 (households with preschool children). It appears that men of this stage are less mobile ($a < 0.01$), while women are more mobile in 1974 than they were in 1963 ($a < 0.05$). Males of two-worker households of lifecycle stages 2 and 5 also show decreased person trip generation in 1974 ($a < 0.1$, and $a < 0.05$, respectively), while females from households with children show increases. As noted earlier, gender differences are very small in 1974 among those of two-worker households with children.

The patterns shown in Figure 2 indicate that similar changes have taken place in the number of trip chains made by adult individuals between 1963 and 1974. The overall means indicate that the number of chains increased slightly from 1.32 in 1963 to 1.37 in 1974 ($a < 0.1$). The gender differences appear to be decreasing, especially among those women who have children. Table 4 summarizes the differences for the number of trips and number of trip chains.

Figures 1 and 2 suggest that the lifecycle stage affects behavior of adult women, whether employed or not, more significantly than that of adult men. To examine this, the between-group variances of the number of trips and trip chians are evaluated and presented in Table 5. The result confirms this conjecture. Childcare needs, which vary greatly across lifecycle stages and which are typically undertaken by women, are perhaps the main reason for this result. Presumably intra-household task assignment is based strongly on gender. Note that this differential effect appears to be increasing in 1974.

It is evident from the results that the lifecycle stage, an attribute defined at the household level, significantly influences individuals travel behavior. Household characteristics remain to be important factors even when the unit of analysis is the individual. Finally, it is notable that the lifecycle stage exhibits similar pattern between 1963 and 1974.

Activity Engagement by Activity Type: One would expect that participation in obligatory activities such as work and school would be stable over time. The observation of this study is against such expectation as it has been seen in Table 3. The Rochester sample further indicates that this instability is associated particularly with women from households with preschool or school-age children (Stages 3 and 4). The work trip rates has especially increased among women from these two stages ($a < 0.01$ for both stages).

The women's trip rates for personal business and shopping are higher than those of the men in the same lifecycle stage (Figures 3 and 4). This is especially the case among women from one-worker households. Intra-household division of labor can be seen in Figure 5 which shows the fraction of household maintenance activities performed by each of the two adults in the sample households. Clearly, these chores tended to be performed by women ($a < 0.01$ for shopping, both 1963 and 1974), an observation frequently found in previous analyses [e.g., 39]. Furthermore, this tendency is more apparent in 1974, contraty to the overall statistics given in Table 4. Intra-household division of labor continued to be gender-based between the two time points.

Participation in social-recreation activity declines as one progresses through the lifecycle stages, both for 1963 and 1974 (Figure 6). This result confirms the previous finding on the 1965 Detroit data [30]. The only exceptions are the depression found for 1974 two-worker households with preschool children (Stage 3), and the very distinct pattern found for women from 1963 one-worker households. Figure 6 also shows that women from one-worker households tended to pursue social-recreation activities more frequently than men ($a < 0.01$ for 1963, $a < 0.05$ for 1974).

Participation in personal business, shopping, and social-recreation is also different between one-worker and two-worker households. Individuals from one-worker households tended to engage in these activities more frequently (personal business: $a < 0.01$ for 1963, $a < 0.05$ for 1974; shopping: $a < 0.01$ for 1963 and 1974; social- recreation: $a < 0.02$ for 1963, $a > 0.1$ for 1974). The difference is mainly due to the higher trip rates found among the women of one-worker households for these purposes.

Effects of lifecycle and sex on serving passengers are shown in Figure 7. Quite notable are the distinct patterns found between men and women, most dramatically among 1963 two-worker households. In 1963, males of two-worker households frequently undertook the task of chauffeuring. This no longer is the case in 1974. A conjecture can be developed from this: the serve-passenger trips made by males from two-worker households were mainly to drive spouses to and from work, and the needs to perform such tasks diminished in 1974 when more households owned two cars and more women were licensed to drive. The decreased participation in this activity by males from one-worker households can be explained by the same factor: more multi-car households and more woman drivers have decreased the needs for male adults to chauffeur their spouses and other family members.

Female adults of these households, on the other hand, exhibit almost identical patterns across the stages of lifecycle, whether one-worker households or two-worker households. The pattern is that the rate of participation in this activity is low in the first stage of lifecycle; it keeps increasing as the lifecycle stage progresses and peaks when the youngest child's age is between 5 and 14; then the rate shows a decline as the lifecycle stage matures. This clear pattern suggests that the lifecycle stage is the major determinant of the frequency of women's serve-passenger trips. The strong influence of lifecycle further suggests that these trips are made for the children of the households. The pattern obviously points to the additional travel needs generated by the presence of children. The analysis has also indicated that the additional trips are assigned to women, whether employed or not.

Time Allocation: The patterns of variations in activity par- ticipation and trip making are now studied together with additional findings on daily time allocation. The result is a conjecture as to how individuals adjust their behavior in response to increased or decreased needs and desire for out-of-home activities that may arise as they progress along the lifecycle.

As they progressed through the stages of lifecycle, male adults tended to spend lesser amounts of time out of home and on travel, although some irregularities are present (Figures 8 and 9). This result is consistent with many previous results [e.g., 18,41]. Women, on the other hand, exhibit quite different patterns. It is obvious that lifecycle influences time use patterns of men and women quite differently.

The differences in total out-of-home time and travel time expendi- ture between men and women are quite substantial. The differences seen for individuals from one-worker households are not surprising since most of the women in this group were not employed. The difference clearly seen between men and women of two-worker households, however, cannot be explained by the employment factor ($a < 0.01$ for total out-of-home time and total travel time, both 1963 and 1974). It is well accepted that women's work trips tend to be shorter, indicating that they tend to seek employment in the vicinity of the residence location [38]. This study extends this by offering a new finding that such gender difference also exists in the total amount of time spent outside the home. It may be the case that women with working spouses tend to seek part-time employment, resulting in the observed difference in out-of-home time.

The analysis has clearly shown the significant effects of the lifecycle stage on various aspects of travel behavior. The interaction effect of lifecycle stage and sex has also emerged as a critical factor. Analysis of variance of the travel pattern indicators studied in this section indicated that the lifecycle main effect is significant (at $a = 0.05$) for all the indicators, and also that the lifecycle-sex interaction is significant for total number of trips ($a < 0.0001$), number of serve-passenger trips ($a < 0.0001$), and total out-of-home time ($a < 0.001$).

Adaptation: Further inspection of Figures 8 and 9 indicates a pattern which can be found commonly among women, whether from one-worker households or from two-worker households: There is a consistent and clear tendency that women with preschool children (Stage 3) spend the least amount of time outside home and disproportinally small amount of time on travel compared with the women of the adjacent lifecycle stages. The only exception is the travel time expenditure by 1963 women from two-worker households. This result prompts a conjecture that women are assigned childcare responsibility, which demands the most time before the child enters school; this leads to an increased amount of time allocated for in-home activities and time spent outside home is accordingly decreased.

It is important to examine the presence of such a pattern in other aspects of travel behavior. Re-examination of the results of this study indicates that no other aspects exhibit such a consistent pattern associated with women with preschool children. For example, the total number of trips shown in Figure 1 indicates that women with preschool children from two-worker households made slightly more trips than those from households with no children (stages 2 or 6), both in 1963 and 1974. Figure 9, on the other hand, indicates that women with preschool children spent least time outside home. Further comparison of Figures 1 and 9 indicates that there is no correspondence between the number of trips and out-of-home time expenditure by women.

The conjecture that can be derived from the depression associated with lifecycle stage 3 can be summarized as follows. Women take the primary responsibility for childcare and presence of preschool or shool-age children leads to increased needs for performing various tasks outside home that originate from this childcare responsibility. At the same time, the same childcare responsibility implies that the amount of time that needs to be spent inside home attending children also increases, especially when the children are in the infant stage. The result is a somewhat increased (definitely not decreased) number of trips combined with a decreased amount of time spent outside home by these women with pre-school children (Stage 3) and also those with school-age children (Stage 4) to some extent. Figures 1 and 9 support this conjecture.

The result agrees with the notion that working women face heavy time constraints [42] and tend to minimize travel distance [39]. In addition, it indicates that they also tend to minimize out-of-home activity time, and further that this tendency applies to nonworking women as well. The sometimes sublte, but consistent effects of lifecycle stages on women's travel patterns shown above clearly indicate that lifecycle is a logical and useful predictor of travel expenditure.

Generalization of the result at this point is by no means warranted and more extensive empirical examination is due before the above conjecture can be regarded as a description of the structure that underlies individuals' daily travel behavior. Nevertheless the following generalization is an appealing one.

Individuals have different degrees of commitments and desires for various types of activities. They are at the same time subjected to various constraints, the most absolute one being the time constraint that stems from the fact that a day contains only 24 hours. Organizing their daily itineraries within these constraints involves the choice of activity participation and allocation of time to the activities. Obviously, one can choose to pursue fewer activities but allocate more time to each, or may choose to engage in more activities and allocate less time to each. The current finding, then, suggests that individuals tend not to respond to the constraints by making adjustments in their activity participation choice; rather, adjustments are made in their time allocation decision. Individuals tend to decrease the amount of time spent on activities rather than foregoing the activities. It may be the case that activity choice is less flexible than time allocation. Or it may be that activity choice has a higher priority than time allocation, i.e., time allocation is subordinated under activity choice.

7 Conclusion

This analysis of travel behavior of adult individuals from nuclear-family households has shown that very few aspects of travel patterns remained unchanged over time. The changes in the two metropolitan areas studied can be partially explained by the changes in the behavior of women with respect to driving, working, and bearing fewer children, that have occured in western industrialized societies over that past 20 years. Comparison of the changes also indicated that the economic state of the area has an impact on trip making. Although the fluctuations in the economic environment are not easy to predict, a measure of economic prosperity should at least be carried along in any cross-sectional travel data collected for use in forecasting.

The gender difference in trip rates appears to be declining, especially in the case of employed women. Yet, the travel patterns of women still show a very different picture in the lifecycle stages with small children. Women in these stages exhibit lower mobility in terms of time expenditure than they do in any other lifecycle stages. The analysis has also shown that assignment of tasks among household members is based more on gender than on employment, especially for shopping and chauffeuring. It was further shown that this assignment pattern remained unchanged over time.

The lifecycle stage significantly affects every aspect of travel behavior examined in this paper. Its effects in general exhibit similar patterns between the two time points for which data were available (an important exception is the drastic change in the lifecycle effect found for serve-passenger trips made by male adults, which reflects the motorization process that continued between 1963 and 1974). More importantly, the examination of travel behavior along the lifecycle stages made it possible to develop the many inferences on household activities, task assignment, gender and empolyment effect, impact of children, and changes over time in these elements. In addition, the use of lifecycle together with the role concept and inspection of time allocation to out-of-home activities across lifecycle subgroups, led to the development of the inference on the decision structure underlying travel behavior: Individuals respond to increased activity needs by spending less time in each activity rather than changing their activity engagement patterns.

The changes in travel patterns over the time periods considered in this study are very much a result of the significant social and cultural changes that took place in the past 20 years. Although such dramatic changes may not occur in every 20 year period, a behavioral shift, once identified, should not be ignored in transportation planning and policy development. The next important changes may very well be in the institutional environment, e.g., more liberal work schedule, acceptance of home as work location, or increased childcare availability at the work sites. Another ongoing change that transportation planners cannot ignore is in telecommunications, which can have great impacts on spatial distributions of activity locations and how activities are pursued (especially shopping and personal business). How these changes will affect the demand for travel will have to be determined by how these changes affect activities of household members. What behavioral modifications households will make in pursuit of their activities can be speculated by lifecycle better than by any other classifier. While not an absolute answer to the transportation planners' prayer, this classifier offers a reasonable focus for viewing travel behavior in the ever-changing environment.

References

1. Jones, P.M., M.C. Dix, M.I. Clarke, and I.G. Heggie (1980) Understanding Travel Behavior. Transport Studies Unit, Oxford University, Oxford.
2. Carpenter, S. and P. Jones, eds. (1983) Recent Advances in Travel Demand Analysis. Gower Press, Aldershot.
3. Brog, W., and E. Erl (1983) Application of a model of individual behaviour (situational approach) to explain household activity

patterns in an urban area and to forecast behavioural changes. In Recent Advances in Travel Demand Analysis (S. Carpenter and P. Jones, eds.), Gower Press, Aldershot, pp. 350-370.

4. Clarke, M., C. van Knippenberg, and J. Splinter (1984) Analysing the effect of rural bus service reductions on life styles. Paper presented at the 63rd Annual Meeting of the Transportation Research Board, Washington, D.C.
5. Heggie, I., and P. Jones (1978) Defining domains for models of household behavior. Transportation, 7, 655-672.
6. Burnett, K.P. (1978) Markovian models of movement within spatial structures. Geographic Analysis, 10, 142-153.
7. Hanson, S. (1979) Urban travel linkages: review. In Behavioral Travel Modeling (D.A. Hensher and P.S. Stopher, eds.), Croom Helm, London, pp. 81-100.
8. Kitamura, R., L.P. Kostyniuk, and M.J. Uyeno (1981) Basic properties of urban time-space paths: empirical tests. Transportation Research Record 794, 8-19.
9. Recker, W.W., M.G. McNally, and G.S. Root (1983) A methodology for activity-based travel analysis: the STARCHILD model. Paper presented at the 30th North American Meetings of the Regional Science Association, Chicago, IL.
10. Koppelman, F.S., and E.I. Pas (1984) Travel-activity beahavior in time and space: methods for representation and analysis. In Measuring the Unmeasurable (P. Nijkamp, ed.), Martinus Nijhoff, The Hague (forthcoming).
11. Federal Highway Administration (1967) Guidelines for Trip Generation Analysis. Washington, D.C.
12. Sosslau, A.B., A.B. Hassam, M.M. Carter, and G.V. Wickstrom (1978) Quick Response Urban Travel Estimation Techniques and Transferable Parameters. National Cooperative Highway Research Program Report 187, Transportation Research Board, Washington, D.C.
13. Institute of Transportation Engineers (1977) Trip Generation Manual. Washington, D.C.
14. Hanson, S., and P. Hanson (1978) The impact of women's employment on household travel patterns: a Swedish example. In Women's Travel Issues: Research Needs and Priorities (S. Rosenbloom, ed.), U.S. Department of Transportation, Washington, D.C., pp. 127-169.
15. Van der Hoorn, T. (1979) Travel behavior and the total activity pattern. Transportation, 8, 309-328.
16. Rosenbloom, S. (1983) Demographic shifts: the changing organization of society and the implications for the transportation planning process. In Transportation and Stagnation: Challenges for Planning and Research (P.H.L. Bovy, ed.), C.V.S., Delft, pp. 189-211.
17. Boyle, D.K., and J.E. Chicoine (1982) Application of the Life Cycle Concept to Trip Generation Procedures. New York State Department of Transportation, Albany, NY
18. Kostyniuk, L.P., and R. Kitamura (1982) Life cycle and household time-space paths: empirical investigation. Transportation Research Record 879, 28-37.
19. Clarke, M., and M. Dix (1983) Stage in lifecycle--a classificatory variable with dynamic properties. In Recent Advances in Travel Demand Analysis (S. Carpenter and P. Jones, eds.), Gower Press, Aldershot, pp. 215-231.
20. Kostyniuk, L.P., and R. Kitamura (1984) Temporal stability of urban travel patterns. Transportation Policy and Decision Making, 2, 481-500.
21. Ladd, W.M., R.H. Larson, and K. Boka (1969) A Trip Generation Model for the Detroit Region. The Detroit Regional Transportation and Land Use Study, Detroit, MI.

22. Hill, D., and N. Dodd (1966) Studies of trends of travel between 1954 and 1964 in large metropolitan areas. Highway Research Record 141, 1-23.
23. Smith, R.L. and D.E. Cleveland (1967) Time stability of trip generation and predistribution modal choice models. Transportation Research Record 569, 76-86.
24. Downes, J.D., S. Johnsen, and D. Morrell (1978) Household and Person Trip Generation Models. Supplementary Report 401, Transport and Road Research Laboratory, Crowthorn.
25. Gunn, H.F. (1981) Travel budgets--a review of evidence and modeling implications. Transportation Research, 15A, 7-23.
26. Cullen, I., and V. Godson (1975) Urban Networks: The Structure of Activity Patterns. Progress in Planning Vol. 4, Pergamon Press, Oxford.
27. Chapin, S.F. (1974) Human Activity Patterns in the City. John Wiley, New York, NY.
28. Burns, L.D. (1979) Transportation, Temporal, and Spatial Components of Accessibility. D.C. Heath, Lexington, MA.
29. Zimmerman, C. (1982) The life cycle concept as a tool for travel research. Transportation, 11, 51-69.
30. Salomon, I., and M. Ben-Akiva (1982) The use of life style concept in travel demand models. Transportation Research Record 879, 37-45.
31. Skinner, L.E. (1984) Family Composition and Life Style Variable Use in Urban Transportation Planning. Draft Report, Urban Planning and Transportation Management Division, Federal Highway Administration, Washington, D.C.
32. McDonald, K.G., and P.R. Stopher (1983) Some contrary indication for using household structure in trip generation analysis. Transportation Research Record 944, 92-100.
33. Kitamura, R. (1983) Sequential, history-dependent approach to trip-chaining behavior. Transportation Research Record 944, 13-22.
34. Kostyniuk, L.P., and R. Kitamura (1984) Trip chains and activity sequences: test of temporal stability. Transportation Research Record (forthcoming).
35. Kitamura, R., and L.P. Kostyniuk (1985) Maturing motorization and household travel: the case of nuclear-family households (in preparation).
36. Baanders, A., J. Kremer-Nass, and C.J. Ruijgrok (1983) Income decline and travel behaviour--some recent Dutch findings and research directions. In Transportation and Stagnation: Challenges for Planning and Research (P.H.L. Bovy, ed.), C.V.S., Delft, pp. 25-44.
37. Horton, F.E., and W.E. Wagner (1969) A Markovian analysis of urban travel behavior: pattern responses by socioeconomic and occupational groups. Highway Research Record 283, 19-29.
38. Madden, J.F., and M.J. White (1978) Women's work trips: an empirical and theoretical overview. In Women's Travel Issues: Research Needs and Priorities (S. Rosenbloom ed.), U.S. Department of Transportation, Washington, D.C.
39. Hanson, S. and P. Hanson (1980) Gender and urban activity patterns in Uppsala, Sweden. Geographical Review, 291-299.
40. Kostyniuk, L.P., and D.E. Cleveland (1978) Gender-role identification in the methodology of transportation planning. In Women's Travel Issues: Research Needs and Priorities (S. Rosenbloom, ed.) U.S. Department of Transportation, Washington, D.C.
41. Wilbur Smith and Associates, and Alan Voorhees and Associates (1966). Traffic Planning for the North Central Freeway. Washington, D.C.
42. Michaelson, W. (1983) The Impact of Changing Women's Roles on Transportation Needs and Usage. Institute of Transportation Studies, University of California, Irvine, CA.

TABLE 1.

DEFINITION OF LIFECYCLE STAGES USED WITH 1963 AND 1974 ROCHESTER DATA

LIFECYCLE STAGE	AGE OF MALE ADULT	PRESENCE OF CHILDREN	AGE OF YOUNGEST CHILD
2	<45	No	
3	-	Yes	<5
4	-	Yes	5-14
5	-	Yes	≥15
6	≥45	No	

- : Not used to define lifecycle stage

Note: Lifeycycle stages 1 and 7 are for unmarried individuals and not used in this study.

TABLE 2.

TEMPORAL CHANGES IN SOCIOECONOMIC SAMPLE CHARACTERISTICS

a. LIFECYCLE STAGE AND NUMBER OF WORKERS (ROCHESTER)

NO. OF WORKERS	YEAR	LIFECYCLE STAGE 2	3	4	5	6	TOTAL
1 Worker	1963	4.0	43.9	28.8	5.2	18.1	100.0
	1974	6.9	34.0	26.0	5.7	27.4	100.0
2 Workers	1963	16.2	11.7	29.1	13.6	29.4	100.0
	1974	20.3	15.1	34.4	9.4	20.8	100.0
Total	1963	7.3	35.1	28.9	7.5	21.2	100.0
	1974	13.3	24.9	30.1	7.5	24.2	100.0

YEAR	NO. OF WORKERS	LIFECYCLE STAGE 2	3	4	5	6	TOTAL
1963	1 Worker	39.6	90.9	72.4	50.3	62.2	72.7
	2 Workers	60.4	9.1	27.6	49.7	37.8	27.3
	Total	100.0	100.0	100.0	100.0	100.0	100.0
1974	1 Worker	26.8	70.8	44.9	39.4	58.8	51.9
	2 Workers	73.2	29.2	55.1	60.6	41.2	48.1
	Total	100.0	100.0	100.0	100.0	100.0	100.0

b. SEX AND EMPLOYMENT (DETROIT AND ROCHESTER)

		SEX AND EMPLOYMENT				
		MALE		FEMALE		
AREA	YEAR	EMPLOYED	NOT EMPLOYED	EMPLOYED	NOT EMPLOYED	TOTAL
Detroit	1965	46.3	6.5	19.2	28.0	100.0
	1980	34.8	15.4	23.1	26.7	100.0
Rochester	1963	39.2	7.2	16.6	37.0	100.0
	1974	36.8	8.8	24.1	30.3	100.0

c. PERCENTAGE OF DRIVER'S LICENSE HOLDERS (ROCHESTER)

	NO. OF WORKER AND SEX			
	1-WORKER HOUSEHOLDS		2-WORKER HOUSEHOLDS	
YEAR	MALE	FEMALE	MALE	FEMALE
1963	94.9	70.3	94.4	70.5
1974	98.5	87.5	99.8	92.7

TABLE 2.

TEMPORAL CHANGES IN SOCIOECONOMIC
SAMPLE CHARACTERISTICS
(cont.)

d. PERCENTAGE OF HOUSEHOLDS BY CAR OWNERSHIP (ROCHESTER)

NO. OF WORKERS AND CAR OWNERSHIP

YEAR	1-WORKER HOUSEHOLDS 0	1	≥2	TOTAL	2-WORKER HOUSEHOLDS 0	1	≥2	TOTAL
1963	6.1	73.5	20.4	100.0	6.2	68.7	25.1	100.0
1974	0.4	40.5	59.1	100.0	0.2	31.4	68.4	100.0

e. PERCENTAGE OF TRIPMAKERS (ROCHESTER)

NO. OF WORKERS AND SEX

YEAR	1-WORKER HOUSEHOLDS Male	Female	2-WORKER HOUSEHOLDS Male	Female
1963	95.6	62.3	97.3	93.1
1974	92.8	70.8	94.6	87.0

NOTE: See Appendix Table for sample size.

Table 3

Temporal Changes in Activity and Travel Pattern Indicators
(Detroit and Rochester)*

Travel Pattern Indicators

	Detroit 1965	1980	Difference (%)	Rochester 1963	1974	Difference (%)
No. of Trips	4.05	3.44	-15.1	4.21	4.14	-1.7
No. of Trip Chains	1.51	1.34	-11.1	1.55	1.61	-3.9
No. of Sojourns per Chain	1.68	1.56	-7.1	1.72	1.57	-8.7
Total Travel Time	75.5	73.1	-3.1	69.1	71.9	4.1

Mean Frequency of Activity Engagement

	Detroit 1965	1980	Difference (%)	Rochester 1963	1974	Difference (%)
Work	0.81	0.72	-11.4	0.89	0.98	9.7
School	0.04	0.14	+	0.01	0.03	+
Eat Meal	0.11	0.13	23.6	0.11	0.13	18.9
Personal Business	0.35	0.34	2.8	0.39	0.37	-6.0
Shopping	0.51	0.38	-25.0	0.46	0.44	-5.4
Social-Recreation	0.38	0.23	-40.2	0.38	0.31	-18.4
Serve Passengers	0.35	0.16	-54.2	0.42	0.27	-35.1
Total	2.54	2.10	-17.5	2.66	2.53	-4.9
Sample Size	53,928	2,221		6,626	1,515	

* : Statistics are average per tripmaker
+ : Greater than 200%

TABLE 4.

GENDER EFFECTS ON NUMBER OF TRIPS AND TRIP CHAINS (ROCHESTER)

TOTAL NUMBER OF TRIPS

	1-WORKER HOUSEHOLDS			2-WORKER HOUSEHOLDS		
	Male	Female	Difference	Male	Female	Difference
1963	4.02	2.73	1.29	4.23	3.50	0.73
1974	3.71	2.84	0.87	4.02	3.69	0.33

NUMBER OF TRIP CHAINS

	1-WORKER HOUSEHOLDS			2-WORKER HOUSEHOLDS		
	Male	Female	Difference	Male	Female	Difference
1963	1.53	1.02	0.51	1.59	1.38	0.21
1974	1.45	1.09	0.36	1.59	1.41	0.18

NOTE: See Appendix Table for sample size.

TABLE 5.

RATIO OF VARIANCE BETWEEN LIFECYCLE STAGE SUBGROUPS TO TOTAL VARIANCE (IN PERCENT)

		HOUSEHOLD TYPE AND SEX			
		1-WORKER HOUSEHOLDS		2-WORKER HOUSEHOLDS	
	Year	Male	Female	Male	Female
No. of Trip Chains	1963	2.00	2.89	0.81	1.67
	1974	2.05	5.46	0.73	4.94
No. of Trip Chains	1963	1.72	3.15	1.96	2.53
	1974	2.31	6.06	1.28	7.28

NOTE: See Appendix Table for sample size.

APPENDIX TABLE

Numbered of Individual in Rochester Sample by Sex, Number of Workers in Household, Lifecycle Stage, and Year*

		LIFECYCLE STAGE AND SEX											
No. of Workers	Year	2 M	2 F	3 M	3 F	4 M	4 F	5 M	5 F	6 M	6 F	Total M	Total F
1 Worker	1963	118	113	1268	1270	826	837	149	150	521	527	2882	2897
	1974	33	30	156	155	118	120	26	26	125	126	458	457
2 Worker	1963	178	174	126	129	315	318	147	148	317	321	1083	1090
	1974	87	85	64	64	146	146	40	40	88	88	425	423

*The number of males and females within each category of lifecycle stage and number of workers are not identical because of observations with inconsistent or incomplete trip records.

FIGURE 1. AVERAGE NUMBER OF TRIPS PER PERSON BY NUMBER OF WORKERS, SEX, AND LIFECYCLE STAGE

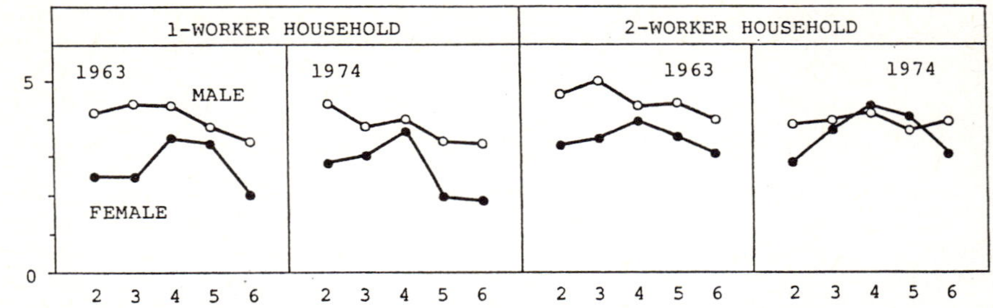

FIGURE 2. AVERAGE NUMBER OF TRIP CHAINS PER PERSON BY NUMBER OF WORKERS, SEX, AND LIFECYCLE STAGE

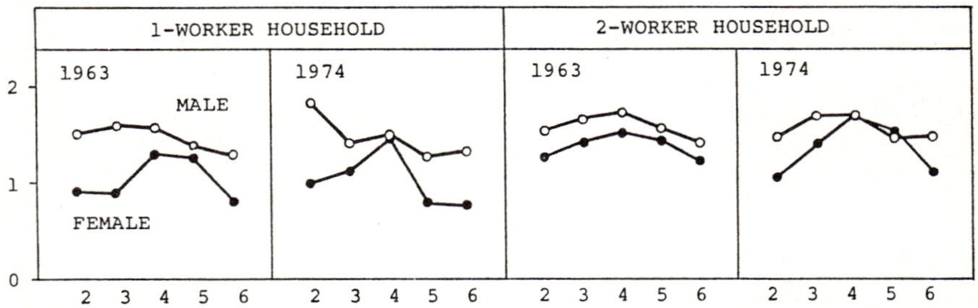

FIGURE 3. AVERAGE NUMBER OF PERSONAL-BUSINESS TRIPS PER PERSON BY NUMBER OF WORKERS, SEX, AND LIFECYCLE STAGE

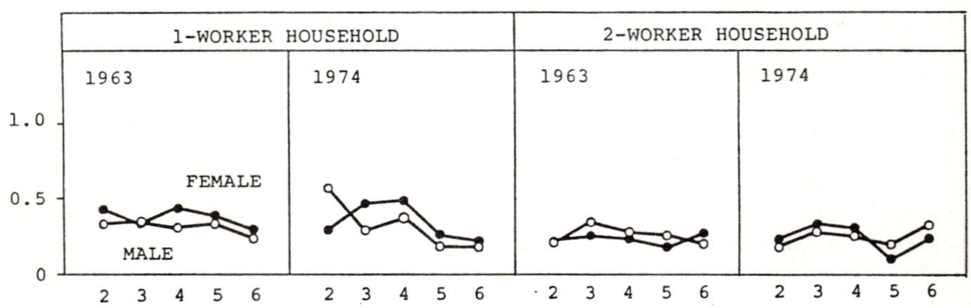

FIGURE 4. AVERAGE NUMBER OF SHOPPING TRIPS PER PERSON BY NUMBER OF WORKERS, SEX, AND LIFECYCLE STAGE

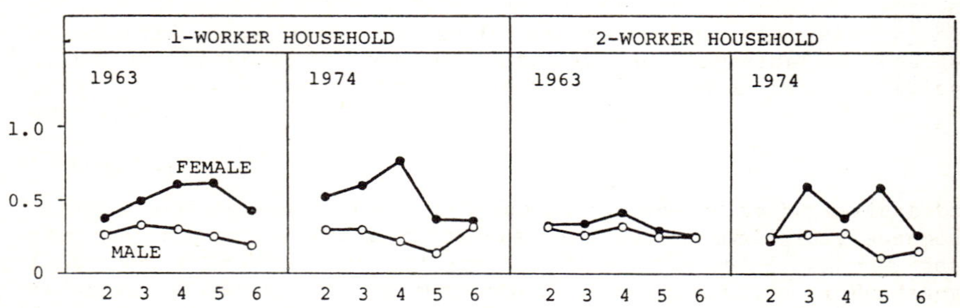

L I F E C Y C L E   S T A G E

FIGURE 5. ACTIVITY ASSIGNMENT BETWEEN MALE AND FEMALE MEMBERS OF HOUSEHOLD

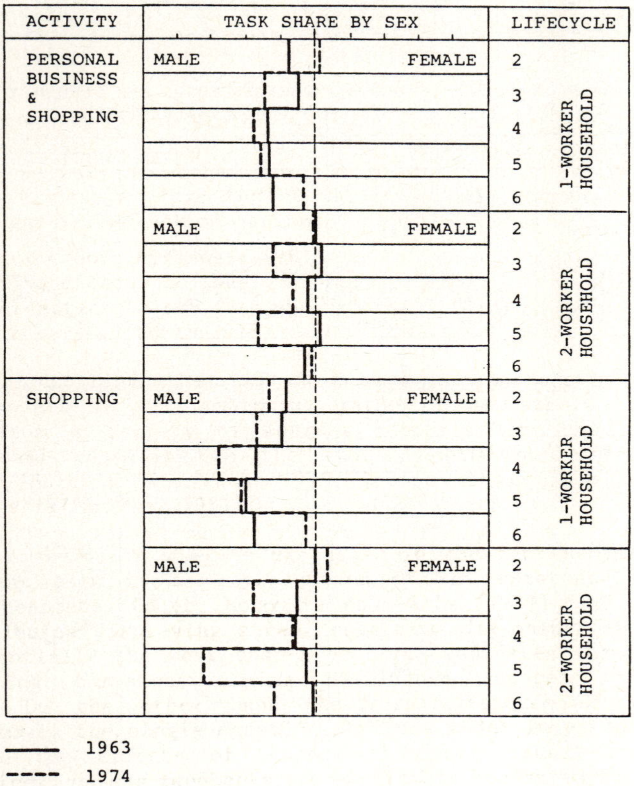

FIGURE 6. AVERAGE NUMBER OF SOCIAL-RECREATION TRIPS PER PERSON BY NUMBER OF WORKERS, SEX, AND LIFECYCLE STAGE

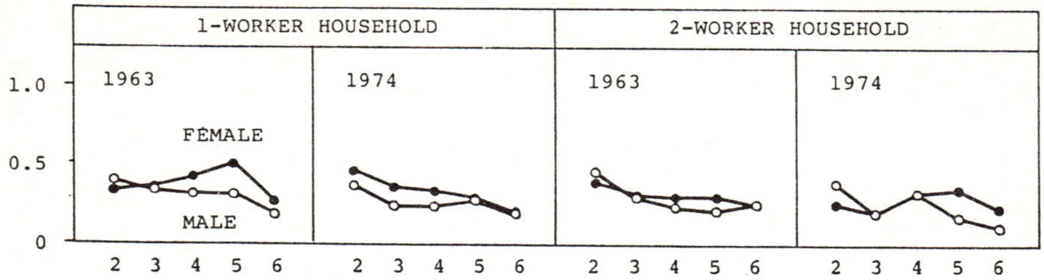

FIGURE 7. AVERAGE NUMBER OF SERVE-PASSENGER TRIPS PER PERSON BY NUMBER OF WORKERS, SEX, AND LIFECYCLE STAGE

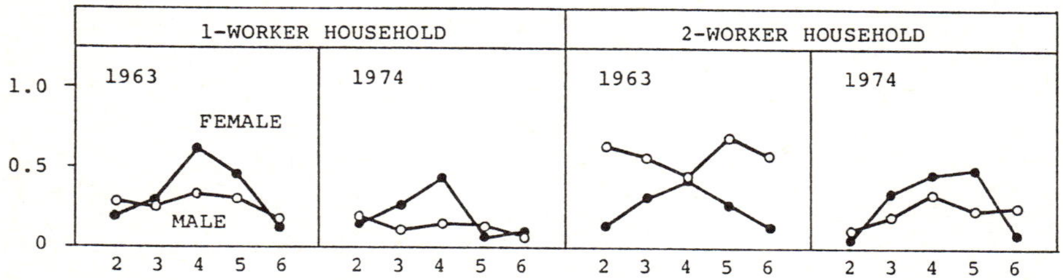

FIGURE 8. AVERAGE TOTAL TRAVEL TIME PER PERSON BY NUMBER OF WORKERS, SEX, AND LIFECYCLE STAGE (IN HOURS)

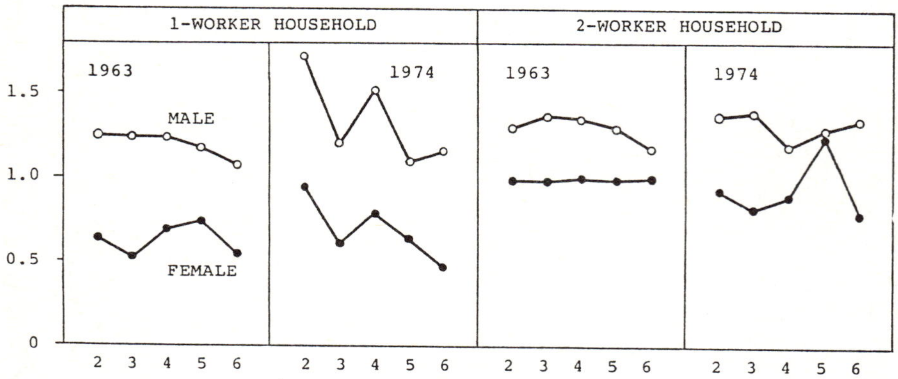

FIGURE 9. AVERAGE TOTAL OUT-OF-HOME TIME PER PERSON BY NUMBER OF WORKERS, SEX AND LIFECYCLE STAGE (IN HOURS)

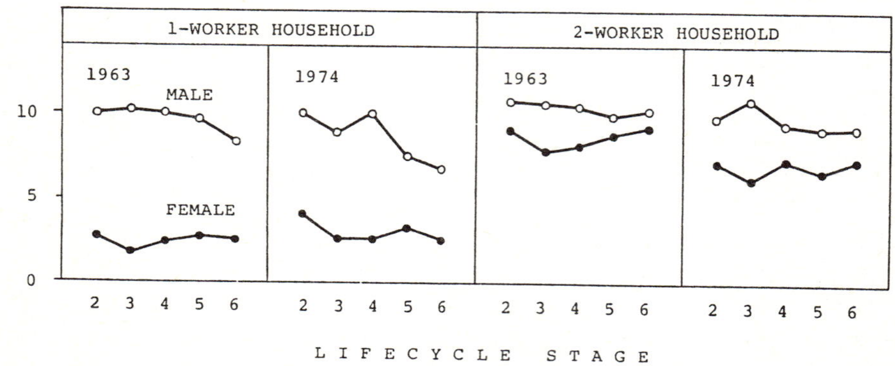

Behavioural Research for Transport Policy
pp. 363–380.

EMPLOYMENT, CHILDCARE AND TRAVEL BEHAVIOR: FRANCE, THE NETHERLANDS, THE UNITED STATES

Charles Raux
Ingenieur C.N.R.S.
Laboratoire d'Economie des Transport
Lyon, France

Dr. Sandra Rosenbloom
Professor of Planning
The University of Texas at Austin

1.0 Introduction

There are significant demographic trends taking place throughout the world which are altering the ways in which families interact with their environment and among themselves. In the years since the Second World War, and indeed in the last decade, the United States and most of Europe have seen an unprecedented increase in the labor force participation of women, and particularly married women with young children. In 1950 the average female labor force participation rate for the 24 OECD countries was 35.7%; it had grown to 44.2% by 1975 with 13 OECD countries, including France and the United States, having rates in excess of 50%. (OECD, 1979)

The growth of female labor force participation is driven by two complementary trends - the increasing activity of married women, and the increasing participation of women with children under six. In the United States from 1970 to 1978 the participation of married women with children five and under, whose spouse was present in the household, increased from 37% to just under 48%. By 1983 almost 38% of all women with spouse present, whose children were under three, were in the labor force. (U.S. Census, 1978)

In 1975 the labor force participation rate of French women having two children, the youngest of whom was under two, was almost 48%; over 65% of all French women having only one child under two were employed. The Netherlands which has had traditionally lower rates of female participation has also seen rapid changes; in 1975 only 15% of married Dutch women with children under six worked but in just four years this figure had increased to almost 26%.

And while women make up a significant percentage of the part-time labor force - 64.8% in the U.S., 82.0% in France and The Netherlands, most women are not employed part-time. In 1977, 87 % of all employed French women, 81% of employed Dutch women, and almost 74% of all employed U.S. women worked full-time. (OECD,1980)

Moreover there are changes taking place in the structure of the traditional household which create even greater incentives for women with children to work. In 1970 there were 5.5

million U.S. households headed by a women alone; that number grew 72% by 1983. Today families headed by women comprise 11.3% of all U.S. families, 9.7% of all French families and 4.3% of all Dutch families. In all three countries female heads of household (with children) are far more likely to be employed than other women; sixty-four percent of U.S. women who have children three to five and 55.5% of those whose youngest child is under three are employed outside the home.

While there is a growing body of inter-disciplinary research which recognizes that individual and family differences generate a variety of household activities which in turn create very different transportation patterns and needs, we have little empirical information on the patterns of these non-traditional households. The research presented in this paper was aimed at filling a significant gap in existing knowledge of the transportation patterns of non-traditional families and their individual members.

This paper presents some of the results of a comparative international small-scale but in-depth study of the travel patterns of a variety of households. The paper focuses on the activity patterns of the increasing number of households with working women in the United States, France and the United States. The paper specifically examines how members of a household organize their travel activities, and particularly their work-related travel, to respond to their family and household responsibilities.

2.0 Policy Relevance

Conventional transport planning approaches implicitly assume either the continued dominance in society of traditional families or of the travel patterns generated by traditional families. Such an assumption ignores the travel pressures created by household and childcare responsibilities, particularly in two worker and single parent households. This oversight restricts the ability of transport policymakers to provide meaningful transport services and facilties appropriate to the changing needs of the population.

Data and findings from a small-scale study such as this can be used by policymakers in several ways. First we can analyze if non-traditional families fall into well-defined market segments. Second we can see if the basic elements used for forecasting travel behavior and assessing transport service modifications are open to serious question.

If the answer to either of these questions is <u>yes</u> - as indeed our data show, we can develop surveys large enough to obtain reliable statistical relationships. If the relationships reported on here were found to be statistically significant they would be of great use in forecasting travel behavior from fairly simple and easy-to-collect socio-demographic indicators of household groups.

In addition, reliable predictions of the travel needs and behaviors of different types of families would allow policymakers to assess the impacts of any proposed policy on specific family groups.

3.0 Study Background

In 1982 a tri-country study of the transportation implications of the striking demographic changes described above was begun by Sandra Rosenbloom with the support of the German Marshall Fund of the United States. With additional support from the University of Texas and the Dutch Ministry of Transport in-depth surveys of families in different life situations were conducted in 1982 in Rotterdam and Austin. Charles Raux undertook a similar study in Lyon in 1984 with the support of the Centre National de La Recherche Scientifique through the Laboratoire d'Economie des Transports of the Universite de Lyon II.

The survey was designed to give initial indications of the differences between and among families in different life situations. Families to be interviewed were categorized by the employment status of the wife in two adult households, the presence or absence of both parents, and the number and age of children. Families were also categorized by the occupation of the head of household in France.

Eight types of traditional and non-traditional families were identified:

Both parents working, the youngest child under six;

Both parents working, the youngest child over six but under eighteen (and living at home);

One parent working, the youngest child under six;

One parent working, the youngest child over six but under eighteen;

Two adult workers, no children;

One adult worker, no children;

Working parent with no partner, the youngest child under 6; and

Working parent with no partner, the youngest child over six but under eighteen.

Both an attitudinal survey and an activity diary were administered to selected households, in the categories identified above. Households were selected for interviews based on a random, quota sample; this method is useful in obtaining sufficient interviews from groups that may not be well represented in either the society as a whole or in a completely random sample.

Ultimately eighty-one families, or 160 individuals were interviewed in Austin; 204 families, including 199 women and

117 men were interviewed in Rotterdam. One hundred-fifty households, including 151 women and 97 men, were interviewed in Lyon. The French data set contains no control families, that is, those without children, and the U.S. data sets contains no single heads-of-household.

Because the U.S. and Dutch data were collected earlier than the French, many of the basic descriptions of the data have already been published (Rosenbloom, 1985, Ruhl, Rosenbloom, and Kremer,1983). Therefore this paper will center largely on the inital French findings as reported by Charles Raux. A brief summary of comparable U.S. and Dutch findings will follow.

4.0 French Data Analysis

4.1 Work Trip and Family Responsibilities

This section of the paper asks:

> How complex are adult work travel patterns?
>
> Which work schedules are preferred by adults in different family situations?

4.11 Combining trips to and from work

Tables 1 and 2 show data on the trip linkages of those adults who report combining their work trips with other trip purposes.

On the home-to-work link, trip-tours are more frequent for two-worker households with young children than for any other group except single heads of household. Adults in this group are equally likely to combine trips; 39% percent of the women and 38% of the men in such families combine their trip to work with some other travel link.

The overwhelming reason for linking the trip to work is to take children to day-care facilities; between three-fourths and ninety percent of adults combining trips in two-worker families with young children are doing so. And, in fact, this is the single most frequently cited trip-link for all respondents who combine trips.

Eighty-six percent of single parents with children under six combine trips from home to work; 84% of those reporting linking trips take their children to childcare on their way to work.

From work-to-home many more adults link their travel with some other trip purpose, and again there are marked differences between the sexes; almost 70% of all women but only 36% of all men link trips from work. Shopping is the major trip purpose for all adults, but adults in two worker families are also making a large number of trips to pick up children at

Table 1

COMBINATION OF HOME TO WORK TRIPS
French Respondents

	Responding Yes		Trip type*										Sample	
			Day-Care school		Shopping		Other Serve-passenger trip		Visits		Leisure			
Family Group	%F	%M	%F	%M	%F	%M	%F	%M	%F	%M	%F	%M	F	M
2 workers - child<6	39	38	89	75	11	25	0	13	0	0	0	0	23	21
2 workers - child>6	23	23	50	50	50	17	0	17	0	17	0	17	26	26
1 worker - child<6	0	22	0	20	0	80	0	0	0	0	0	0	1	23
1 worker - child>6	0	11	0	50	0	50	0	0	0	0	0	0	3	19
Single working parents child<6	86	-	84	-	37	-	5	-	5	-	0	-	22	1
Single working parents child>6	38	-	80	-	40	-	0	-	0	-	10	-	26	0
Total	44	24	80	55	34	41	5	9	5	5	5	5	101	90

* 2 answers accepted

Table 2

COMBINATION OF WORK TO HOME TRIPS
French respondents

	Responding Yes		Trip type*										Sample	
			Day-Care school		Shopping		Other Serve-passenger trip		Visits		Leisure			
Family Group	%F	%M	%F	%M	%F	%M	%F	%M	%F	%M	%F	%M	F	M
2 workers - child<6	65	43	53	67	80	78	0	11	0	22	0	0	23	21
2 workers - child>6	62	31	31	38	94	75	6	13	0	13	0	0	26	26
1 worker - child<6	0	29	0	0	0	86	0	0	0	0	0	14	1	24
1 worker - child>6	67	40	0	0	100	88	0	0	0	13	0	13	3	20
Single working parents child<6	91	-	60	-	70	-	5	-	10	-	0	-	22	1
Single working parents child>6	65	-	40	-	100	-	7	-	7	-	0	-	26	0
Total	69	36	46	30	84	82	4	6	6	12	0	6	101	92

* 2 answers accepted

school or day-care facilities.

It appears that single women with young children have a very complex travel pattern: home-day-care-work-day-care-shopping home.

The presence of children under six seems to create complexity in the work travel of all working adults; on the way to work families roughly equally distribute these trip links between the two workers. On the way from work women bear more of the responsibility for picking up children and for shopping.

4.12 Changing Work Hours

Table Three shows which respondents would like to change their work hours if they could and exactly how they would make those changes. Women in almost all groups are more likely to want to modify their schedules with 41% of single mothers of young children desiring changes. Most women would prefer to start their jobs earlier or have continuous work hours. In data not shown here, women and men report they would like to have more free time; the need for more time for childcare is cited by women.
This preliminary analysis of the influence of employment on travel behavior and childcare confirms that the presence of young children creates growing complexity in the travel patterns of adults in the household, particularly in two worker families. Women display the most complex travel patterns because of their greater responsibility for taking children to school or day-care; not surprisingly they have the greatest wish to change their work schedule even though men in two worker families share more of these responsibilites than men in other households.

4.2 Travel Mode and Transport System Characteristics

This section examines how adults in families in different situations choose their travel mode to work and what consequences potential changes in the transport system would have for them.

4.21 Work-trip Modal Choice

In data not shown here, 91% of all adults said that they had a convenient bus stop less than ten minutes walk form their home. Almost all male respondents said that they had a car for private use, while only 55% of all women did.
Non-salaried women were less likely to have the use of a car (33-44%) while single parents were almost as likely as men (86%) to have access to a car.

The mode choice for work trips are shown in Table 4; the data reflect the relative availability of autos. Seventy-nine percent of all men but only 54% of all women drive to work or combine car and public transit. Working women in intact households actually use a car for their work trip mode <u>less</u>

Table 3

WILLINGESS TO CHANGE WORK HOURS
French respondents

	Responding Yes		Hour Changes*														Sample	
			Earlier		Continued		Variable		Later		Regular, Fixed		Grouped in the week		Other			
Family Group	%F	%M	%F	%M	%F	%M	%F	%M	%F	%M	%F	%M	%F	%M	%F	%M	F	M
2 workers - child<6	35	32	13	29	50	29	0	14	13	0	13	14	13	29	50	14	23	22
2 workers - child>6	35	27	33	43	33	14	0	0	22	14	0	0	33	14	11	29	26	26
1 worker - child<6	0	25	-	17	-	17	-	0	-	17	-	0	-	0	-	50	1	24
1 worker - child>6	33	20	-	25	-	75	-	25	-	0	0	0	-	0	-	0	3	20
Single working parents child<6	41	0	22	-	11	-	11	-	11	-	11	-	11	-	33	-	22	1
Single working parents child>6	15	-	0	-	25	-	25	-	25	0	0	0	25	-	0	-	26	0
Total	31	26	19	29	29	29	6	8	16	8	6	4	19	12	23	25	101	93

* 2 answers accepted

Table 4

TRAVEL MODE TO WORK
French respondents

	Public transport		Car		Car & public transport		Walk		Other		Sample	
Family Group	%F	%M	%F	%M	%F	%M	%F	%M	%F	%M	F	M
2 workers - child<6	30	14	43	59	9	14	9	0	9	14	23	22
2 workers - child>6	27	8	38	69	15	12	15	8	4	4	26	26
1 worker - child<6	-	8	-	73	-	8	-	12	-	0	1	26
1 worker - child>6	-	5	-	76	-	5	-	14	-	0	3	21
Single working parents child<6	18	-	64	-	0	-	14	-	5	-	22	1
Single working parents child>6	27	-	46	-	0	-	27	-	0	-	26	0
Total	26	8	48	69	6	10	16	8	5	4	101	96

often than all women on average; the average is raised by single parents with young children, 64% of whom drive to work.

Walking as a work-trip mode increases as children get older; 9% to 15% for women in two worker households and 16% to 27% for single women.

4.22 Changes in Transportation Characteristics

Respondents were asked their reactions to hypothetical situations; Table 5 shows data on how respondents would react to the doubling of their work-trip costs. Two-thirds of both men and women would do nothing; that is, they would continue to follow their usual travel patterns.

Single women with children over six were the most likely to change their schedules (48%). The women most "resistant" to change were women in two worker households with older children; 68% of these women would change nothing.

Men in two worker households with older children were also the most resistant to changing their schedules; 77% of these men would change nothing while 48% of men in two worker households with young children would change some aspect of their schedule to respond to new travel costs; 34% of them would change their travel mode (eg. go by bus or walk); almost only women would change some aspect of their job.

Table 6 gives data on how respondents would react if their work-trip took one hour longer each day. Here again most men would change nothing while such changes seem to impact women in all groups far more significantly; 58% of all women would change something about their schedule. The women most likely to change their schedules are those in two worker households with young children; most would change something about their job(48%) although few (4%) would stop working.

It is striking that 81% of men in two worker households with young children but only 26% of women in those households would change nothing. This can be perhaps be attributed to the distribution of work and household responsibilities in these families.

Over half of all single parents would change their schedules to respond to increases in travel time; most would move their homes closer to their jobs. It is probably easier for these women to change their home location than any aspect of their jobs, compared to women in couples.

In summary, women do seem the most sensitive to changes in travel time. In two worker families, men have the option of changing either their home or their job while women only seem to have the option of changing their job. Single parents, the group most responsive to travel time changes, appear to have fewer constraints to changing their home than married women, and so act more like male heads of household.

Table 5

DOUBLING OF THE COSTS OF TRANSPORTATION
French Respondents

	Nothing		Change travel mode		Stop or change work		Go by bus		Change home		Ask for compensation		Walk		Other		Sample	
Family Group	%F	%M	%F	%M	%F	%M	%F	%M	%F	%M	%F	%M	%F	%M	%F	%M	F	M
2 workers - child<6	57	52	0	14	10	0	5	10	5	5	10	10	10	10	5	0	21	21
2 workers - child>6	68	77	4	4	8	0	4	12	0	0	4	8	12	0	0	0	25	26
1 worker - child<6	-	58	-	13	-	0	-	8	-	17	-	0	-	0	-	4	1	24
1 worker - child>6	-	80	-	0	-	7	-	0	-	7	-	0	-	0	-	7	2	15
Single working parents child<6	65	-	5	-	0	-	5	-	5	-	0	-	10	-	10	-	20	1
Single working parents child>6	52	-	0	-	4	-	13	-	17	-	9	-	0	-	4	-	23	0
Total	62	66	2	8	5	1	7	8	7	7	5	5	8	2	4	3	92	87

Table 6

TRIP TO WORK ONE HOUR LONGER EACH DAY
French respondents

	Nothing		Change work		Change home		Work part-time		Change travel mode		Stop work		Other		Sample	
Family Group	%F	%M	%F	%M	%F	%M	%F	%M	%F	%M	%F	%M	%F	%M	F	M
2 workers - child<6	26	81	48	0	9	14	9	0	0	0	4	0	4	5	23	21
2 workers - child>6	50	69	31	15	8	12	0	0	0	4	12	0	0	0	26	26
1 worker - child<6	-	52	-	12	-	24	-	0	-	4	-	0	-	8	1	25
1 worker - child>6	-	65	-	10	-	10	-	0	-	0	-	0	-	15	3	20
Single working parents child<6	45	-	5	-	23	-	0	-	14	-	0	-	14	-	22	1
Single working parents child>6	36	-	20	-	40	-	0	-	4	-	0	-	0	-	25	0
Total	42	66	25	10	19	15	2	0	4	2	4	0	4	7	100	93

4.3 Childcare and Adult Responsibilities

This section addresses three specific questions about how adults view their responsibilities to their children and how these responsibilities affect their travel and employment decisions.

4.31 Interaction between Work and Childcare

Table 7 shows respondents answers to the questions of <u>if</u> and <u>how</u> family childcare needs affect employment choices and scheduling. Very few men report a relationship between childcare needs and their employment choices. Fifty-two percent of all women report a positive relationship; the highest response was among working women in two worker households with young children (65%).

Women who do report a positive relationship betwen childcare needs and aspects of their job frequently choose jobs with variable work hours, or take either all day Wednesday or the afternoon off (when most French children have no school). Chosing part-time work is a more common response for both married and single women with children <u>under six</u>; only 7% of married women and none of the single parents with children over six choose part-time employment as a reponse to childcare needs.

Single working mothers seem to have fewer options than married women in two worker households; it appears that single mothers have less financial freedom to chose those options which reduce their income.

Data not shown here indicate that, overall, women seem to arrange their work schedules to fit their childcare needs rather than the opposite. This finding may reflect the fact that women have difficulties in finding childcare arrangements, especially for younger children, and so they first arrange childcare and <u>then</u> find a complementary job. Employment scheduling for mothers of older children may be dictated by the inflexible hours of public schools.

4.32 Children's activities and adult travel patterns

Table 8 displays data on whether respondents make special trips for their children (other than school or daycare), whether they combine trips with their trips to take children somewhere, and whether they postpone trips for times when the children are not around.

Roughly two-thirds of both men and women make special trips for the children other than school or daycare. Non-salaried married women with older children are the most likely to make such special trips; interestingly men in such households are also the most likely of all men to make special trips for children. The impact of the age of the children is not clear; two worker and single parent households with younger children

Table 7

JOB WITH REGARD TO CHILDCARE NEEDS
French respondents

	Responding Yes		How ?										Sample	
			Variable Hours		Dont'work on wednesday		Both		Part-time		Other			
Family Group	%F	%M	%F	%M	%F	%M	%F	%M	%F	%M	%F	%M	F	M
2 workers - child<6	65	9	27	-	13	-	27	-	20	-	13	-	23	22
2 workers - child>6	56	8	43	-	29	-	14	-	7	-	7	-	25	26
1 worker - child<6	-	8	-	-	-	-	-	-	-	-	-	-	1	24
1 worker - child>6	-	6	-	-	-	-	-	-	-	-	-	-	3	17
Single working parents child<6	38	-	38	-	0	-	25	-	25	-	13	-	21	1
Single working parents child>6	50	-	38	-	15	-	23	-	0	-	23	-	26	-
Total	52	9	35	-	16	-	21	-	12	-	16	-	99	90

Table 8

CHILDREN ACTIVITIES AND ADULTS TRAVEL
French respondents

	Taking children somewhere		Combining trips				Postponing trips	
			Yes		Sample			
Family Group	%F	%M	%F	%M	F	M	%F	%M
2 workers - child<6	65	62	33	23	15	13	65	47
2 workers - child>6	58	65	20	24	15	17	35	23
1 worker - child<6	73	69	47	19	19	16	58	36
1 worker - child>6	80	74	50	14	22	14	56	27
Single working parents child<6	76	-	44	-	16	-	59	-
Single working parents child>6	62	-	47	-	15	-	50	-
Total	66	67	41	20	102	60	53	34

are more likely to make special trips than their counterpart families with older children. However the reverse is true of the remaining family groups.

Do adults who take their children somewhere combine these trips with other trip purposes? Forty-one percent of all women but only 20% of all men do so, although these trips are made in roughly equal proportion by men and women in two worker families.

Single parents and non-salaried married women are the most likely to combine such trips while men in one-worker families are the least likely to combine trips. Combining trips may be far more necessary for non-salaried women who have the major role in child development and for single women who have this role and the additional obligation of making a living. Single women, especially, must make optinal use of all travel.

Far more women (53%) than men (34%) postpone trips to times when the children are away. Men and women in households with young children are slightly more likely to postpone trips, although the differences are only really obvious for married working women; 65% of all such women with young children postpone trips while only 35% of married working women with older children do so.

4.4 French Conclusions

The results of the French survey do validate the primary study hypotheses and show that three major factors combine to influence the activity and travel patterns of adults in the French socio-economic and cultural context;

> the age of the children in the household;
>
> the sex of the adult;
>
> the number and employment status of adults in a household.

Younger children appear to pose constaints on adult activities in several ways; for working mothers with young children there is a significant relationship between both childcare methods and work trip travel mode. Such women display more complex travel patterns, more often linking trips with both their work and non-work trips to accomodate children's needs. Moreover working women with young children seem very sensitive to time constraints, generally modifying their employment schedules to meet their childcare needs.

Women seem more constrained than comparable men in their travel choices. Non-salaried women specialize in taking children somewhere. Working women simultaneously perform several activities; their activity patterns are the most sensitive to time and other constraints and their travel mode depends strongly on their childcare needs. When both parents work there is some sharing of childcare responsibilities.

Being a single parent poses a number of constraints since it is not possible for such a working women to share household and childcare responsibilities. The economic and temporal constraints faced by these women lead to very complex travel patterns.

5.0 Dutch Analyses

The following sections will summarize analyses comparable to the French issues reported on in the first section of this paper; fuller details are available in Rosenbloom, 1985.

5.1 Combining Trips to and from Work

As in the French survey, Dutch working women in two worker and in single parent households were the most likely to report combining trips <u>to</u> work with other trips. Women with children under six were the group most likely to combine trips; 43% of all women in this group reported trip-chaining behavior. Working women in two parent households with children of any age were the next most likely to report combining trips; between 23 and 28% of these women answered affirmatively.

Men in two worker households with children were the most likely to report combining trips; 12-15% of men in these households combined their trip <u>to</u> work with some other trip purpose. Unlike the French study, however, dropping children at daycare facilities was not even a minor trip-purpose on the trip linked to work. Further analysis of the type of childcare used by working families in the three countries is needed to understand these differences.

As in the French study, a much larger number of those questioned combined trips <u>from</u> work with other trips; overall, 48% of all women but only 16% of all men combined work to home trips with other trips. While fewer Dutch of either sex link trips than their French counterparts, the division between men and women is roughly the same.

The women most likely to combine trips were working women either with or without partners whose youngest child was over six. Seventy percent of single parents and 68% of women with partners with children over six reported combining trips.

Although men did not combine trips to the extent that women did, men with working wives and with children under six were the most likely to report trip-chaining behavior. Twenty-four percent of men with working wives and with children under six and 12% of men similarly situated but with older children reported such behavior.

Unlike the French study, shopping was the single largest trip purpose of both men and women who did link trips from work. Ninety-three percent of all women and 77% of all men who reported a trip purpose reported linking shopping with their work trip.

The Dutch results are consistent with the French findings although there are differences in the actual choices made by families in the two countries. Overall, families with two workers and young children have more complicated activity patterns than one-worker families or those with older children; these activity patterns and the constraints faced by individual members generate more complex travel behavior.

5.2 Changing Work Hours

Dutch respondents who said they would change their work hours were asked why they would do so. Differences between men and women are strong; 52% of all women but only 17% of all men said that they would change their job schedule to better meet family obligations. Most men, 32%, said that they would change their schedule so that other business and personal activities could be more easily accomplished. Roughly equal proportions of men (13%) and women (16%) said that they would change their job schedule to make more free time during the week.

There are differences between and among groups of men and women with regard to this question complement the French study. Working women with small children, with or without partners, were the most likely to want to change their work schedule because of family obligations; 100% of all fulltime working women with partners and 67% of single parents with young children wanted to change for family reasons.

Forty percent of working women with partners and 83% of working women without partners both with children over six wanted to change their job schedules to meet family obligations.

While most men did not advance family reasons for changing job schedules, there are pronounced differences between men with working wives and small children and those without. Fifty percent of those men with working wives and children under six who wanted to change their work schedule wanted to do so to meet family obligations. Twenty-five percent of similarly situated men with slightly older children wanted to change their employment hours for family reasons.

5.3 Work-Trips and Childcare

Most Dutch adults did not report a relationship between their work-trip and their childcare responsibilities; only 18% of fulltime working women and 4% of fulltime working men did. And most of those that answered in the affirmative did not provide a reason for that relationship.

Of those reporting both a relationship between worktrip mode and childcare responsibilities and the reason for that relationship, 67% of the women but only 25% of the men reported chosing their worktrip mode to allow them to drop-off children at school. The few men that reported a reason said that they had chosen the worktrip mode which they had to save time to be with their children (25%) and for other, unspecified purposes (50%).

5.4 Dutch Summary

As in the French study the variables that seem to have the most impact on adult travel behavior are sex of the adult, age of the children, and the employment status of the wife. Working mothers display the most complicated travel patterns.

6.0 Summary of U.S. Findings

6.1 Combining Trips to and from Work

Almost double the number of U.S. residents link trips <u>to</u> work than do Dutch workers; the U.S. figure is more comparable to the French. Forty-seven percent of all women and 33% of all men report linking trips to work with other purposes.

There are interesting differences in the kinds of trips which are linked. Sixty-one percent of all men who link their trip to work do so for <u>recreational</u> purposes. Men without working wives were the most likely to report recreational trip links; 100% of men without working wives who reported trip chaining behavior reported this trip linkage even if they had children.

Most women (57%) link their trip to work with recurring medical trips; these percentages also hold for married working women with children of all ages. It is striking that 75% of women with no children link work trips with recurring medical trips. Men report only sporadic medical-work trip linkages; this trip purpose is the second largest trip link made by all men as a group.

Equal percentages of men and women combine shopping trips with their trips to work - 8%, while 21% of all women report combining trips to banks, and administrative offices (personal business) with their trips to work. Perhaps most striking is the fact that 13% of all working women with children under six link some sort of volunteer or charity activity with their trip to work.

Like the Dutch data but unlike the French, U.S. respondents <u>do not</u> make trips to take children to care facilities; again this may be a function of differences in the childcare options available in the three societies.

Many more Austin respondents link trips on the return home trip than on the original trip to work and many more thanboth French and Dutch respondents. Eighty-five percent of all women and 73% of all men link their trip from work with some other purpose. Shopping and recurring medical trips are the most common trip linkages for both men and women.

Working women are the most likely to make either shopping or personal business trips on the way home from work and they are more likely to make these trips than comparably situated men. Fifty-three percent of working women with children under six but only 31% of comparably situated men make shopping trips, although only the men in this group link personal business trips with trips from work.

Working women with older children are more likely to make shopping and personal business trips than comparably situated men; 70% of working women with children over six link shopping trips, and 22% link personal business trips with their trips home from work. Fifty percent of all men with working wives who reported linking trips linked shopping trips to home trips; only 9% of this group made personal business trip linkages.

6.2 Changing Work Hours

Thirty-five percent of all women and forty percent of all men want to change their work schedules although variance among the family groups is high. Men in households with young children were by far the most likely to want to change their schedules: 75% of such men with salaried wives and 50% of those with non-salaried wives. Strikingly none of the women in households with young children wanted to change their schedules whether they had salaried employment or not.

The change most demanded was a flexible work day; exactly the same number of men and women -65%, wanted that change. Of those that specific specific hours, more men and women wanted to arrive later than either arriving or leaving later.

The largest number of both men and women would change their schedules to reduce their childcare costs; 94% of all women and 58% of all men gave childcare cost problems as the reason for changing their schedules.

These data suggest that while women retain most of the childcare responsibility men are advancing some willingness to change their schedules to reduce the cost of caring for their children while the wife works. One-quarter of both men and women wanted to change their job schedule to better match store opening and closing hours and for personal convenience.

7.0 Conclusions

The data from all three countries give overwhelming support to the hypotheses that non-traditional families have activity and travel patterns different from the traditional families on which we currently base transportation planning methodologies and ultimately policymaking. Dutch, French, and American data show pronounced differences between women and comparably situated men, and among groups of women. Women in all groups seem to be more directed in their activities by the family and childcare responsibilities they accept, with those having young children showing the most complex travel adjustments. Family and household responsibilities seem to be strong motivators of women's travel and employment behavior.

There are important planning and policy implications to be drawn from both these findings and from the basic empirical data on non-traditional families which we have developed. Planners can use such empirical data to link identifiable transportation patterns and needs to specific non-traditional households, and to their individual members. This in turn gives planners the ability to better predict the dimensions and direction of future household trip-making and to better assess traveller response to both government transport policies and to modifications in transport services.

Policymakers must be prepared to respond to the different needs of the growing number of non-traditional families in all developed societies. Among the issues which policymakers can address with the information we present here are how the structuring and restructing of employment opportunities could facilitate the employment of women, how childcare facilities and policies could be made more responsive to the constraints of two-worker families, and how transportation policies could be more sensitive to the needs of the rapidly growing number of single-parent households.

ACKNOWLEDGEMENTS

Both authors wish to thank Constance Rottlander-Myer and Judith Kremer of the Instituut voor Social Beleidsonderzoek of the Hague for their assistance in the administration of the Dutch study and for their guidance in formulating the survey instrument in Lyon. Both authors gratefully acknowledge the assistance of their funding sources.

REFERENCES

Rosenbloom, Sandra, "The Growth of Non-Traditional Families; A Challenge to Traditional Planning Approahces," in G. Jansen, R.M. Gijsbertus, et al. (ed.), Transportation and Mobility in an Era of Transition, Amsterdam: Elsevier North Holland, forthcoming.

Ruhl, Aad, Rosenbloom, Sandra, and Kremer, Judith, "The Influence of the Combination of Employment and Childcare Needs on Transport," in the Proceedings, PTRC Summer Annual Meetings, University of Sussex, England, July 1983.

Raux, Charles, "Les relations entre le travail, les charges familiales et les transports dans le contexte francais", Etudes et Recherches, Laboratorie d'Economie des Transports, Universite' Lyon II, France, forthcoming.

7
Measurement of users' valuations

Behavioural Research for Transport Policy
pp. 383–403.

Utilization of Attitudinal Measurement Techniques to Analyze Demand for Transportation: Methods, Applications and New Directions

Julian Benjamin, Ph.D.
Economics Department
North Carolina A&T State University
Greensboro, North Carolina 27411

ABSTRACT

Over the past decade there has been substantial progress in the development and application of techniques which measure and analyze consumer attitudes. These techniques have varied from initial efforts at unidimensional evaluation of existing facilities to multidimensional scaling of consumer preferences to complex decomposition techniques which estimate travel demand.

The paper classifies uses of a wide range of attitude measurement and analysis techniques according to policy, planning, operations or facilities design, reviews in detail the full spectrum of available techniques, and then presents a match of recommended techniques for each application.

The paper pinpoints gaps in methodology as well as applications. Recommendations are based on continuing research by the author along with research reported in the literature. The paper relies on comparisons of practical aspects of each methodology, includes examples of questionnaires, and refers to analytical results in selected references. Explanations of algorithms are intuitive and conceptual rather than mathematical.

The paper gives a comprehensive, realistic and when necessary, critical view of the appropriate use of these techniques.

Attitudinal measures have been used to understand the motivations and behaviors in replicable studies of individuals for whom transport systems are planned. Attitudes also provide demand estimates in cases where econometric extrapolations are invalid because of large system changes. This paper surveys both uses of attitudinal measurements and discusses transport management and planning applications.

WHY MEASURE CONSUMER ATTITUDES?

Human behavior results from the emotions and feelings that we all have. Attitudes represent those feelings, specifically as they relate to the elements of our environment. It is the relationship of attitudes and behavior that has led researchers to investigate attitudes in the hopes of improving forecasting techniques.

The Development of Attitudinal Concepts

Psychologists (e.g. Anderson, 1974 and 1977), economists (e.g. Lancaster, 1966), and market researchers (e.g. Engle, Kollat, and Blackwell, 1973, Green and Wind, 1973, and Wind, 1982) have long recognized that the emotional appeal of objects and the personal values of individuals to a great extent influence our behaviors. The formation of emotions are complex and often not rational including subconscious adaptive reactions to the symbolic value of events and objects (Valiant, 1977). Psychologists have considered methods of measuring attitudes since Thurstone (Torgerson, 1958).

Early transportation attitudinal studies assessed satisfactions with transport services (e.g. McMillan and Assael, 1968). These studies were followed by evaluations of consumer viewpoints toward transit attributes in the hope of unravelling the complexity of attitudes and of developing market segments of consumers with similar viewpoints (e.g. Dobson and Nicolaidis, 1974, Golob and Nicolaidis, 1974, and Stopher, 1977). Subsequently, relationships were investigated between behavior (mode choice) and attitudes in the hope of establishing a better set of predictors of mode choice (e.g. Recker and Golob, 1976 and Recker and Stevens, 1977). Finally, techniques were refined for assessing preferences for and forecasting utilization of new means of transport (e.g. Lerman and Louviere, 1979, Louviere, et al, 1981, Louviere, 1981, Benjamin and Sen, 1982). Most recent work has focused on incorporating techniques in the decision-making process of various agencies (e.g. Benjamin, 1983).

A Conceptual Framework of Attitude Formation

Attitudes are the set of motivations, emotions and perceptions held by a person about his world. There are complex processes by which attitudes are formed. Green and Wind (1974) outline a conceptual framework of attitude formation. Based on that, a transport decision model was suggested by Benjamin and Sen (1982b) and is presented in Figure 1. Briefly, the background of any consumer provides the basis for the formation of attitudes towards transport. Preferences are mitigated by financial and environmental constraints. Attitudes, however, are not unchangeable. Experience and changing values affect attitudes, and attitudes and preferences are therefore influenced by behaviors.

Cross-sectional analyses by Recker and Golob (1976), Recker and Stevens (1977), and Koppelman and Pas (1980) establish the association of attitudes as predictors of behavior. Longitudinal work by Tischer and Phillips (1979), Louviere, et al (1982), and Benjamin and Sen (1982a) confirm causal relationships. However, work by Horowitz (1978) and more recent work by Lyon (1984) confirm that causal relationships between the formation of attitudes and behaviors can occur in either direction. For example, a man decides to ride the bus because it seems convenient (attitudes cause change in behavior). After riding for a while, he finds it is not convenient (behavior causes changes in attitudes) and he finally decides to travel by car (attitudes change behaviors). This complex set of evaluative steps becomes a learning process by which attitudes, values, income constraints, preferences, and finally behaviors influence each other. It is, however, convenient from a decision-making viewpoint to study influences of environment on attitudes and then attitudes on behaviors since transport system variables are generally under the control of the decision-maker.

APPLICATIONS OF ATTITUDINAL MEASURES

Attitudinal measures have application in four major areas: 1) assessment of community goals and objectives to assist in establishing community policies; 2) assessment of consumer viewpoints and the formation of consumer groups with similar viewpoints to assist in planning services or facilities; 3) forecasts of demand for consumer products or services to help evaluate alternative plans; and 4) evaluation of existing services to help monitor transport systems performance.

Assessment of Community Goals

Studies of community goals and objectives assist planners and policy makers in setting goals that are congruent with community wants and perceived needs. Studies of the opinions of residents are particularly useful in differentiating the objective needs of the community from the pressures of the variety of political interest groups which frequently dominate the decision processes of local government agencies. On a practical level, the weight of evidence of community backing has the potential of providing powerful leverage for planners and managers with policy makers (e.g. Benjamin, 1983).

Assessment of Consumer Preferences

In developing transportation plans, it is necessary to know what factors influence the views of consumers and whether there exist consumer groups with distinct viewpoints and travel requirements. Identification of key decision factors and groups with similar viewpoints aids the decision-maker in proposing a set of different services, each tailored to the travel requirements of a group that was so identified.

Assessment of consumer preferences and identification of groups with similar viewpoints to form market segments revolve first around a valid and reliable means of assessing preferences and then an open-ended approach to develop a posteriori groups. There are numerous examples where consumer preferences for services, service attributes or satisfaction with service characteristics have been assessed and similar groups were then clustered together (e.g. Dobson and Nicolaidis, 1974, Dobson and Tischer, 1978 and Benjamin and Sen, 1981). Group membership was related to background variables by using statistical techniques. This identification of groups makes it easier to assess the characteristics of group members and to tailor services to their needs. The practical outcome of these approaches was to complement the general trend in the transit industry to investigate multi-service mixtures. Identification was made of consumer groups who were transportation disadvantaged, carless, or who had other similar travel requirements. The results coincided with a redefinition of the role of the transit manager more as a broker, matching providers to consumer groups. More recently, the role of the manager or planner has evolved into a wider role affecting broader community policies of land use and zoning regulations to help meet transportation goals.

Preference Analysis and Demand Forecasting

Design of transport services, facilities or analysis of alternative plans requires a more detailed analysis of consumer preferences for specific levels of factors. For example, instead of determining if fare is important in choosing a means of transportation, questions are asked about consumer response to a fare increase from $.50 to $1.00 when an express service is introduced. If the decision-maker is interested in designing a service or facility, preference information is sufficient.

Information about consumer preferences will indicate which set of attributes is most attractive to the consumer group for which the service is intended. However, plans that rely on revenues require a forecast of utilization for proposed new services.

Assessment of preferences for specific factor levels, forecasts of market shares or demand for services requires more exacting techniques. Initial attempts at forecasting related mode choice to attitudes. Logit models were used with attitudinal predictors. It was found that these variables were expensive and provided little additional predictive power (Recker and Golob, 1976) beyond system variables. Further, it was found that prediction of individual behaviors was even more difficult. Finally, under the introduction of new modes or service changes, it is difficult to forecast the attitudes which predict behaviors.

In response to those more exacting needs, a set of experimental techniques has been developed. Included are trade-off analysis (Johnson, 1974), conjoint measurement (Green and Wind, 1973), the information integration approach (Anderson, 1974), direct utility assessment (Louviere, et al, 1981), and experimental choice modelling (Louviere and Hensher, 1982). In each case, the subject is required to indicate his preference or intended utilization of a mode or service option. These hypothetical choices provide two sets of information. First, they can be decomposed to find what factors and levels of factors are most important in forming preferences. Second, market simulations can be made to forecast market response to proposed changes, new services, or modes. A discussion of market simulation techniques is provided by Neveu and Koeppel (1978), Kocur, et al (1982) and Louviere and Hensher (1983). Experimental techniques have been shown to give similar results to econometric forecasts (Louviere, et al, 1981) and are particularly useful when major changes or new modes are introduced and econometric extrapolations are not valid.

Applications of experimental techniques have been made to mode choice (Meyer, et al, 1981, Benjamin and Sen, 1982, Kocur, et al, 1982), residential locational choice (Benjamin, 1977), destination choice (Louviere and Hensher, 1983, and Louviere and Meyer, 1980), electric vehicle design (Kidder and Benjamin, 1982), carpooling (Levin and Gray, 1982), paratransit services (Louviere and Kocur, 1982), demand for air travel (Hensher and Louviere, 1983), and design of services for the elderly and handicapped (Benjamin, 1984).

Service Evaluation

The final application is evaluation of existing services. This also includes evaluation of pseudo-experiments under which new services are introduced or old services are changed. Substantial work has been carried out in the evaluation of the effects of demonstration projects (e.g. Heaton, McCull and Waksman, 1976). Evaluation of ongoing service, although not new, has been carried out as a regular part of the decision process very rarely. Benjamin (1983) illustrates this application and highlights many of the difficulties in using attitudinal information in conjunction with system and environmental information to make planning decisions. Figure 2 illustrates the configuration of a computer information system which was designed to combine this variety of information in a decision-making format.

A REVIEW OF ATTITUDINAL TECHNIQUES

For the purposes of this paper, all techniques are classified into three basic methods of measuring attitudes that have been employed to date. They are: unidimensional scaling techniques, multidimensional scaling techniques and experimental techniques. Within each category, there are differences among specific procedures and their application. Techniques vary with respect to their ability to analyze responses to either simple or complex stimuli, procedures for collecting data, assumptions concerning the integrity of data as either rank order, categorical or interval data and the ability of techniques to forecast behavior or to analyze belief constructs. In each case, a brief discussion of these assumptions and descriptions will be made. More complete discussions of each technique will be found in suggested references.

Unidimensional Scaling Techniques

These are techniques that compare stimuli on one single scale at a time. Either simple or complex stimuli can be compared. An example would be to compare preference for different modes of transportation. One example of this approach is Thurstone scaling. Input to Thurstone scales are either pared comparisons or rankings. Outputs provide the relative preference on an interval scale. Individual characteristics or attributes (simple stimuli) can be analyzed using comparisons based on the importance of each attribute in selecting a mode or some other relevant scale. Transport services (complex stimuli) can be compared in a similar manner using preference as a criteria. Output again is the relative standing of attributes on an interval scale. A good discussion of this technique is found in Green and Tull (1978).

There are other ways of obtaining information about items one attribute at a time. These include categorical indices such as satisfaction with or importance of transit attributes. There is a body of empirical evidence that indicates that these categorical responses can be analyzed as though they are interval scaled so that both input and output provide interval measures using conventional statistical measures such as the mean and standard deviation (Green and Wind, 1973, Benjamin, 1977).

Another use of categorical variables is the systematic evaluation of user values. Questions require subjects to agree or disagree with statements that relate to basic beliefs such as "I like to travel with people just like myself." Responses are useful in setting public policy.

Semantic differential questions are also categorical and elicit attitudes toward transport on bi-polar attributes. These enable direct comparisons of perceived quality of service such as expensive service versus inexpensive service. The advantage of these scales is that they relate directly to attributes of the service without relating the evaluations to consumer expectations as do satisfactions questions.

Overall, unidimensional questions are easiest to use and have had the widest application. During the period of time around 1980, they were used as part of many transit development plans. In developing a consumer information system, Benjamin (1983) found that simple measures of satisfaction could be best understood and, therefore best utilized. Although the measurement of unidimensional scales is not state of the art, they have high potential for utilization in planners' and managers' decision-making processes on a regular basis.

Multidimensional Scaling Techniques

Multidimensional scaling techniques analyze attitudes toward stimuli in several dimensions simultaneously. The most widely used of these techniques is principal components analysis. The techniques are applicable to both simple and complex stimuli. Although they have been proven as a research tool for both market research as well as transportation analysis, aside from principal components analysis, they have found little practical application. This is no doubt due to their complexity and their need for sophisticated special-purpose programs.

Principal components analysis is well known. It is generally used to reduce a large set of unidimensional attribute ratings to a smaller set of underlying unrelated factors. Even this obviously useful application is not widely used outside of research studies.

Carrol (1972) describes a set of scaling techniques that permit the analysis of preferences for different modes of transportation (e.g. Dobson and Tischer, 1976 and Dobson and Tischer, 1977), evaluations of different modal attributes (e.g. Dobson, Golob and Gustafson, 1974 and Benjamin and Sen, 1981) and consideration of individual differences in evaluation procedures by individual subjects (e.g. Stopher, 1977). These techniques are useful for both identifying underlying decision variables as well as grouping individuals according to their viewpoints. One advantage of multidimensional preference analysis is that existing services can be compared without the development of experimental designs. In fact, multidimensional preference mapping procedures are closely related to the experimental procedures that are discussed in the next section of this paper. However, despite their utility in research, applications to routine decisions seem remote.

Experimental Techniques

Perhaps the techniques with the most untapped potential are experimental techniques. These include a variety of different techniques in which consumers reveal their preferences for existing and proposed new services by performing hypothetical choices or indicating preferences for services or modes under controlled conditions. Included here are multi-dimensional preference analysis (unfolding analysis), tradeoff analysis, conjoint measurement, direct utility assessment, and experimental choice modelling. The effectiveness and wide applicability to transport problems of these techniques have been illustrated for unfolding analysis (Benjamin and Paaswell, 1977 and 1981), tradeoff analysis (Koeppel, et al, 1979), conjoint measurement (Benjamin and Sen, 1982a), direct utility assessment (e.g. Louviere, et al, 1981, Kocur, et al, 1982, and Meyer, et al, 1978) and experimental choice (e.g. Louviere, 1981, Hensher and Louviere, 1983).

The techniques all assume that subjects are capable of making comparisons and indicating preferences under controlled hypothetical conditions. There is evidence that these hypothetical choices are closely related to the behaviors exhibited by individuals in actuality (Lerman and Louviere, 1979, Benjamin and Sen, 1982, and Louviere, et al, 1982).

Examples of surveys designed for handicapped paratransit services and mode choice are presented in Figure 3 and Figure 4. The survey in Figure 3 was administered to handicapped residents of Greensboro, North Carolina. The major objective of the survey was to assess the preferences of persons with different handicaps for dial-a-ride services with various levels of round trip cost, lead time, and scheduled availability. Results are summarized in Benjamin (1984).

Figure 4 was designed to assess the importance of potential factors that influence the choice between private automobile and carpooling during a gasoline crisis. The questionnaire is designed to measure binary choices as part of a more complex multiple choice questionnaire. Results are capable of direct application in market forecasting models and are summarized by Kocur, et al (1982).

Direct utility assessment has shown promise because of the ease of obtaining assessments from subjects by using intended usage as the criterion variable and the ability to use standard statistical techniques such as regression analysis to analyze results. The extension of these methods to choice models also uses techniques familiar to planners.

One of the results of each analysis are a set of part-utilities, which, when combined according to some utility function (usually linear additive or polynomial) indicate the utility for any of the hypothetical cases under study. The analysis may be cross-sectional, in which case average part-utilities are found for the entire sample. On the other hand, analysis may be individual resulting in different part-utilities for each individual. This permits clustering and market segmentation according to service preference.

Another use of these techniques is for forecasting. Neveu and Koeppel (1978) provide a good discussion of the elements to be considered in using market simulation techniques to develop forecasts. Essentially, total utilities are calculated for a series of competing modes or services for each subject. The mode or service with the highest utility is considered the mode or service of choice. Other variations of these methods consider the strength of utility in each choice. Benjamin and Sen (1982a) found that full factor presentations such as those utilized for direct utility assessment provide the best forecasts. Louviere (1981) extended these methods by introducing choice as a criterion variable and adapting cross-sectional econometric forecasting models directly to the experimental format. Forecasts are developed from choice models as part of the UTPS process, and market shares and elasticities are obtained directly from the models.

One major difficulty with these approaches is the total number of stimuli that can be constructed by permitting all factors and levels. For example, three factors of four levels each result in $4^3 = 64$ possible stimuli. This is an impossible task to incorporate in any questionnaire. There are two ways to reduce these tasks. First, orthoganal experimental designs make it possible to greatly reduce the required number of stimuli for a set number of factors and levels. Second, the number of factors in any comparison can be reduced by using block designs (using comparable sets of smaller factors). The smallest blocks are two-factor pairs which is the identifying characteristic of tradeoff analysis. Winer (1971) explains experimental design in detail. Kocur, et al (1982) present a good set of applications to survey research techniques.

In sum, the development of experimental techniques started with the use of hypothetical choices in paired comparisons of complex stimuli in multidimensional scaling. In this case, stimuli were not necessarily selected systematically. Conjoint measurement introduced complex stimuli that were selected systematically to contrast factors and levels of factors. At the same time, it was found that assumptions of transitivity were reasonable and that ranking of stimuli provided sufficient information to perform an analysis. The techniques used for conjoint measurement were related to regression analysis and analysis of variance except for the rank-order integrity of the criterion variable. Special-purpose algorithms were developed to perform this task. Tradeoff analysis was introduced as a block design to simplify the tasks of the respondents. It was then found that the subject ratings used in information integration theory could be adapted to assess subject utilities making possible the use of the more familiar statistical methods such as regression analysis and ANOVA which require interval criterion variables and provide a basis for testing functional forms of utility functions. It was further found that cross-sectional analysis was possible directly with this data. This is used in functional analysis (direct utility assessment). Finally, overlapping experimental designs, cross-sectional random choice econometric models (the logit model) and randomized factors and factor levels were incorporated together in experimental choice models. Simulated choices are easier to obtain and interpret and make possible the integration of experimental results directly into the most familiar forecasting process (the UTPS process). This development has taken place over the last decade. All of these techniques assume utility maximizing behavior and all require some choice or rating under hypothetical conditions.

When first conceived, the experimental methods were recognized as having potential if they could be demonstrated to be cost effective. As a result of extensive research over a decade, techniques have been developed which do not depend on special purpose algorithms, manuals have been written to assist in questionnaire design and forecasts are possible directly from choice models. Despite these advances, there is still a lack of general understanding by planners and managers of the usefulness of these techniques as part of the planning process. It is the experience of this researcher that many planners and most management personnel do not have the technical background sufficient to allow for an understanding of techniques that differ from accepted practice. Few managers are competent statisticians and few planners have had courses in design of experiments. Further, these techniques are presented in modern planning texts only as a footnote. Despite advances in technique, the need for technical understanding of what are still a set of complex procedures will continue to hamper their regular use.

Suitability of Techniques for Different Applications

The question of which technique is most suitable for an application depends to a great extent on the match between the technique, the decision to be made, and the background of the decision maker. The applicability of a technique depends on the use and how detailed and precise the information needs are, the cost of the technique including all aspects of implementation from survey design and sampling procedure to analysis, the availability of analytical procedures and the congruence with existing decision processes. Table 1 analyzes each major technique classification according to these categories. Entries in Table 1 are based on the experience of the author and the synthesis of the literature presented in this paper.

NEW DIRECTIONS FOR ATTITUDES

There are three areas of potential investigation: 1) basic theoretical models of attitude formation and behavior; 2) methods of assessing or measuring attitudes, preferences or behavioral intentions; 3) application and incorporation of attitudinal measures into all levels of the planning and decision processes.

New Directions for Theoretical Models

The purpose of theoretical models is to gain a more thorough understanding of how people make decisions. Despite the emphasis in recent years on refining methodologies for analyzing behavior intentions (e.g. conjoint analysis and experimental choice) further investigation is needed to find how and why decisions are made. There are five major gaps in theoretical models: first is the relation of basic psychological structures, values, and personality in formation of preferences. Although basic work on structural models has been completed by Charles Rivers Associates (1977), Koppelman and Lyon (1981), and Lyon (1984), more detailed work is needed especially in light of newly refined experimental methodologies.

The second area that is not clearly understood is the dynamic nature of attitude formation. Although it is most convenient for planning applications to think of attitudes as predictors of behavior, this is really a dynamic process. Despite the pioneering work by Lyon (1984), analytical tools need to be refined and data sets gathered which take advantage of experimental procedures and more thoroughly relate personality constructs to intended and observed behaviors.

The third area is to relate recent work in activity analysis to travel preferences. The relation of preferences to different activity patterns is not understood. Clearly, preferences for mode are related to trip purpose (e.g. Benjamin, 1984), but most work has been done for mode choice for the journey to work. In fact, one may measure behaviors (i.e. activity patterns) to gain insights into personality constructs.

The fourth area in need of further investigation is the interpersonal relationships that occur within a group or particularly a family in deciding on the various dimensions of travel. Investigations are usually of one subject per household. However, questions arise as to the roles of family members in making decisions. For example, Benjamin (1977) found that husbands and wives had distinctly different preferences and, in general, dominated on different dimensions in experimental choice of residential location. The implications of these results for household surveys relating to all travel dimensions must be investigated.

Finally, although family life cycle has been shown to be a predictor of travel behavior, little attention has been given to personal life cycle. To what extent do maturity and adaptability relate to travel preferences?

There is still much to be understood before a thorough conceptual model can be formed of total travel behavior. Implications of these findings for modelling methods will yet be seen.

New Directions for Assessing Attitudes

Over the last few years, much work has been spent on refining experimental techniques. There are three major areas that should be investigated as part of this process.

First, virtually all methods assume utility maximization. Combs (1955), however, lists many ways that choices may be made. Lexicographic and satisficing models both do not assume tradeoffs and are not necessarily utility maximizing. Benjamin (1979) demonstrates that polynomial models are not powerful enough to distinguish between decision processes that include tradeoffs and those that do not. Therefore, there is a need for techniques that test choice processes.

The second area is the dynamic nature of attitude formation. What influences the formation of preferences and behavioral intentions over time? One untouched new area is the use of microcomputer simulations and video graphics to inform subjects of the possible consequences of their decisions. The applicability of microcomputers as part of simulations and experiments has barely been touched.

Third, techniques and materials must be refined with the objective of applications in mind. In general, the simpler the technique and the more ease in understanding, the better its chances of acceptance. This need for simplicity of understanding by non-technical decision makers must be weighed against the complexities of the world and of consumer decision processes. This makes the task of application challenging.

Applications must be made in light of financial, physical, and institutional limitations. Currently, decision-makers are particularly interested in supplying service within budget limitations. Techniques that will help forecast demand or provide demand elasticities accurately and inexpensively are of most utility to them now.

It is this understanding of the need to adapt techniques to existing well-known methodologies that has led to the development of continually simpler techniques that are easy to use for forecasting and are compatible with the well known UTPS approach. Choice modelling has been the basis of the most recent experimental works (e.g. Louviere, 1981 and Louviere and Hensher, 1983).

These works still, however, focus on demand alone. Measures of productivity, efficiency and system performance are part of emerging operations planning techniques. A project currently in progress by the author (Benjamin, 1983) examines the practicality of using satisfactions measures in an information system. The system used comparisons of satisfactions with measures of system performance and costs. A schematic of the system is presented in Figure 2. One unique aspect is its use of microcomputers. Simpler measures of effectiveness were initially seen as most useful by the management team. However, as the management team learned the meaning of more complex measures, their confidence and willingness to incorporate these measures in decision making has increased dramatically. This was accompanied by increased service efficiency and a willingness and budget to try new ideas.

New applications must take the most cost effective techniques for any situation and use them in a decision framework. Market researchers have been wrestling with these ideas for a decade. One analogy with market research that shows promise is market positioning (Wind, 1982). The relative position of services already provided in a community would be compared on key characteristics. Decisions can then be made as to the market position of any new service (the relative perceived characteristics). Another idea that has served marketing specialists is the idea of a portfolio analysis. Used as part of strategic planning, the technique examines the perceived tradeoffs made by decision makers of the risks and payoffs of different product groups. Applied to transport, these would be evaluations of different service combinations and city policies offered by a transportation management organization. Finally, another approach, the matrix technique, has also been shown to be one way to analyze policy priorities by decision makers (Wind, 1982).

Following the hypothesis that techniques will be most useful which fit most easily into existing procedures, new applications for attitudes would examine how attitudinal or experimental methods can improve each stage of the UTPS procedure or would help with existing marketing and operations analyses. New experimental methods should mimick existing procedures but take advantage of their ability to investigate responses to new untested situations for which econometric extrapolations are invalid. Such situations include the introduction of new modes, service configurations or regulatory policies. Assumptions that it is possible for new situations to be evaluated on old dimensions is one cause of error in forecasting. Consumer input is a must wherever these situations arise.

SOME CONCLUDING COMMENTS

The problem in the application of these techniques is that researchers have not applied the market research techniques they develop to their own product--the techniques themselves. Although these problems are not well documented, they are part of the experience of the author.

On the other hand, the need for understanding attitudes is no less important today than it ever has been. Evaluating elderly and handicapped systems must account for the willingness of handicapped persons to travel. Estimates of mode split for automobiles and buses must take into account the stigma of riding in a bus and the sexiness of sporty cars. These complex and symbolic processes govern travel behavior and should never be ignored.

Trends in transit mirror trends elsewhere in government. Increased cost consciousness, a willingness to tailor services to different segments of the community and deregulation of the private sector are occurring in all areas of the transportation industry. The challenge of new techniques will be to fit into existing management and planning processes, to help clarify the complexities of human behavior and to help chart the future for a crucial element of our nation's infrastructure.

REFERENCES

Anderson, Norman H., (1974), "Information Integration Theory: A Brief Survey," in D.H. Drante, R.D. Atkinson, R.D. Luce, and P. Suppes (eds.), Contemporary Developments in Mathematical Psychology, Vol. 2, Freeman, San Francisco.

Anderson, Norman H., (1977), "Note on Functional Measurement and Data Analysis," Perception and Psychophysics, Vol. 21, No. 3, pp. 201-215.

Bauer, Herbert J., (1972), User Attitude Surveys and Transportation System Development, General Motors Research Report, GMR-1138, Warren, MI.

Benjamin, J., (1977), "An Assessment of User Preferences for New Apartments: An Application of Multidimensional Scaling," Doctoral Dissertation, State University of New York at Buffalo, available from Xerox Microfilms.

Benjamin, J., (1981), "An Empirical Comparison of Four Multiattribute Preference Models," ORSA-TIMS Joint National Meeting, Houston, TX, October.

Benjamin, J., (1983), "A Consumer-Oriented Transit Information System," UMTA/USDOT and NTIS, Washington, DC.

Benjamin, J., (1984), "A Conjoint Analysis of Preference for Service by Handicapped Persons," paper presented to the Third International Conference on Problems of Mobility of the Handicapped, Orlando, FL.

Benjamin, J. and R. Paaswell, (1977), "A User Oriented Housing Choice Model," Journal of Urban Systems, Vol. 1.

Benjamin, J. and R. Paaswell, (1981), "A Psychometric Analysis of Residential Location," Journal of Socio-Economic Planning Sciences, Vol. 15, No. 6, pp. 305-319.

Benjamin, J. and L. Sen, (1981), "An Evaluation of Public Transportation Using Market Segmentation Techniques," Journal of Advanced Transportation, Vol. 15, No. 2, pp. 143-161.

Benjamin, J. and L. Sen, (1982a), "A Comparison of the Predictive Ability of Four Multiattribute Approaches," in the Transportation Research Record, No. 890, pp. 1-6.

Benjamin, J. and L. Sen, (1982b), "A Manual of Procedures to Analyze Attitude Towards Transportation," USDOT and NTIS, Washington, DC.

Carroll, J.D., (1972), "Individual Differences and Multidimensional Scaling," in R.N. Shepard, A.K. Rommey and S. Nerlove, eds., Multidimensional Scaling: Theory and Applications in the Behavioral Sciences, Seminar Press, New York.

Charles Rivers Associates (1978), On the Development of a Theory of Travel-Attitude-Behavior Interrelationships, Report No. DOT-TSC-RSRA-78-14,USDOT, Washington, DC.

Combs, (1964), A Theory of Data, Wiley, New York.

Dobson, R., F. Dunbar, C. Smith, D. Reibstein, and C. Lovelock, (1978), "Structural Models for the Analysis of Traveler Attitude-Behavior Relationships," Transportation, Vol. 7, pp. 351-363.

Dobson, R., T.F. Golob, and R.L. Gustafson, (1974), "Multi-Dimensional Scaling of Consumer Preferences for a Public Transportation System: An Application of Two Approaches," Socio-Economic Planning Sciences, Vol. 8, pp. 23-26.

Dobson, R., and G.C. Nicolaidis, (1974), "Preferences for Transit Service by Homogeneous Groups of Individuals," Transportation Research Forum, Vol. 15, pp. 325-335.

Dobson, R. and M.L. Tischer, (1976), "Beliefs About Buses, Carpools, and Single-Occupant Autos: A Market Segmentation Approach," Proceedings of the Transportation Research Forum, Vol. 17, pp. 200-209.

Dobson, R. and M.L. Tischer, (1977), "Comparative Analysis of Determinants of Modal Choices by Central Business District Workers," TRB, Transportation Research Record 649, pp. 7-14.

Dobson, R. and M.L. Tischer, (1978), "Perceptual Market Segmentation Technique for Transportation Analysis," TRB, Transportation Research Record 673, pp. 145-152.

Engle, James, David Kollat, and Roger Blackwell (1973), Consumer Behavior, Dryden Press, Hinsdale, IL.

Green, P.E., (1974), "On the Design of Choice Experiments Involving Multifactor Alternatives," Journal of Consumer Research, Vol. 5, pp. 103-123.

Green, Paul E., and V.R. Rao, (1972), Applied Multidimensional Scaling: A Comparison of Approaches and Algorithms, Holt, Rinehart, and Winston, Inc., New York.

Green, Paul E., and Donald S. Tull, (1978), Research for Marketing Decisions (4th Edition), Prentice-Hall, Inc., Englewood Cliffs, NJ.

Green, Paul E., and Yoram Wind (1973), Multiattribute Decisions in Marketing: A Measurement Approach, Dryden Press, Hinsdale, IL.

Hartgen, D.T. and G. Tanner, (1970), "Individual Attitudes and Family Activities," High Speed Ground Transportation Journal, Vol. 4, pp. 439-467.

Hartgen, D.T., (1974), "Attitudinal and Situational Variables Influencing Urban Mode Choice: Some Empirical Findings," Transportation, Vol. 4, pp. 377-392.

Heaton, C., C. McCall, and R. Waksman (1976), Evaluation Guidelines for Service and Methods Demonstrations Projects, Report No. UMTA-MA-06-0049-76-16.

Hensher, D. and J. Louviere, (1983), "Individual Preferences for International Air Fares," Transport Economics and Policy, September, pp. 225-245.

Horowitz, A.D., (1978), "A Cognitive Dissonance Approach to Attitudinal Modeling in Travel Behavior Research," presented at 57th Annual Meeting, TRB.

Johnson, R.M., (1974), "Trade-off Analysis of Consumer Values," Journal of Marketing Research, Vol. 11, pp. 121-127.

Kidder, A., J. Benjamin, and R. Shuman, (1982), "Application of Conjoint Analysis in Assessing the U.S. Market for Electric Cars," Proceedings of the Annual Meeting of the Academy of Marketing Sciences, Las Vegas, NV, May.

Kocur, G., T. Adler, W. Hyman, and B. Aunet, (1982), Guide to Forecasting Travel Demand with Direct Utility Assessment, Washington, DC: UMTA/USDOT and NTIS.

Koeppel, K.W.P. and A.J. Neveu, (1979), "An Acid Test of the Trade-off Method of Attitude Measurement," Preliminary Research Report 160, NYSDOT.

Koppelman, F.S. and E.I. Pas, (1980), "Travel-Choice Behavior: Models of Perceptions, Feelings, Preference and Choice," Transportation Research Record 765, pp. 26-33.

Lancaster, K.J. (1966), "A New Approach to Consumer Theory," Journal of Political Economy, Vol. 14, pp. 132-157.

Lerman, S.R. and J.J. Louviere, (1979), "Using Functional Measurement to Identify the Form of Utility Functions in Travel Demand Models," TRB, Transportation Research Record 673, pp. 78-86.

Levin, I., and M. Gray, (1979), Evaluation of Interpersonal Influences in the Formation and Promotion of Carpools, Technical Report 101, University of Iowa, Institute of Urban and Regional Research.

Louviere, J.J. (1981), "On the Identification of the Functional Form of the Utility Expression and Its Relationship to Discrete Choice," Appendix B in Applied Discrete-Choice Modelling by D. Hensher and L. Johnson, New York: John Wiley, pp. 385-415.

Louviere, J.J., D.H. Henley, G. Woodworth, R.J. Meyer, I.P. Levin, J.W. Stoner, D. Curry, and D.A. Anderson, (1981), "Laboratory-Stimulation Versus Revealed-Preference Methods for Estimating Travel Demand Models," TRB, Transportation Research Record 794, pp. 42-51.

Louviere, J.J. and D. Hensher (1983), "Using Discrete Choice Models with Experimental Design Data to Forecast Consumer Demand for a Unique Cultural Event," The Journal of Consumer Research, Vol. 10, No. 4, pp. 348-361.

Louviere, J.J. and R.J. Meyer, (1980), "Behavior Analysis of Destination Choice: Theory and Empirical Evidence," Transportation Research.

Lovelock, C.H., (1975), "A Market Segmentation Approach to Transit Planning, Modeling, and Management," Proceedings, Transportation Research Forum, Vol. 16, pp. 247-258.

Lyon, P. (1984), "Time-Dependent Structural Equations Modelling: A Methodology for Analyzing the Dynamic Attitude-Behavior Relationship," Transportation Science, Vol. 8, No. 4, pp. 395-414.

McMillan, Robert K. and Henry Assael (1968), National Survey of Transportation Attitudes and Behavior, NCHRP Report No. 49, Washington, DC.

Meyer, R.J., I.P. Levin, and J.J. Louviere (1978), "Functional Analysis of Mode Choice," TRB, Transportation Research Record 673, pp. 1-7.

Neveu, A.J. and K.W.P. Koeppel, (1978), "Trade-off Model: Documentation of Computer Programs," Preliminary Research Report No. 139, Planning Research Unit, NYSDOT.

Recker, W.W. and T.F. Golob, (1976), "An Attitudinal Mode Choice Model," Transportation Research, Vol. 10, pp. 299-310.

Recker, W.W. and T.F. Golob, (1979), "A Non-Compensatory Model of Transportation Behavior Based on Sequential Consideration of Attributes," Transportation Research, Vol. 13B, pp. 269-280.

Recker, W.W. and R.F. Stevens, (1977), "An Attitudinal Travel Demand Model for Non-Work Trips of Homogeneously Constrained Segments of a Population," Transportation Research, Vol. 11, pp. 167-176.

Stopher, P.R., (1977), "The Development of Market Segments of Desgination Choice," TRB, Transportation Research Record 649, pp. 14-22.

Tischer, M.L. and R.V. Phillips, (1979), "The Relationships Between Transportation Perceptions and Behavior Over Time," Transportation, Vol. 8, pp. 21-36.

Torgerson, W.S. (1958), Theory and Methods of Scaling, John Wiley and Sons, Inc., New York.

Tversky, A., (1972) "Elimination by Aspects: A Theory of Choice," Psychological Review, Vol. 79, pp. 281-299.

Valiant, G., (1977), Adaptation to Life, New York: Wesley.

Wind, Y., (1982), Product Policy: Concepts, Methods and Strategy, Reading, MA: Addison-Wesley Publishing Company.

Winer, B.J. (1971), "Statistical Principles," Experimental Design, McGraw-Hill, New York.

Table 1

Tabular Guide to Use of Techniques

	(1)	(2)	(3)	(4)	(5)	(6)	(7)
I. Uses							
1. Provides high degree of detailed information				X	X	X	X
2. Provides general information about individual transport attributes	X						
3. Provides overview of many general transport attributes		X	X				
4. Can be used to group market segments		X	X	X	X	X	X
5. Easily used to detect interactions						X	X
II. Costs							
6. Easily self-administered	X	X	(X)				
7. Requires personal interview format (usually)			(X)	X	X	X	X
8. Easily adopted to telephone	X	X					
III. Computer Analysis							
9. Easily analyzed on popular packages (SPSS)	X	X					
10. Can be analyzed on popular packages after proper preparation						X	
11. Requires special computer packages				X	X		X

---

1. Unidimensional scales
2. Factor analysis
3. Multidimensional scaling
4. Trade-off analysis
5. Conjoint measurement
6. Functional analysis
7. Experimental choice analysis

	(1)	(2)	(3)	(4)	(5)	(6)	(7)
12. Requires little computer time	X						
13. Requires greater computer time		X	X	X	X	X	X
14. Requires little statistical background	X						
15. Requires some statistical background		X					
16. Requires special knowledge of programs and procedures			X	X	X	X	X
IV. <u>Congruence with Existing Processes</u>							
17. Operations evaluation	X						
18. Marketing	X						
19. Service and facilities design				X	X	X	X
20. Demand forecasting						X	X

FIGURE 1*

ELEMENTS OF DECISION MODEL

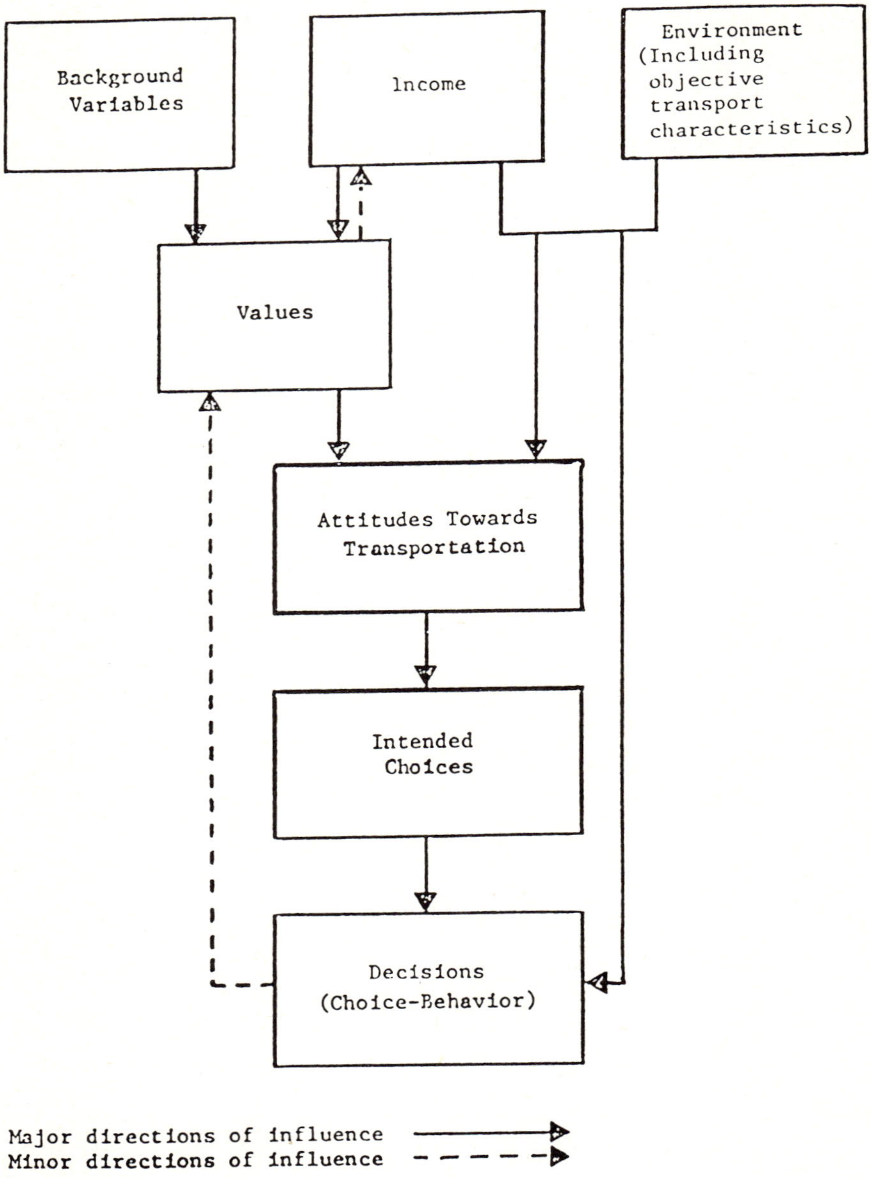

*Originally published in Benjamin and Sen (1982b)

FIGURE 2

INFORMATION REQUIREMENTS OF A CONSUMER-ORIENTED INFORMATION SYSTEM *

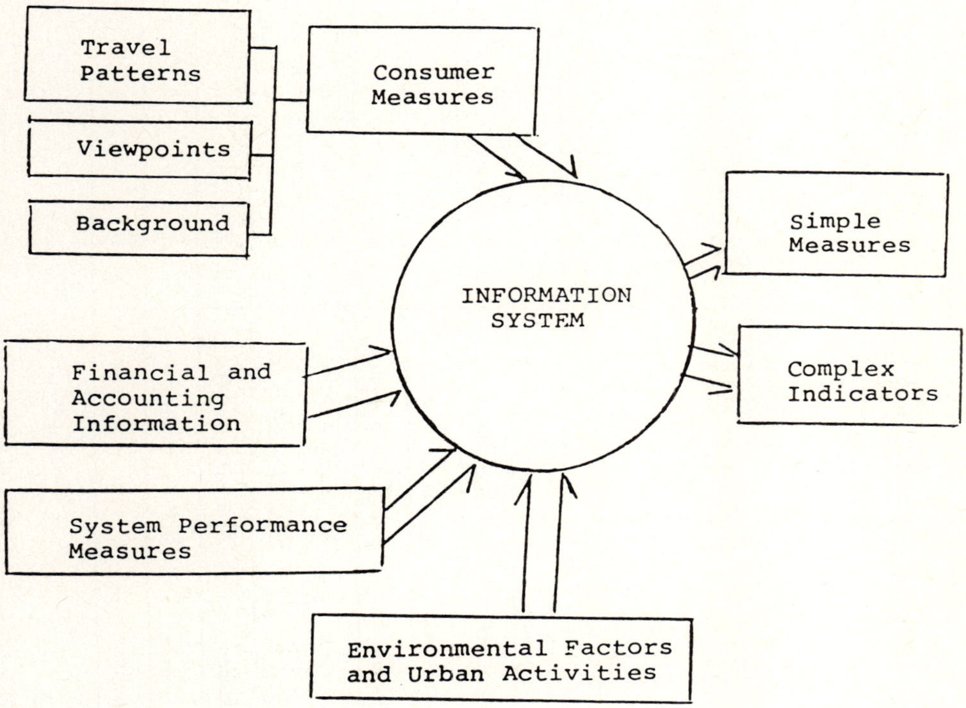

* Originally published in Benjamin(1983)

FIGURE 3. An Example of Conjoint Questionnaire*

Part A - Work Trips or Shopping Trips
Personal Preferences

Questionnaire I.D.

Check one: ☐ Work ☐ Shop

	Round Trip Cost	Lead Time (Hours)	Time when Service is available	How Frequently You Might Use the Service (Never, Rarely, Half the Time, Usually, Always)
1	**Free**	On demand	6 a.m. - 10 p.m.	1 2 3 4 5 6 7 8 9 10 11
2	$4.00	On demand	24 hours a day	1 2 3 4 5 6 7 8 9 10 11
3	$1.00	On demand	8 a.m. - 5 p.m.	1 2 3 4 5 6 7 8 9 10 11
4	$4.00	2 hours	8 a.m. - 5 p.m.	1 2 3 4 5 6 7 8 9 10 11
5	$1.00	24 hours	24 hours a day	1 2 3 4 5 6 7 8 9 10 11
6	$4.00	24 hours	6 a.m. - 10 p.m.	1 2 3 4 5 6 7 8 9 10 11
7	Free	2 hours	24 hours a day	1 2 3 4 5 6 7 8 9 10 11
8	Free	24 hours	8 a.m. - 5 p.m.	1 2 3 4 5 6 7 8 9 10 11
9	$1.00	2 hours	6 a.m. - 10 p.m.	1 2 3 4 5 6 7 8 9 10 11

* Originally published in Benjamin(1984)

FIGURE 4. An Example of a Mode Choice Questionnaire*

UNDER WHAT SITUATIONS WOULD YOU DRIVE ALONE OR SHARE A RIDE (CAR POOL/VAN POOL) TO WORK?

Consider that you are going to work and that driving alone or sharing a ride in a car pool or van pool are your only choices.

Below are a number of factors describing eight different situations where you are faced with choosing whether to drive alone or share a ride to work.

Look at each situation across the entire line and please answer in the last column to the right how likely you are to drive alone or share a ride to work.

	AUTO FACTORS			CAR POOL/VAN POOL FACTORS		PLEASE– ANSWER IN THIS COLUMN: HOW LIKELY ARE YOU TO DRIVE ALONE OR SHARE A RIDE? (CIRCLE A NUMBER)				
	Gas Availability	Gas Price	Parking Cost to Drive Alone	People You Share A Ride With	Employee Work Schedule	Always Drive Alone	Probably Drive Alone	In-different	Probably Share A Ride	Always Share A Ride
SITUATION 1	Ample Supply	$1.30/gallon	Free	Co-Worker/ Neighbor	Flexi-time (hours can vary daily)	1	2	3	4	5
SITUATION 2	Ration of 10 gallons/week*	$2.60/gallon	Free	General Public (Carpool Matching)	Flexi-time (hours can vary daily)	1	2	3	4	5
SITUATION 3	Ration of 10 gallons/week*	$2.00/gallon	$30/month	Co-Worker/ Neighbor	Flexi-time (hours can vary daily)	1	2	3	4	5
SITUATION 4	Ample Supply	$2.60/gallon	$30/month	Co-Worker/ Neighbor	Fixed 8 hour day	1	2	3	4	5
SITUATION 5	Ration of 10 gallons/week*	$1.70/gallon	Free	Co-Worker/ Neighbor	Fixed 8 hour day	1	2	3	4	5
SITUATION 6	Ample Supply	$2.00/gallon	Free	General Public (Carpool Matching)	Fixed 8 hour day	1	2	3	4	5
SITUATION 7	Ample Supply	$1.70/gallon	$30/month	General Public (Carpool Matching)	Flexi-time (hours can vary daily)	1	2	3	4	5
SITUATION 8	Ration of 10 gallons/week*	$1.30/gallon	$30/month	General Public (Carpool Matching)	Fixed 8 hour day	1	2	3	4	5

*If your car gets 15 miles per gallon, you can travel 150 miles per week.

T-277D-80

*Adopted from Kocur et al (1982)

OVER →

Behavioural Research for Transport Policy
pp. 405–422.

THE APPLICATION OF ATTITUDINAL RESEARCH TO THE MANAGEMENT OF URBAN PUBLIC TRANSPORT

Elaine Forsyth and Austin Smyth

INTRODUCTION

The recently completed Tyne and Wear Metro forms the backbone of what is the first fully integrated public transport system in the United Kingdom. Nevertheless, users of the Metro form (and will continue to form) a relatively small proportion of the total public transport market in the conurbation; the vast majority of public transport passengers making use of the bus services provided under the umbrella of Tyne and Wear Passenger Transport Executive (Table 1). The PTE is responsible for the provision of bus, Metro and ferry services within the Metropolitan County of Tyne and Wear. The services provided are operated by the Executive itself and two subsidiary companies of the National Bus Company - United Automobile Services Limited and Northern General Transport Company.

RECOGNITION OF THE NEED TO MONITOR 'QUALITY OF SERVICE' IN TYNE AND WEAR

Over the past five years the public transport network has undergone substantial revision in conjunction with the opening of the Metro. Integration has been a key feature including systemwide through-ticketing in addition to a harmonising of routes and services operated. The PTE is aware that track-based public transport frequently commands a superior 'image' to conventional bus services in the minds of potential passengers, and has attempted to portray a high quality image for the public transport system as a whole through efforts at 'integration' and a common approach to marketing.

According to the executive 'the requirement is now to consolidate on the gains of the past few years'. In spite of these attempts, the PTE is aware that the Metro component continues to be generally perceived as superior to the bus-based elements. Faced with the fact that the period of major capital investment is now at an end and with indications that revenue support may well be cut in the near future, how can the authority achieve its objective of 'sustaining and enhancing the image of public transport, and continuing to improve the overall quality of service'?

It is recognised that with resources now severely restricted, management effort and the available expenditure should be targeted on those areas which will yield the highest return. To enable a programme of low cost and yet 'effective' action to be developed the transport operator has to be aware of how passengers perceive the 'quality of service' offered. For Tyne and Wear 'information on certain critical areas of the passengers' perception of overall service quality is lacking' according to the PTE, who therefore propose to monitor 'perceptions' of services on a regular basis. However, also recognised are the difficulties involved in obtaining 'qualitative data' and the lack of a readily applicable method of data collection and interpretation.

THE ROLE OF THE UNIVERSITY IN MONITORING 'QUALITY OF SERVICE'

Following invitations to a number of academic institutions the Transport Operations Research Group (TORG) at the University of Newcastle upon Tyne was commissioned to report on the various techniques for determining peoples' perceptions of the service quality and to recommend a survey programme, **to be undertaken subsequently by the PTE** on a regular basis, enabling the Executive to monitor the effects of various measures taken to improve the quality of public transport in Tyne and Wear. It was also envisaged that the data collected should be capable of being analysed in a straightforward way.

In drawing up its study brief Tyne and Wear PTE had identified a need to answer a series of questions on the basis of the information collected:-

(1) How important is any given attribute of service?

(2) What is any individual passenger's perception of service quality with respect to a given attribute?

(3) What impact to (a) service frequency; (b) fare levels; and (c) the provision of information have on the attitudes expressed?

(4) To what extent are attitudes influenced by media comment and other external events?

(5) What is the passengers' overall perception of service quality? and

(6) What changes have there been to attitudes over time which are related to changes in the system?

The Executive saw a need for attitudinal information from their customers relating to:

(a) Driving standards
(b) Service punctuality/headway variations
(c) Vehicle appearance
(d) Interior/exterior cleanliness
(e) Staff attitudes, particularly those of drivers, inspectors and information staff
(f) Other factors which they feel important.

According to the PTE there was also some evidence to suggest that 'the staff's perception of passenger behaviour plays a part in determining attitudes on both sides'.

In TORG's proposal (February 1984) a number of comments were made with reference to the PTE's brief indicating; (1) the apparent shortness of the timescale planned for the study (the PTE envisaged the first set of fieldwork and results being completed by 31st March, 1984); (2) the need to identify the salient 'attributes' of the services perceived by public transport users rather than prejudging the relevant aspects of quality of service; (3) that valuable information could be revealed by means of conventional attitudinal survey techniques, but that a 'model' for attitude or behaviour prediction should not be expected.

The overall aim of the study undertaken by the University team and described in this paper is, then, to assist the transport operator in making practical decisions to improve the quality of services by providing a clearer understanding of how passengers with different levels of dependence on public transport perceive the quality of service offered. This paper draws heavily on the team's report (Forsyth et al 1984) and we would acknowledge the contributions of our colleagues in this respect.

The specific objectives to be met included:

(1) to compare and evaluate various attitudinal survey methods and analytical techniques;

(2) to review the data available from other sources, for direct use and for control of samples;

(3) to select preferred methods and sampling techniques;

(4) to test and refine the preferred methods in pilot surveys;

(5) to design surveys and methods of analysis.

The focus of much of the remainder of this paper is on attempts to achieve objectives (3), (4) and (5).

THE MEASUREMENT OF ATTITUDES CONCERNING QUALITY OF SERVICE

With respect to objective (1), a comprehensive review of definitions of attitude, the availability and appropriateness of various methods for identifying salient attributes and the use of various attitude scaling techniques is provided in the team's report. In brief, attitude has been variously defined encompassing ideas about inclinations, feelings, biases and concern so we were clearly involved with the collection of subjective data. For our purposes the definition offered by Drever (1952) was appropriate and we accpeted that the inquiry concerned 'that more or less stable set or disposition of opinion or purpose, involving expectancy of a certain kind of experience and readiness with an appropriate response'.

We were concerned with very individual reactions, and preferences, both social and psychological in origin, according to the individuals interaction with the physical and social environment. Attitudes are not merely a response to objects or events encountered, but are coloured by the experiences and attitudes of peers and relevant others. They are also able to be broken down into three component parts: (1) 'cognition' - those beliefs we hold concerning the object or event; (2) 'affect' - our feelings for or against the object or event; and (3) 'conation' - our tendencies or intentions to act with respect to an object or event.

PRELIMINARY QUALITATIVE RESEARCH

Having defined the problem more clearly, it was necessary to establish what comprises the public transport service in the minds of the relevant people; i.e. residents within Tyne and Wear County. This was crucial to ensure that attitudes to appropriate aspects of service be monitored and to avoid asking individuals to respond on attributes neither previously nor spontaneously considered. It was also important that service attributes be labelled in the respondents' own terms and therefore be meaningful, suggesting a benefit in having potential respondents generate the components for later consideration.

This qualitative work has the objective of securing insight into the problem and there are many techniques in popular use; each tailored to fit slightly different techniques in popular use; each tailored to fit slightly different study goals. Observation, introspection and group discussion are three such methods, but for our purposes (to establish those attributes **normally salient** to individuals) a part-structured interview incorporating open-ended questions was considered most appropriate.

A questionnaire was designed on the basis of past experience to categorise respondents prior to the open-ended questions. Locational (district of residence), personal (e.g. age and gender) and journey-specific (e.g. frequency of use and for what purpose) details were collected. Next, respondents were required to think about using the bus or Metro service and to report the comments which came to mind with respect to likes and dislikes, advantages and disadvantages associated with use of the mode. In this way, a generous list of service attributes salient to the respondents was collected. A frequency table of service aspects elicited was then drawn up, and those to be used in the quantitative surveys were selected on the basis of their frequency of elicitation.

It was considered of utmost importance that this preliminary work be thorough to ensure a sound basis on which to build the subsequent survey. Nearly 500 interviews were therefore conducted at varying times over 4 days. Five diverse locations were selected as survey sites to include geographical variation and also levels of car availability.

Content analyses were carried out in order that the frequency table could be completed and there was a general finding that lack of experience of a particular mode led to a diminished flow of attributes. Metro and bus differed slightly in their most salient features and Tables 2 and 3 illustrate the percentage of respondents offering the listed attributes where this was greater than or equal to 4. For Metro, speed came out on top with almost 50% giving mention to this. Twenty four per cent mentioned the 'good service' provided and slightly less its 'frequency' and its 'accessibility'. For bus, the 'frequency/regularity', 'reliability' and 'goodness of service' shared the same popularity with 23% of the sample eliciting each. This work did largely confirm the validity of those attributes of service of interest to the PTE but also others and in the respondents' own terms.

It is important to note here that the items listed in Tables 2 and 3 represent concepts, and their favourableness or otherwise should not be attended to. For example, from Table 2 we see that 15% of respondents referred to the ease of access to the system: 'not near me' was one response given for this category, but respondents who described Metro to be 'near me' are also included within this 15%.

This brings us to objectives (3) and (4) concerning the selection and testing of quantitative research methods. Equipped with a list of those most salient attributes and those potentially important in facilitating later intermodal comparison, several tools were designed for application as follows.

SELF-COMPLETION CARDS

The first of these tools, cards designed for distribution, self-completion and subsequent collection on-vehicle on selected routes, were tested for their practicality. This also allowed testing of two alternative rating scales. As illustrated in Figures 1 and 2, they comprised basic stratification questions with two sets of rating scale; in one, respondents indicated the extent to which they considered each of the listed attributes important and in the second they gauged the favourableness or otherwise of attitudes concerning these attributes. Card A used the conventional semantic-differential rating scales (Osgood et al 1957), allowing a choice from seven intermediate positions, while card B allowed a choice from only three positions in each case.

On examining the responses, there was found a clear trade-off between the higher level of information obtained with the 7-point scale and the relative ease of completion and slightly higher response rate achieved with card B. The importance rating unfortunately failed to distinguish between attributes satisfactorily as nearly all respondents indicated all aspects as being very important. This may have been due to the perceived social desirability of attributing importance to components of service but, in any case, the reliability of the information gained was questionable.

SELF-COMPLETION QUESTIONNAIRES

Next, questionnaires were designed for both on-vehicle distribution and household drop; each with a pre-paid reply envelope for its later return. It was considered that the increased time and care possible in the absence of distractions on-board the vehicle would result in better quality information. Three lengths of questionnaire were tested, differing in the number of attributes included in the semantic-differential rating scale, and these were distributed on selected bus routes and to houses in selected streets in Tyne and Wear. The questionnaire requested information on car ownership and district of residence and contained a section concerned specifically with bus services. A semantic-differential scale was employed to obtain information on the considered importance of service attributes. Those routes used most often by respondents were determined and a rating scale on which to indicate the level of satisfaction with the various attributes was also included. Comparable importance and satisfaction scales for the Metro service and the opportunity for free comment completed the questionnaire.

The principal finding was that the on-bus distribution achieved a higher response rate than the conventional postal surveys although substantial differences were evident according to the area in which questionnaires were delivered. It was felt that the personal contact in on-bus distribution aids willingness to contribute, but the higher response rate may simply have resulted from non-users having less interest or feeling that they did not have anything worthwhile to offer given their lack of experience. It is interesting to note that the length of the attribute list seemed not to influence response rates.

A scan of responses revealed once more that the importance measure had attracted unanimously high ratings for all attributes thereby denying the possibility of weighting the satisfaction scores for attributes accordingly. It was decided at this stage to replace this scale with a request that each respondent state the 3 attributes from the list which he or she held to be most important.

Because of the nature of the pre-pilot work, it would not be appropriate to draw specific inferences from the limited analysis of responses. There did seem, however, to be differentiation in the scores according to survey site and Metro was generally rated more favourably than the buses.

On-bus distribution with postal reply is cheaper per valid return than the self-completion cards collected on-vehicle, in part because of the extra manpower required for administration of the latter. Overall, on-bus distribution achieves the most cost-effective and satisfactory set of returns, in part because it enables more time to be taken in assimilating the arguably complex requirements.

ON-STREET INTERVIEWS

For completeness, on-street interviews were also considered and it was with this procedure that attempts were made to investigate the PTE's fourth question of the extent to which attitudes are influenced by media comment.

The questionnaire had basically six sections which were used in various combinations at various sites in Tyne and Wear County. The introduction allowed categorisation of the respondent according to personal and journey-specific factors. Section A asked interviewees what they thought of each attribute in turn, and the response was a number from 1 to 7 read from a flash-card according to the position between the 'favourable' and 'unfavourable' poles considered representative of the respondent's attitude. The 3 most important attributes were established at the end of this section. Section B required respondents to think about media coverage and to rate, in turn, television, press and local radio from 1 to 7 according to their perceived accuracy. The respondent then rated the 3 again according to their perceived adequacy, in each case using a flash-card for guidance.

Section C sought the same information as A, but here instead of a full flash-card with each question displayed only one differential was used; a 1 to 7 scale of satisfaction. Section D made a final attempt at eliciting importance ratings using a similar flash card with a 1 to 7 scale of importance.

Section E was directed at the non-user to investigate: (1) why public transport was seldom used; (2) the travel patterns of peers; and (3) which mode of transport peers would expect the respondent to use for selected journey purposes, as relevant others have been found previously to influence such behaviour quite considerably. Lastly, section F was an open request for comment on media coverage of local public transport.

Initially the semantic-differential rating scales were presented with mixed polarity, i.e. sometimes with the positive pole on the left and sometimes on the right. This is generally accepted to be proper conduct to avoid systematic bias. However, this meant that 7 would correspond in some cases to a favourable position and sometimes to an unfavourable one which caused sufficient confusion for the scales subsequently to be altered to show consistent poles throughout.

The importance rating again failed to differentiate and infrequent users found section E problematical. Also opinion concerning media comment proved very difficult to collect, due in part to lack of awareness on the part of respondents. In any case, it must be appreciated that attempting to attribute causality or establish a clear link between the nature and extent of media coverage and people's attitudes to transport services would be extremely tenuous.

In the presence of an interviewer, there seemed to be some resistance to assimilating the instructions and providing responses in numerical terms. Rather, verbal responses were offered for the interviewer to translate

PILOTING OF PROPOSED QUESTIONNAIRES

In the light of all survey work undertaken to find the optimum quantitative method and with due consideration of the practicality and appropriateness of each, a pilot questionnaire was developed. The 7-point differential was retained for attitude measurement but respondents were required to consider the importance of factors relative to one another rather than to rate each independently. A more extensive enquiry was made concerning travel details on this questionnaire. There was a separate section for bus and Metro, but each contained the 7-point scale to measure satisfaction with a list of attributes. Using the same scale, respondents were also asked to indicate their opinion about the 'goodness' or 'badness' of the service as a whole. This general comment was frequently elicited in the course of the preliminary work, but was included principally to address the fifth question posed by the PTE; that of the passengers' overall perception of service quality. It is appreciated that this means of measurement is not ideal, but an overall impression will also be gained from the attribute ratings in aggregate. The three most important attributes were once more requested.

An example was included with the instructions for use of the rating scale illustrating a mark at the favourable extreme and the scale was presented with a boxed grid in place of the series of segmented lines used previously.

An acceptable response rate of 22% resulted using on-bus distribution with postal return but the responses proved less than satisfactory, with most respondents using only the extremes of the scales. There was clearly a serious misunderstanding of the requirements attributable, in view of past success to the altered presentation.

The boxed grid may have been interpreted as a 2-dimensional 'matrix' and the provision of an extreme example may have restricted the spread of responses. This clearly illustrates the susceptability of such social research to minor changes in presentation and emphasises the need for thorough testing of all survey tools used.

A second pilot study was necessitated and this time two questionnaires were developed. Only the presentation of the attribute rating scales differed from that used in pilot study. In both cases, the example for guidance was at point 6, i.e. not at the extreme. The first of these questionnaires (see Figure 3) incorporated the conventional semantic-differential with the segmented line displays and the second offered seven labelled spaces for each attribute from which the most appropriate was to be selected.

Distribution was as for the first pilot study for comparison, and despite a slightly diminished response rate 93% of responses were valid. Both questionnaire types had the same return and success rates leading to the decision to recommend that the PTE implement the first of these 2 questionnaires described with the semantic-differential presented in its traditional form. It is the accepted method of enquiry and has been demonstrated by its exponents to yield reliable and valid data.

The use of labelled response options requires extensive qualitative preparation to ensure that labels appropriate to the population in question are used and to avoid interpretational bias. Without this, equal intervals between positions cannot be assumed. As there were no difficulties with the spatial choice, this questionnaire type was discarded.

THE RECOMMENDED APPROACH TO MONITORING QUALITY OF SERVICE

Having selected the optimum means of quantification of the attitudes of people in Tyne and Wear County to the services provided, there remained the determination of the best strategy for interpretation.

In developing a preferred strategy a variety of alternative procedures were examined (Figure 4). Table 4 illustrates suitable locations for obtaining various types of sample, and this question of where to contact respondents also has implications for the choice of an appropriate data-collection technique as highlighted in Table 5. It is clear from this research that in recommending a preferred method there are no 'right' or 'wrong' approaches, but rather each has its advantages and disadvantages according to circumstances (Table 5). In coming to a decision on the choice of a technique appropriate to this study, however, three questions must be addressed: who to ask? where to sample? what to ask?

We concluded that the envisaged budget (£10,000 per annum for quality of service monitoring) would be spent most cost-effectively by concentrating on users of the system, surveyed on-board vehicles, about aspects of the transport service which they were experiencing directly. It can be claimed that this would provide a valid link between survey responses and the actual quality of services on offer.

It was clear from the preliminary and pilot surveys that a set of semantic-differential scales should form the core of the questionnaire. This is intended to yield quantifiable responses on satisfaction with the selection of the three most important attributes providing useful data on relative importance. The inclusion of open-ended questions on media coverage was also recommended.

The preliminary surveys suggested that reply-paid self-completion questionnaires, distributed on-vehicle would be the most cost-effective survey method. Taking into account the likely budget available, the strategy envisaged for the monitoring exercise would involve the following:

(a) standard questionnaires to be administered twice yearly on each of a set of 'core' routes selected for their stability and reflecting the pattern of public transport operation in Tyne and Wear. The core surveys would cover three routes from each of the seven marketing areas, inter alia covering each operating company and all the main depots, i.e 42 route surveys in all. 100-150 completed questionnaires per route are expected; and

(b) the same questionnaires - lengthened if necessary to cover specific issues - to be administered twice yearly on roughly a 3-year cycle to cover all the major routes and depots, i.e. 120 routes over 3 years.

In addition, to establish the views of non-users and to probe more deeply into factors helping to form attitudes, 400 household interviews per annum were recommended.

In parallel with these surveys, it was recommended that management information should be collected in a systematic way to develop reliable indicators for each salient attribute in order to maximise benefits obtained from the additional surveys. Crucially, the qualitative responses from the attitude surveys should be compared with 'hard' data obtained by such indicators.

The framework for analysis (Figure 5) envisaged use of simple and widely used parametric (or where appropriate, non-parametric) statistical tests to establish whether significant differences exist over time, between the responses of users of different groups of services.

It is also important to establish whether respondents who differ in respect of their personal characteristics or car-availability will hold different attitudes about the quality of public transport. In the light of this possibility two-part statistical tests were recommended. Where differences exist on the basis of personal characteristics, appropriate weightings will need to be applied, derived from MICSS surveys, to ensure the representativeness of scores.

The original PTE brief referred to the development of an overall measure of attitude. The research undertaken confirmed that such a measure based on sound psychological theory is not appropriate in this case. Nevertheless, the proposed questionnaire did include a request for respondents to rate bus and Metro services overall. It was suggested that it might be useful to correlate this overall measure with an 'index' of a respondent's attitude based on individual attribute scores. As already mentioned, respondents were also to be asked to select the three most important items from the attribute set as they saw it. A frequency analysis of the complete list of all attributes would yield useful information on the relative importance of each.

In developing any framework for analysis the temporal dimension is highly significant. The case for 'control' routes is based on a need to establish whether attitudes are changing over time - either due to secular effects or to exogenous improvements in the quality of service. Once several rounds of surveys have been completed, it should be possible to establish a statistically valid trend against which current survey results may be compared.

THE PTE'S REACTION TO THE PROPOSALS

The extent to which our proposals may assist the transport operator in making practical decisions to improve the 'quality of service' can be considered by referring back to the original objectives outlined above.

In general terms the five objectives described have been achieved. Tyne and Wear Passenger Transport Executive has been provided with a flexible survey programme which it can implement using its own staff. Data analysis is intended to be straightforward and, providing the PTE collect data on objective indicators of quality of service it should be possible for the Executive to see clearly where improvements are needed and to monitor the effects of any changes implemented on the attitudes of its customers.

These observations are reinforced by the initial reaction of the transport operator to our proposals, and by the fact that our suggestions were braodly accepted in spite of the uncertainty in the public transport industry over deregulation. Indeed, the research carried out is timely and the envisaged programme of surveys to be undertaken by Tyne and Wear PTE staff should be seen as part of an overall market responsive strategy being adopted by the operator. This will also include customer care campaigns encouraging staff to be more attentive to passengers' needs.

'Quality of service' monitoring will take place under the slogan 'Tell us what you think', and a single-fold leaflet has been designed following TORG's recommendations. Two/three month 'blitzes' biannually are envisaged, with on-vehicle distribution of questionnaires. The PTE plan to commence the survey programme in the Spring of 1985.

Initially it is not proposed to develop new objective measures or indicators of service attributes, but other readily available information will be taken into consideration in discussion of potential remedial measures. Furthermore, it is proposed to involve district managers in assessing survey results. This will also permit geographical differences in quality of service to be followed up. An important issue is reliability and it is anticipated that further research will be focussed on this aspect of service provision.

The Passenger Transport Executive also intend to use the survey instrument as a method for monitoring the effects of deregulation of the bus industry in Tyne and Wear. In particular, the regular sequence of surveys will allow the PTE to

(i) relate changes in the 'quality of service' to patronage data;

(ii) monitor the performance of 'contractors' on tendered services;

(iii) provide comparisons of passenger reactions on 'contractor' provided services with those operated by the PTE; and

(iv) permit more effective competition with other operators in a deregulated operating environment through greater awareness of the market's requirements.

The survey and analytical techniques outlined in this paper are not new, nor do they represent 'state of the art' approaches. They have, however, been incorporated into a practical survey instrument and framework for analysis, the use of which will enable a major UK transport operator to be substantially more market responsive in conditions likely to produce the most radical changes seen in the UK public transport industry in the last fifty years.

REFERENCES

Drever, J. (1952) The Penguin Dictionary of Psychology. Penguin Books Ltd. UK.

Forsyth, E., Hills, P. J., James, N., Silcock, D. T., and Smyth, A. W. (1984). Monitoring the Quality of Public Transport Services in Tyne and Wear. Transport Operations Research Group. University of Newcastle upon Tyne. (Unpublished)

Osgood, C. E., Suci, G. V. and Tannenbaum, P. H. (1957). The Measurement of Meaning. University of Illinois Press: Urbana.

Tyne and Wear Transport (1984) Tyne and Wear Public Transport Plan 1985/86-1987/88

Table 1

TYNE AND WEAR TRAVEL MARKET

- PASSENGER BOARDINGS BY MODE

BUS	86%
METRO	13%
BR/FERRY	1%

(Tyne and Wear Transport 1984)

Table 2 Quality of Service Attributes for Metro by Frequency of Elicitation and by Location

Attributes of Metro service elicited	All locations %	Monument %	Regent Centre %	Heworth %	West Rd %	Sunderland %
quick/slow	47	56	55	46	35	31
good service/bad service	24	8	33	27	31	20
frequent/infrequent	17	28	20	17	9	9
near me/not near me	15	19	15	10	17	15
clean/dirty	9	18	8	4	9	4
reliable/unrealiable	9	17	26	5	-	-
comfortable/uncomfortable	9	10	10	8	11	4
convenient/inconvenient	9	11	18	6	10	-
efficient/inefficient	8	15	10	4	7	2
plenty of room/crowded	8	11	8	7	8	-
no need to change/need to change	7	13	5	8	1	6
station facilities good/station facilities bad	7	-	12	10	11	2
cheap/expensive	6	7	8	5	4	4
direct service/no direct service	5	3	20	6	1	4
personal safety good/personal safety bad	5	8	13	-	5	-

Table 3 Quality of Service Attributes for Buses by Frequency of Elicitation and by Location

Attributes of Bus service elicited	All locations	Monument	Regent Centre	Heworth	West Rd	Sunderland
frequent/infrequent	23	28	28	16	22	26
reliable/unreliable	23	29	30	15	20	28
good service/bad service	23	14	20	14	33	44
shelter good/shelter bad	10	13	2	10	10	11
considerate drivers/inconsiderate drivers	9	16	6	1	13	17
able to get a seat/not able to get a seat	8	12	5	5	10	7
near me/not near me	8	10	10	7	5	9
cheap/expensive	8	11	8	4	9	9
smoking on buses	7	13	2	4	8	2
clean/dirty	6	3	10	7	5	9
convenient/inconvenient	6	6	10	3	8	4
comfortable/uncomfortable	5	3	5	6	6	4
good driving/bad driving	4	1	8	1	7	3
smooth ride/bumpy ride	4	4	10	2	2	2

N.B. Attributes elicited by less than 4% of respondents have been left out of Tables 2 and 3.

Table. 4: Practical combinations of sampling approaches and survey locations for Tyne and Wear Quality of Service Monitoring

Location of survey / Sampling approach	At home	On-board vehicle	At stops/ stations	Activity centres	Special locations
Quota (non-random)	✓	✓	✓	✓	✓
Random/ probability	✓				
Choice-based		✓	✓		
Panel study	✓				

Table 5 :Practical combinations of survey locations and data collection techniques for Tyne and Wear Quality of Service Monitoring

Data collection technique / Location of survey	Personal interview	Self-completion questionnaires		Group discussion
		'collect'	postal	
At home	✓	✓	✓	
On-board vehicles	(metro only)	✓	✓	
At stops/ stations	(bus only)	✓	✓	
Activity centres	✓	✓	✓	
Special locations				✓

Table 6 The advantages and disadvantages of selected data collection techniques in various situations for Tyne and Wear Quality of Service Monitoring

Location of survey / Data collection technique	In general	At home	On-board vehicles	At stops/stations	Activity centres	Special locations
Personal interview	Minimum non-response Spontaneity of interview obtained Complex questionnaire design possible Most expensive option - given limited budget of £10,000 not possible to obtain sample large enough to derive population estimates. Reluctance to give personal information.	Difficult to obtain large enough sample of 'choice' public transport users.	With small number of interviewers cannot contact more than a small proportion of intending passengers (not feasible for bus)	As for 'on-board' (not feasible for Metro at peak times)	Criticisms as for 'on-board' and 'at stop' and more general limitations of quota sampling/ location choice	Not appropriate
Self-completion questionnaire 'collect'/postal	Non-response tends to be higher/more variable Spontaneity of information reduced. Questionnaire design must be simpler. Much less expensive (except for follow-up home interviews)	Benefits of self-completion in terms of cost advantage over personal interview significantly reduced	Much more cost-effective/practical than personal interviews - with a small budget, the only way of contacting substantial numbers of potential respondents	As for 'on-board'	Criticisms as for 'on-board' and 'at stop' and more general limitations of quota sampling/ location choice	Not appropriate
Group discussion	Not suitable for obtaining attitude scale data - possible role for qualitative research/elicitation of salient attributes of Quality of Service. Expensive and time consuming to set up/ logistical problems Group conditioning bias in responses and among participants - 'axe to grind'. Does not provide a representative sample population estimates not possible.	Not suitable	Not suitable	Not suitable	Not suitable	See under 'in general' heading

Figure 1

TYNE AND WEAR TRANSPORT

TO HELP US PLAN TRANSPORT SERVICES IN TYNE AND WEAR, WE WOULD LIKE TO KNOW WHAT YOU THINK OF PUBLIC TRANSPORT IN THE COUNTY. PLEASE FILL IN THIS SURVEY FORM AND HAND IT TO ONE OF OUR SURVEY STAFF WHEN YOU GET OFF.
THANK YOU FOR YOUR HELP.

1. WHAT IS YOUR AGE?
   - LESS THAN 16 ☐
   - 16 - 64 ☐
   - 65 + ☐

2. WHAT IS YOUR SEX?
   - FEMALE ☐
   - MALE ☐

3. HOW OFTEN DO YOU USE A BUS IN TYNE AND WEAR?
   - ONCE A WEEK OR LESS ☐
   - MORE THAN ONCE A WEEK ☐

4. HOW OFTEN DO YOU USE METRO?
   - ONCE A WEEK OR LESS ☐
   - MORE THAN ONCE A WEEK ☐

5. LISTED BELOW ARE SOME FEATURES WHICH MAKE UP A BUS SERVICE. THINK OF THE BUS SERVICE YOU WOULD LIKE AND INDICATE THE IMPORTANCE OF EACH OF THESE FEATURES TO YOU BY PLACING ONE CROSS (X) AT THE APPROPRIATE PLACE ON EACH LINE:

e.g. IMPORTANT — UNIMPORTANT

HOT : ___ : ___ : ___ : _x_ : ___ : ___ : ___ :

... WOULD MEAN THAT YOU CONSIDER IT NEITHER IMPORTANT NOR UNIMPORTANT THAT BUSES SHOULD BE HOT INSIDE

IMPORTANT — UNIMPORTANT

HELPFUL CREW : ___ : ___ : ___ : ___ : ___ : ___ : ___ :
RELIABILITY : ___ : ___ : ___ : ___ : ___ : ___ : ___ :
QUICK JOURNEY : ___ : ___ : ___ : ___ : ___ : ___ : ___ :
GOOD DRIVING : ___ : ___ : ___ : ___ : ___ : ___ : ___ :
REGULAR SERVICE : ___ : ___ : ___ : ___ : ___ : ___ : ___ :
CHEAP FARES : ___ : ___ : ___ : ___ : ___ : ___ : ___ :
CLEAN INSIDE : ___ : ___ : ___ : ___ : ___ : ___ : ___ :
BUS SHELTERS : ___ : ___ : ___ : ___ : ___ : ___ : ___ :

6. WE WOULD NOW LIKE YOU TO THINK OF THE BUS SERVICES YOU USE IN TYNE AND WEAR. PUT ONE (X) AT THE APPROPRIATE PLACE ON EACH LINE TO SHOW HOW YOU THINK TYNE AND WEAR BUSES PERFORM:-

RELIABLE :___:___:___:___:___:___:___: UNRELIABLE
UNHELPFUL CREW :___:___:___:___:___:___:___: HELPFUL CREW
GOOD DRIVING :___:___:___:___:___:___:___: BAD DRIVING
DIRTY INSIDE :___:___:___:___:___:___:___: CLEAN INSIDE
GOOD BUS SHELTERS :___:___:___:___:___:___:___: BAD BUS SHELTERS
EXPENSIVE FARES :___:___:___:___:___:___:___: CHEAP FARES
QUICK JOURNEY :___:___:___:___:___:___:___: SLOW JOURNEY
ENOUGH BUSES :___:___:___:___:___:___:___: TOO FEW BUSES
BAD SERVICE :___:___:___:___:___:___:___: GOOD SERVICE

7. ANY OTHER COMMENTS?

Figure 2

TO HELP US PLAN TRANSPORT SERVICES IN TYNE AND WEAR, WE WOULD LIKE TO KNOW WHAT YOU THINK OF PUBLIC TRANSPORT IN THE COUNTY. PLEASE FILL IN THIS SURVEY FORM AND HAND IT TO OUR SURVEY STAFF WHEN YOU GET OFF.

TYNE AND WEAR TRANSPORT

THANK YOU FOR YOUR HELP

1. WHAT IS YOUR AGE?
   - LESS THAN 16 ☐
   - 16 - 64 ☐
   - 65 + ☐

2. WHAT IS YOUR SEX?
   - FEMALE ☐
   - MALE ☐

3. HOW OFTEN DO YOU USE A BUS IN TYNE AND WEAR?
   - ONCE A WEEK OR LESS ☐
   - MORE THAN ONCE A WEEK ☐

4. HOW OFTEN DO YOU USE METRO?
   - ONCE A WEEK OR LESS ☐
   - MORE THAN ONCE A WEEK ☐

5. LISTED BELOW ARE SOME FEATURES WHICH MAKE UP A BUS SERVICE. THINK OF THE BUS SERVICE YOU WOULD LIKE AND TICK ONE BOX ALONGSIDE EACH FEATURE ACCORDING TO HOW YOU RATE IT.

VERY IMPORTANT	IMPORTANT	DON'T MIND	
☐	☐	☐	HELPFUL CREW
☐	☐	☐	RELIABILITY
☐	☐	☐	QUICK JOURNEY
☐	☐	☐	GOOD DRIVING
☐	☐	☐	REGULAR SERVICE
☐	☐	☐	CHEAP FARES
☐	☐	☐	CLEAN INSIDE
☐	☐	☐	BUS SHELTERS

6. WE WOULD NOW LIKE YOU TO THINK OF THE BUS SERVICES YOU USE IN TYNE AND WEAR. TICK ONE BOX ON EACH LINE WHICH BEST EXPRESSES YOUR OPINION OF TYNE AND WEAR BUSES:

RELIABLE	☐	NEITHER	☐	UNRELIABLE	☐
UNHELPFUL CREW	☐	NEITHER	☐	HELPFUL CREW	☐
GOOD DRIVING	☐	NEITHER	☐	BAD DRIVING	☐
DIRTY INSIDE	☐	NEITHER	☐	CLEAN INSIDE	☐
GOOD BUS SHELTERS	☐	NEITHER	☐	BAD BUS SHELTERS	☐
EXPENSIVE FARES	☐	NEITHER	☐	CHEAP FARES	☐
QUICK JOURNEY	☐	NEITHER	☐	SLOW JOURNEY	☐
ENOUGH BUSES	☐	NEITHER	☐	TOO FEW BUSES	☐
GOOD SERVICE	☐	NEITHER	☐	BAD SERVICE	☐

7. ANY OTHER COMMENTS? ....
....

Figure 3

TYNE AND WEAR PTE - TRANSPORT SURVEY

PROPOSED QUESTIONNAIRE

SECTION A

Please place a tick in the correct box

1. What is your age? Under 16 ☐ 16 - 64 ☐ 65 + ☐

2. What is your sex? Male ☐ Female ☐

3. How often do you use a bus? More than once a week ☐ Once a week or less ☐

4. How often do you use Metro? More than once a week ☐ Once a week or less ☐ Not at all ☐

5. How often do you make journeys requiring interchange between bus and Metro? More than once a week ☐ Once a week or less ☐ Not at all ☐

6. Do you own a car? Yes ☐ No ☐

7. Do you have regular access to a car owned by another member of your household? Yes ☐ No ☐

8. Do you hold a current Tyne and Wear Travel Ticket/Travel Permit? Yes - all day (blue) ☐ Yes - off peak (green) ☐ Yes - OAP (brown) ☐ Yes - other ☐ No ☐

8A. If no, have you held one in the last 12 months? Yes ☐ No ☐

9. Whereabouts do you live? (Please give street and name of District) ________________ ________________

TYNE AND WEAR PTE - TRANSPORT SURVEY

SECTION B

10. Which bus routes do you use most often? Route No. ________ Route No. ________ Route No. ________

We would now like you to consider these bus services and then tell us what you think of them.

Place a cross (X) on each line in the place which best expresses your opinions of these bus services in the terms described.

Example: 1 2 3 4 5 6 7

Unpleasant |__|__|__|__|__|_X_|__| Pleasant

... would mean that you consider Tyne and Wear buses to be fairly pleasant.

Now place one cross (X) in each of the lines below.

	1 2 3 4 5 6 7	
11. Unhelpful bus crew	|__|__|__|__|__|__|__|	Helpful bus crew
Dirty inside bus	|__|__|__|__|__|__|__|	Clean inside bus
Slow journey	|__|__|__|__|__|__|__|	Quick journey
Expensive fares	|__|__|__|__|__|__|__|	Cheap fares
Unreliable	|__|__|__|__|__|__|__|	Reliable
Inadequate bus shelters	|__|__|__|__|__|__|__|	Adequate bus shelters
Poor driving standards	|__|__|__|__|__|__|__|	Good driving standards
Too few buses	|__|__|__|__|__|__|__|	Enough buses
Inadequate information	|__|__|__|__|__|__|__|	Adequate information
Dirty exterior	|__|__|__|__|__|__|__|	Clean exterior
Never get a seat	|__|__|__|__|__|__|__|	Always get a seat
Uncomfortable seats	|__|__|__|__|__|__|__|	Comfortable seats

12. Using the same scale, what do you think of these services overall?

1 2 3 4 5 6 7

Bad service |__|__|__|__|__|__|__| Good service

13. Which three of the above factors are most important to you in a bus service?

1. ________________
2. ________________
3. ________________

TYNE AND WEAR PTE - TRANSPORT SURVEY

SECTION C

We would now like you to consider the Tyne and Wear Metro and tell us what you think of it.

Place a cross (X) in each line in the place which best expresses your opinions of Metro.

14.

	1	2	3	4	5	6	7	
Slow journey								Quick journey
Unreliable service								Reliable service
Poor bus links								Good bus links
Infrequent								Frequent
Dirty inside								Clean inside
Expensive fares								Cheap fares
Poor access to Metro								Good access to Metro
Unhelpful drivers								Helpful drivers

15. Using the same scale, what do you think of Metro services overall?

	1	2	3	4	5	6	7	
Bad service								Good service

16. Which three of the above features are most important to you in a Metro service?

1. ______________________________
2. ______________________________
3. ______________________________

17. Any other comments you would like to make?

THANK YOU VERY MUCH FOR YOUR HELP

PLEASE RETURN YOUR COMPLETED FORM TO US IN THE REPLY-PAID ENVELOPE

Figure 4 Research design options for monitoring Quality of Service in Tyne and Wear

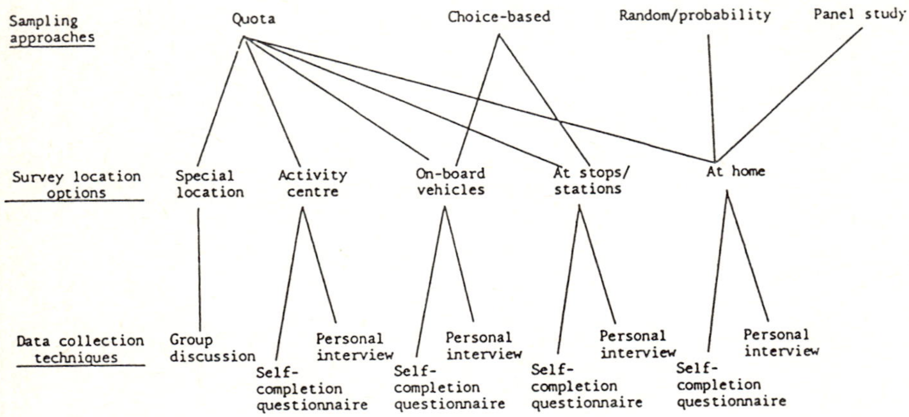

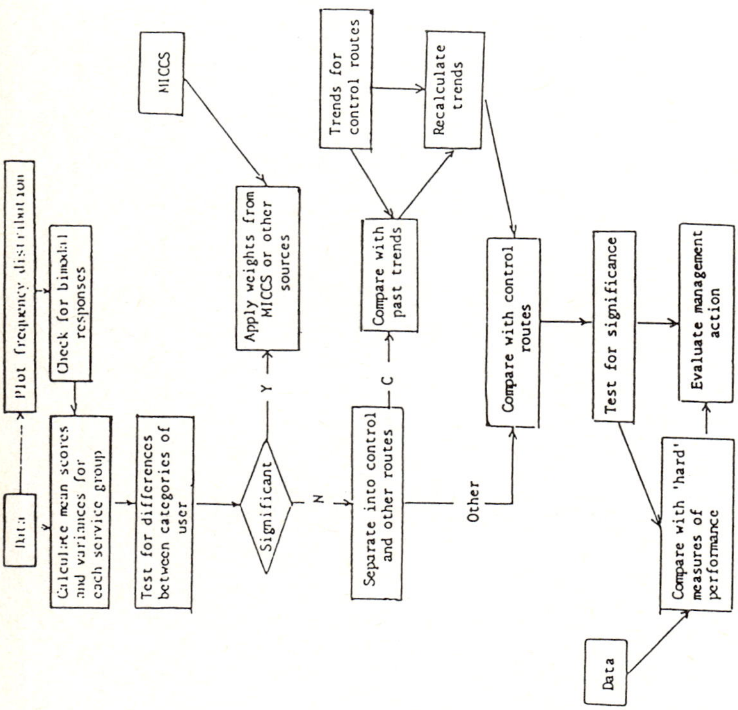

Figure 5 Procedure for Analysis

Behavioural Research for Transport Policy
pp. 423–441.

Using Quality to Predict Demand for Special Transportation

Claire E. McKnight
Urban Transportation Center
University of Illinois at Chicago

Anthony M. Pagano
Urban Transportation Center
University of Illinois at Chicago

Robert E. Paaswell
Urban Transportation Center
University of Illinois at Chicago

Introduction

There are a wide variety of special transportation services targeted to the elderly and handicapped. The demand for trips on these services vary greatly. In expanding these services or designing new ones, planners need to have accurate estimates of demand. However, the usual method of demand analysis for special services is based on estimating the size of the target population and multiplying it by an assumed trip rate taken either from a survey of the target population or from a similar system in another locale. An implicit assumption is that the characteristics of the actual service to be provided have little impact on its use. Not surprisingly, the demand that develops for the actual system often has little relation to the pre-implementation estimate. In order to improve demand analysis for special services, the effects of the pertinent causal factors need to be incorporated into the analysis. These include the characteristics of the target population, fare, marketing, other available transportation, and the quality of the service.

Many characteristics of service contribute to quality, such as reliability, safety, and comfort. Because including them all is not feasible in demand analysis, finding a way to reduce the number of separate variables is necessary. One approach is to develop a comprehensive measure to represent quality. The purpose of this paper is to describe such a measure or index of quality and how it was derived from a ranking by users of special service.

The approach that we took was first to develop a comprehensive list of service attributes that contribute to the quality of service. The second step was to survey users of special transportation to determine the relative importance of each attribute to the users. In the third step, an index was constructed from the responses to the survey using psychometric methods. Finally, a method of measuring the attributes was devised.

The next section of the paper will briefly discuss previous work on 1) measuring quality of service for transportation and 2) demand for special transportation. The third section will describe how we developed the index of quality. The fourth section presents and discusses the index. The final section discusses its uses and further research.

Previous Work

Most demand analysis for the elderly and handicapped has dealt with describing existing travel patterns (e.g., Grey, 1978; Paaswell, 1977) or estimating latent demand. Some of the descriptive studies have looked at the relationships of economic, demographic, or life style characteristics to trip patterns (e.g., Paaswell and Recker, 1978; Wachs, 1980). The few traditional models of demand that have been developed for special services have not been reliable (Hoel and Davis, 1984).

Latent demand analysis has generally been either based on gap analysis or surveying the target population. In gap analysis (e.g., Falcocchio, 1976), the trip rates of elderly or, more frequently, handicapped people are compared with the trip rates of able-bodied people in the 20-55 year range. The difference in trip rates is attributed to travel barriers; the implicit assumption is that travel barriers are the only or major reason for a reduced trip rate. This type of analysis is not useful for estimating demand for a particular service both because of the weakness of the implied assumption and because no transportation service can ever remove all the travel barriers.

In survey methods, (e.g., Michaels and Weiler, 1974; Abt, 1972) the elderly and/or handicapped person is asked how many more trips he or she would take if a transportation mode, which is briefly described in the question, was available. One problem with this method is that the descriptions of the transportation mode are, of necessity, overly simple. A more serious problem is that of non-commitment bias, i.e., the tendancy for people to over estimate their demand for a hypothetical service because they are not faced with the actual problems of use or alternative choices. Hartgen, Pasko, and Howe (1976) estimated that the actual trip making rate would be from one third to one half of the rate projected based on a survey.

Paaswell and Recker (1978) did seminal work in combining modal representations by attributes with service characteristics desired by an elderly and handicapped population. Their introduction of multi-dimensional scaling techniques into this problem clarified the heterogeneous nature of the target group in terms of modal attributes (safety, timeliness, etc.) as a function of preferred mode, available mode, and trip purpose.

Lewis (1984) has made one of the few attempts to model demand for a special transportation service on the characteristics of the service and of the users. He also attempted to incorporate quality variables in the model. The final model incorporated fare, service area population, and four quality variables, all as dummy or binomial variables: use of lift-equipped vehicles, restrictive eligibility requirements, requirements for a subscription, and 24 hour or greater advance reservation. A surprising implication of the model is that changing from lift-equipped vehicles to vehicles without lifts would reduce demand by almost 50 percent. This is surprising in that a small percent of the mobility limited population is wheelchair bound. One explanation for this, which will be supported later in the paper, is that the use of lift-equipped vehicles is correlated with and, in the model, acts as a proxy for high quality of service.

There is a fairly large body of research on quality of service for conventional transit. Some of these deal with definition and measurement (e.g., Tomazinis, 1975; Allen and Grimm, 1980; Allen and DiCesare, 1976). Others deal with mode preference or mode choice (e.g., Foerster, Young, and Gilbert, 1977; Recker and Stevens, 1977; Recker and Golob, 1976). We used these previous studies to help develop a comprehensive list of the attributes of quality of service in transportation. The most frequently mentioned attributes were:

- Delays en route
- Traffic safety
- Personal security
- Frequency of service or convenience of pick-up times
- Directness of service (no transfers)
- Ride comfort
- Temperature control
- Noisiness
- Seat availability
- Crowdedness
- Walking distance to vehicle
- Sheltered waiting areas

In the literature on special services, there are many references to the need for quality, but few attempts to define it or relate quality to demand. Paaswell (1977), in discussing techniques of demand analysis, suggests that model attributes, which we include under quality (e.g., step height, provisions for wheelchairs), should be variables in demand models.

Falcocchio (1979), while not directly discussing either quality or demand, comes the closest to developing a system of performance measures that might be considered definitions of quality. Using semantic ratings by handicapped users, he developed separate weighting systems for six subgroups of the elderly and handicapped. He then used these systems to evaluate three transportation systems. The attributes he included in his system are:

Convenience: Reliability
Waiting Time
Transfers
Ease of Entry and Exit
Walking Distance

Comfort: Heating and Ventilation
Noise
Sudden Stops and Turns
Having a Seat

Safety: Fear of Falling
Fear of Muggings

Development of the Index

Quality, as is clear from the previous section, is a broad term. It covers such concepts as reliability, safety, and comfort. These concepts are too general to be measured. They need to be defined in specific and measurable terms. We have defined quality in a hierarchical fashion. Quality is the sum of general aspects which in turn are the sums of specific attributes as shown in Exhibit 1.

The first step in developing the index was to compile a comprehensive list of attributes of quality. Our initial list included all the attributes discovered in the literature review. Because most of the literature was concerned with fixed route transportation, we drew from our own experience to add attributes specific to demand responsive and special transportation. This initial list of 64 attributes was then categorized into nine general aspects (e.g., reliability, safety).

In order to ensure that the list covered all important attributes and to eliminate the less important ones, we sent it to a panel of 22 specialists in the field of transportation for the elderly and handicapped. The panel included academics, managers of special transportation systems, government officials in agencies concerned with this issue, and advocates for the elderly and/or handicapped. The members of the panel were asked to add any additional attributes that they felt should be on the list and to rate all the attributes on a five point scale.

Based on their responses, we added a few attributes and

Exhibit 1

Hierarchical Structure of Partial Index

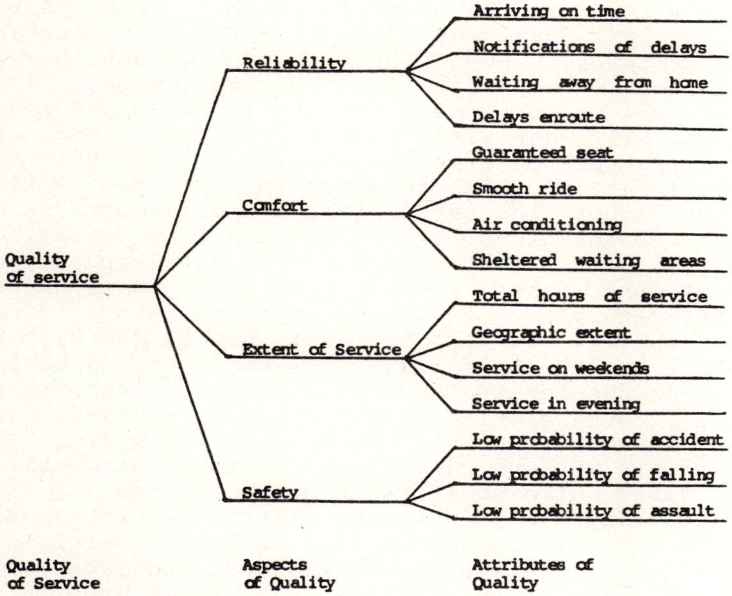

reduced the total number from 64 to 42.

This list, recategorized into eight aspects, was then sent to a sample of elderly and/or handicapped people. The sample was asked to rank each attribute according to its importance to its aspects and then to rank the eight aspects according to their importance to quality of service. They were asked to rank the attributes and aspects, rather than rate them, in order to force distinctions. We were concerned that the majority of the attributes would be rated as extremely important, reducing our ability to determine their relative importance. However, the task of ranking is more difficult than rating, particularly as some of the list of attributes were relatively long. As a result, although 236 of the 659 questionnaires were returned (a 36% response rate), only 148 of these were useable (22% useable rate). Exhibit 2 presents some of the characteristics of the sample.

Although the elderly and handicapped are often discussed as if they were a homogenous group, they have many differences that may contribute to differing valuations of service attributes. Even within the two groups there may be wide variations in how they value the attributes. Thus our next step was to determine if one index could represent quality for all elderly and handicapped people and, if not, what subgroupings should be considered. To do this we used the Kruskal-Wallis test to determine if the rankings of various subgroups indicated that the subgroups were from independent populations. We tried subdividing the overall sample by all combinations of age and handicap. From the results of the test, we concluded that there were four independent groups: Under 65 semiambulatory, under 65 wheelchair users, over 65 wheelchair users, and over 65 non-wheel chair users. (We use the word semiambulatory to describe non-wheelchair users who have a mobility problem such as difficulty walking long distances or climbing steps.) The last group, over 65 non-wheelchair users, includes elderly people with no handicap affecting their mobility as well as those with mobility problems but who do not use wheelchairs. There were only six elderly wheelchair users in the sample and this is too small a number on which to base a separate index. The Kruskal-Wallis test indicated that the rankings of the elderly wheelchair users had more in common with the other elderly than with the under 65 wheelchair users. Thus, we ended up with three groups: the under 65 semiambulatory, the under 65 wheelchair user, and the elderly.

The results of the questionnaires were rank orderings of the aspects of quality of service and rank orderings of the attributes under each aspect. In order to develop weights, cardinal data, rather than ordinal, is required. Therefore, the next task was to transform the information into an implied interval scale using psychometric scaling techniques.

Exhibit 2

CHARACTERISTICS OF USER SAMPLE

	Total Sample	Under 65		65 or older		
		SA	WC	No	SA	WC
Sample Size	148	26	58	17	41	6
Age						
under 55	72	17	55	—	—	—
55 - 65	12	9	3	—	—	—
65 or older	64	—	—	17	41	6
Male	34	9	20	2	2	1
Female	112	16	38	15	38	5
Employment						
Full time	32	6	25	0	1	0
Part time	29	7	15	3	3	1
None	85	13	17	14	36	5
Household income						
under $7,000	55	10	12	7	24	2
$7,000 - $10,000	24	4	7	5	7	1
$10,000 - $15,000	30	2	12	2	3	1
over $15,000	39	10	27	-	1	1
Transportation						
Have used special service	129	20	53	15	36	5
Often use special service	67	8	22	9	27	1
Often drive auto	36	8	23	1	3	1
Often ride in auto	61	15	22	9	13	2
Often use bus or train	35	11	8	4	12	0
Often use taxi	18	2	14	0	1	1

Note: Some subcategories do not add to the sample total because of missing data.
SA = Semiambulatory
WC = Wheelchair bound
No = No handicap

According to psychophysical theory (Guilford, 1954) there is a psychological scale that is cardinal and consistent between individuals. Further, the relation between ranked responses to stimuli (or attributes in this case) and the psychological scale can be determined using psychometric scaling techniques. The theoretical basis for the scaling techniques is the assumption that a stimulus (or attribute) will not elicit the same response from all observers. The distribution or dispersion of responses is assumed to be normally distributed about a mean which represents the true location of the response on the scale. The overlap of the distributions of responses for the attributes is the basis for determining the distances between attributes on the psychological scale.

The method for determining a psychological scale from responses to a questionnaire in which respondents have indicated rank order is derived from the method for comparison of pairs which, in turn, is based on Thurstone's law of comparative judgement which states:

$$R_k - R_j = z_{ij} \, [var(j) + var(k) - 2r_{jk} \, var(j)^{\frac{1}{2}} \, var(k)^{\frac{1}{2}}]^{\frac{1}{2}}$$

where

R_k, R_j = the mean psychological values of stimuli S_j and S_k

z_{jk} = the deviate from the mean of a unit normal distribution corresponding to the proportion of responses where k is judged greater than j

var(j), var(k) = the variances of distributions R_j and R_k

r_{jk} = the coefficient of correlation between the responses to R_j and R_k

The value of z_{jk} can be derived directly from the responses to the questionnaire using normal curve tables. However, there are still three unknowns in the equation. Thurstone suggested simplifying the equation by assuming that r_{jk}, the intercorrelation between responses, is zero (Thurstone's Case III). This is most likely to be true if the stimuli vary in only the aspect in question. In that case the equation simplifies to:

$$R_k - R_j = z_{jk} \, [var(j) + var(k)]^{\frac{1}{2}}$$

The standard deviations of the responses can be estimated using Burros's approximation:

$$var(j)^{\frac{1}{2}} = [n / \sum_j (1/V_{zj})] \; / \; V_{zj}$$

where

n = number of stimuli

V_{zj} = variances of the z_{jk} around their mean

Using this method, cardinal values were assigned to each attribute within an aspect and to each aspect as follows:

1. A table of proportional frequencies was constructed for each aspect. The tables were n by n matrices where n was the number of attributes that were rank ordered. A table was also constructed for the group of eight aspects.

2. The appropriate standardized deviation (or z-value) for each proportion was read from the normal curve table.

3. The standard deviation for each attribute was estimated using Burros's approximation.

4. A new table of intervals between attributes was constructed by multiplying each z_{jk} by $(var(j) + var(k))^{\frac{1}{2}}$

5. The rows of the matrix were averaged to get the final scale intervals for the attributes.

The results of these steps were scale values for each set of attributes under each aspect and for all eight aspects. The attributes were proportional to their relative importance to the aspect (or, for the aspects, to quality) and added to zero. While this met the objectives for the weights in theory, we felt that it would be preferable for all the weights to be positive. So each of the nine scales (one for each of the eight groups of attributes plus one for the group of eight aspects) were linearly transformed so that all values are positive and each scale sums to ten. It should be pointed out that the transformation value is somewhat arbitrary. By choosing a different value, the range of the final weights will change, although the relative distance between weights will remain constant. Thus, given the transformation:

$$y_i = ax_i + ab$$

where

x_i = the initial scale value

y_i = the transformed weight

$a = 10/\sum_i (x_i + b)$

i indicates the attribute or aspect

b = a positive value larger than the absolute value of the smallest x.

While y_i/y_n depends on the value of b that is chosen, the ratio $(y_i - y_j)/(y_i - y_k)$ is independent of b (i, j, and k are three different attributes in the same aspect).

For the final scale of weights, the weight of each attribute is multiplied by the weight of the aspect of which it is a part. The index can be represented as:

$$Q = \sum_j \sum_i W_j \, W_{ij}$$

where

Q = the index

W_{ij} = the relative contribution of attribute i to aspect j and

W_j = the relative contribution of aspect j to quality of service.

Index of Quality

This methodology was used to develop three indices, one for each of the three subgroups: the under 65 semiambulatory, the under 65 wheelchair user, and the elderly. Exhibit 3 shows the weight of each attribute within its aspect group and its final weight. Because the index sums to 100, the final weight for each attribute is also the percentage that the attribute contributes to the index of quality of service.

The under 65 semiambulatory respondents placed the most weight on the reliability aspect, followed by convenience and extent of service, which have equal weight. Reliability is also the most important of the aspects to the under 65 wheelchair users, followed by extent of service and ease of vehicle access. The elderly, on the other hand, value safety most, with reliability second and vehicle access third.

Many special transportation services provide trips for all or a combination of two of the above groups. While the quality of a transportation service may be different for one subgroup than for another, it is convenient for some purposes to have just one index. For a particular special service, a composite index can be constructed by combining the three indices weighed by the percentage of service consumed by the relevant subgroup. Such a composite index was developed using average proportions of users for four services described by Grey (1980). These proportions were 70 percent elderly, 22 percent semiambulatory, and eight percent wheelchair users. This combined index is shown in Exhibit 4.

Exhibit 3

Indices of Quality

	Under 65 Semiambulatory		Under 65 Wheelchair User		Elderly	
ASPECTS	Weight within aspect	% of Final Index	Weight within aspect	% of Final Index	Weight within aspect	% of Final Index
Reliability	2.12	21.2	2.02	20.2	1.60	16.0
Comfort	0.45	4.5	0.61	6.1	0.85	8.5
Convenience	1.51	15.1	1.36	13.6	1.22	12.2
Extent of Service	1.51	15.1	1.49	14.9	0.88	8.8
Vehicle Access	1.18	11.8	1.46	14.6	1.40	14.0
Safety	1.31	13.1	1.35	13.5	1.73	17.3
Driver Characteristics	1.13	11.3	1.09	10.9	1.39	13.9
Ease of Dealing w/Office*	0.79	7.9	0.62	6.2	0.93	9.3
Attributes related to RELIABILITY						
Arriving on time **	4.3	9.1	4.3	8.7	3.7	5.9
Notification	3.2	6.8	3.2	6.5	2.8	4.5
Wait time at home	0.8	1.7	0.8	1.6	1.5	2.4
Wait time away from home	1.0	2.1	1.0	2.0	1.3	2.1
Delays on Vehicle	0.7	1.5	0.7	1.4	0.7	1.1
Attributes related to COMFORT						
Guaranteed seat	2.5	1.1	3.8	2.3	2.3	2.0
Condition & Cleanliness	1.1	0.5	0.9	0.5	1.5	1.3
Smoothness of ride	1.4	0.6	1.7	1.0	1.4	1.2
Air conditioning	1.5	0.7	1.5	0.9	1.4	1.2
Sheltered waiting areas	2.0	0.9	2.1	1.3	1.8	1.5
Seats at waiting areas	1.5	0.7	—	—	1.6	1.4
Attributes related to CONVENIENCE						
Accommodation	2.1	3.2	2.1	2.8	2.1	2.6
Selection of time	3.2	4.8	3.2	4.4	2.7	3.3
Reservation length	2.9	4.4	2.9	3.9	2.5	3.0
Return reservation	1.8	2.7	1.8	2.4	2.7	3.3
Attributes related to EXTENT OF SERVICE						
Few restrictions	2.2	3.3	2.8	4.2	2.8	2.5
Total hours of service	2.7	4.1	2.3	3.4	2.6	2.3
Service in evening	1.8	2.7	1.7	2.5	1.1	1.0
Service on weekend	1.5	2.3	1.6	2.4	1.7	1.5
Capacity	1.8	2.7	1.6	2.4	1.8	1.6
Attributes related to VEHICLE ACCESS						
Aisle width	1.4	1.6	1.6	2.3	1.1	1.5
Step height	2.0	2.4	—	—	2.1	2.9
Number of steps	1.6	1.9	—	—	1.5	2.1
Lift (or ramp)	1.1	1.3	4.4	6.4	0.7	1.0
Driver assistance	1.1	1.3	1.7	2.5	1.4	2.0
Assistance w/parcels	1.0	1.2	0.2	0.3	1.4	2.0
Distance to vehicle	1.8	2.1	2.1	3.1	1.8	2.5
Attributes related to SAFETY						
Low probability of assault	2.8	3.7	0.9	1.2	2.6	4.5
Low probability of falling	3.1	4.1	1.5	2.0	2.7	4.7
Safe driver ***	4.1	5.4	2.7	3.6	4.7	8.1
Tie down	—	—	2.9	3.9	—	—
Position of w.c.	—	—	2.0	2.7	—	—
Attributes related to DRIVER CHARACTERISTICS						
Medical emergency capability	1.6	1.8	1.6	1.7	2.2	3.0
Courtesy & friendliness	2.2	2.5	2.2	2.4	2.8	3.9
Neatness & professionalism	1.7	1.9	1.7	1.8	1.5	2.1
Knowledge of general needs	3.5	4.0	3.5	3.8	2.6	3.6
Familiarity with personal needs	1.0	1.1	1.0	1.1	0.9	1.2
Attributes related to EASE OF DEALING W/OFFICE						
Courtesy & friendliness of telephone operators	3.0	2.4	3.0	1.9	4.3	4.0
Ease of complaining	1.7	1.3	1.7	1.0	2.0	1.9
Follow-up to complaints	1.4	1.1	1.4	0.9	0.1	0.1
Clear information	3.9	3.1	3.9	2.4	3.6	3.3

Exhibit 4

Composite Index of Quality

Aspects	Attribute
1.75** Reliability	3.9* Arriving at destination on time
	2.9 Notification of delays or cancellation of service
	1.3 Wait time for pick-up at home
	1.2 Wait time for pick-up away from home
	0.7 Few delays while on the vehicle
1.61 SAFETY	4.4 Low probability of a traffic accident
	2.7 Low probability of falling
	2.5 Low probability of personal assault
	0.2 Type of tie down
	0.2 Position of the wheelchair in the vehicle
1.36 VEHICLE ACCESS	1.9 Height of first step
	1.8 Short distance from house or destination to vehicle
	1.4 Number of steps
	1.4 Assistance in getting from vehicle to destination
	1.2 Assistance in carrying packages
	1.2 Width of the aisle
	1.1 Presence of a wheelchair lift or ramp
1.31 DRIVER CHARACTERISTICS	2.9 Knowledge of general needs
	2.6 Courtesy and friendliness
	2.0 Ability to handle medical emergencies
	1.6 Neatness and professionalism
	0.9 Familiarity with habits and needs of individual user
1.29 CONVENIENCE OF MAKING RESERVATIONS	2.9 Being picked up at time selected by traveller
	2.6 Shortness of reservation time
	2.3 Convenience of return reservation procedure
	2.1 Accommodation to changes in reservations
1.07 EXTENT OF SERVICE	2.7 No or few restrictions on where vehicle will go
	2.6 Total number of hours of service
	1.8 Low rate of turning down reservations
	1.6 Service on weekends
	1.3 Service on evening
0.87 RESPONSIVENESS	3.9 Courtesy and friendliness of telephone operators
	3.7 Ease of getting clear information on service
	1.9 Receptiveness to complaints and user suggestions
	0.5 Procedures for follow-up on complaints
0.74 COMFORT	2.5 Guaranteed seat or location for wheelchair
	1.9 Sheltered waiting areas for pick-ups away from home
	1.4 Condition and cleanliness of vehicle
	1.4 Smoothness of the ride
	1.4 Air conditioning and good ventilation
	1.4 Seats at waiting areas for pick-ups away from home

* The number in front of each attribute is the weight of that attribute or its relative contribution to the relevant aspect.

** The number in front of the aspect is the weight of that aspect or its relative contribution to quality of service

Before the index can be applied to a special transportation service, a method for measuring the service's achievement of each attribute (i.e., a performance measure) must be developed. We refer to the service's achievement of satisfaction for a particular attribute as its score for that attribute. The method for deriving a score from performance measures is a scoring function.

The scoring function has a form such that the maximum score possible, representing complete satisfaction with that attribute, is one and the minimum score possible, representing no satisfaction, is zero. In between, a marginal change in the score should equal a marginal change in percent of total satisfaction. Performance measures are measurable forms of the attributes; for example, for the attribute "few or no delays on vehicle", the performance measures are expected minutes of delay per trip due to picking up other passengers and due to vehicle breakdowns.

The performance measures were selected by relying on the literature on service or vehicle standards (e.g., North Central Texas Council of Governments, 1977) and through prolonged discussion among the project staff and other professionals in the field utilizing the approach to scoring developed by Miller (1970). In some cases, performance measures are obvious from the nature of the attribute. For instance, for the attribute "guaranteed seat or location for a wheelchair," the measure is a simple yes or no. The measure of the number of steps into the vehicle is a simple count. For other attributes, measures are not obvious and require further definition. Courtesy and friendliness of the driver and condition and cleanliness of the vehicle fall into this area. Another problem is that attributes that have straightforward performance measures may be difficult to obtain reliable data for. For example, there are recognized quantitative measures of arriving at a destination on time or the smoothness of the ride that would be suitable for our purposes. However, they require either a degree of record keeping that is unusual among special transportation services or equipment that is not easily obtained. Therefore, in many cases, we relied on either estimates by the providers as in the case of measuring on-time-performance or surrogate measures such as vehicle type for measuring smoothness of ride. Graphical illustrations of scoring functions are shown in Exhibit 5. Details on all the functions are described in Pagano and McKnight (1983).

The final quality score for a special transportation service is:

$$Q_k = \sum_j \sum_i W_j \; W_{ij} \; S_{ijk}$$

where

Q_k = quality score for provider k and

Exhibit 5

Examples of Scoring Functions

SCORING FOR LENGTH OF RESERVATION

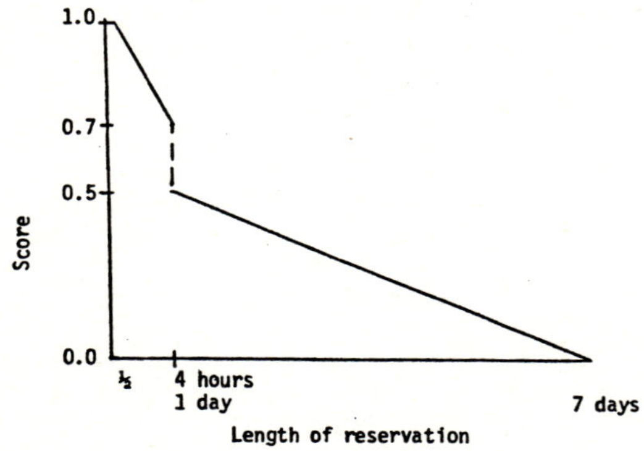

SCORING FOR HOURS OF SERVICE

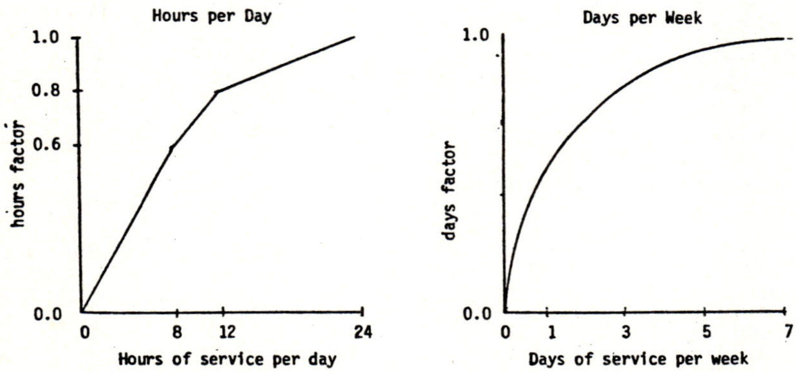

S_{ijk} = score for attribute i under aspect j for provider k

The score for any attribute ranges from zero to one and the index sums to 100; therefore the quality score for a provider falls in the range from zero to 100.

This methodology was applied to a sample of 42 special transportation providers with the following results:

User Subgroup	Quality scores: Mean	Standard deviation	Median	Range
Semiambulatory	67	7	69	45-84
Wheelchair users	66	10	64	39-86
Elderly	69	7	70	47-82
Composite of users	68	7	69	46-83

Uses of the Index

The most obvious use for the index of quality is as an evaluative tool. This could be either for self-evaluation or for comparison of different services. For self-evaluation, the manager of a special tranpsortation service could develop a quality score for the service and compare it with the average scores listed above. He could also use it for analyzing effective service improvements. By noting which attributes have the lowest scores, he knows where the system is weakest. By noting the cumulative weights attached to the attributes and aspects (i.e., the percentages listed next to the attribute weight in Exhibit 3), he can determine which improvements to a particular attribute will result in the greatest improvement in overall quality to the user. Going a step further, if he can assign costs to the various improvements, he can determine the most effective use of expenditures on improving service.

Comparative evaluation of the quality of different services is useful for organizations that want to choose one service with which to contract for service or to provide a subsidy. In some cases an organization that wants to compare several services will not have the staff or time to develop a full score for each special transportation service under consideration. A possible short-cut would be to develop attribute scores for the attributes that contribute the most to the overall quality. Exhibit 6 shows a partial index for a composite service (i.e., a service for the elderly and handicapped including wheelchair users) with suggested performance measures.

Exhibit 6
Partial Index

Attribute	Weigh in Index	Performance Measure(s)
Low probability of traffic accident	7.1	Driver evaluation and training Inspection policies
Arriving at destination on time	6.8	Percentage of time pickups are w/i 5 min.
Notification of delays and cancellations	5.1	Procedures used and delay time that triggers notification
Low probability of falling	4.3	Driver assistance Presence of handrails Driver policies Seat belts and seat positions
Low probability of personal assault	4.0	Pick-up waiting areas
Drivers knowledge of general needs	3.8	Sensitivity training Experience of drivers
Pick-up times chosen by user	3.7	Reservation system Capacity of system
Courtesy and friendliness of drivers	3.4	Judgement of users or of evaluator
Courtesy and friendliness of telephone operators	3.4	Judgement of users or of evaluator
Shortness of reservation time	3.3	Length of reservation Ability to have permanent reservation

Another use for the index of quality is to analyze general policies. For example, the index can be used to determine what types of agencies provide the highest level of service. Pagano and McKnight (1983) used the composite index to look for a relationship between the size and private/public status of an agency and quality. They found that larger agencies and private firms tended to provide a higher quality of service. Another finding was that quality and provision of wheelchair service were significantly correlated. This supports the contention that lift-equipped vehicles acted as a proxy for quality in the Lewis model; see above. There was a negative correlation between vehicle productivity and quality, indicating, not surprisingly, that there is a tradeoff between efficiency and quality.

Finally, the index can be used in demand modeling. The weights of the aspects, which represent the importance of the aspect to the user, are similar to the coefficients for those aspects in a utility function. By using one index rather than a list of aspects or attributes, the model can include all aspects of quality and, at the same time, can be relatively simple. This leaves more room for including other pertinent variables such as income, auto availability, or marketing. It is difficult to collect data for a large sample of special transportation services or even to locate a large number of systems with enough characteristics in common to warrant being included in the same sample. A restricted sample size limits the number of independent variables that can be included in a model. Thus, one variable that can summarize a large amount of information can be a definite asset in demand modeling for special services.

On the other hand, there may be advantages to not trying to summarize too much information in one variable. The decision process of making a choice between modes or between making or not making a trip may be very different from the process of evaluating quality. Even so, the index indicates which aspects and which attributes are most likely to affect demand. Developing a model with the separate aspects or attributes as independent variables would give an indication of the value of the index as a summary of the aspects for demand estimation purposes.

Obviously, additional research is needed on both demand and on the evaluation of quality. Demand models that include variables for quality and other causal factors need to be developed and evaluated by using them to estimate ridership for new and expanded systems and checking the estimates against actual ridership after implementation. Only then will communities and social service agencies be able to realistically project demand and the size of services and budgets that will be needed to meet the demand.

This paper describes a first major attempt at developing a measure of quality and a means of evaluating special services. Similar work is needed to validate the index. Also, the scoring functions need to be checked against how users actually value changes in the level of attributes. In addition, research on how to improve the achievement of some of the most important attributes is needed. Reliability and on-time performance are very highly valued, but there is little knowledge on how to improve them.

Acknowledgement

The research presented in this paper was funded by the University Research and Training Program of the Urban Mass Transportation Administration, U. S. Department of Transportation. The views expressed are those of the authors and not necessarily the views of the sponsor. The participation of Michele Dichter-Figueroa in this project was invaluable.

References

Abt Associates (1974) Accessibility of the Metropolitan Washington, D.C., Public Transportation System to the Handicapped and Elderly, prepared for U.S. Department of Tranpsortation.

Allen, W.G., and DiCesare, F. (1976) "Transit Service Evaluation: Preliminary Identification of Variables Characterizing Level of Service." Transportation Research Record 606, pp. 41-47.

Allen, W.G., and Grimm, L.G. (1980) "Development and Application of Performance Measures for a Medium-sized Transit System." Transportation Research Record 746, pp. 8-13.

Falcocchio, J., et al. (1976) Mobility of People and Goods in the Urban Environment: Mobility of the Handicapped and Elderly, prepared for U.S. Department of Transportation.

Falcocchio, J. (1979) A Methodology for Evaluating the Effectiveness of Transportation Improvements for the Elderly and Handicapped, prepared for U.S. Department of Transportation.

Foerster, J.F., Young, F. W., and Gilbert, G. (1977) "Longitudinal Changes in Public Preferences for Attributes of a New Transit System." Transportation Research 11, pp. 325-336.

Grey Advertising (1978) Summary Report of Data from National Survey of Transportation Handicapped People, prepared for U.S. Department of Transportation.

Grey Advertising (1980) Summary Report of a Survey of Handicapped Users and Non-Users of Special Transportation Systems in Four Cities, prepared for U.S. Department of Transportation.

Guilford, J.P. (1954) Psychometric Methods. McGraw-Hill Book Co., New York.

Hartgen, D.T., Pasko, M., and Howe, S.M. (1976) Forecasting Non-Work Public Transit Demand by the Elderly and Handicapped. New York Department of Transportation.

Hoel, L.A., and Davis, C.J. (1984) "A Market Profile Assessment for Demand Responsive Services." Specialized Transportation Planning and Practice. Vol. 2, No. 1, pp. 73-86.

Lewis, D.L. (1984) "A Model of Paratransit Demand and Supply." Third International Conference on Mobility and Transport of Elderly and Handicapped Persons; Pre-Prints of Papers.

Michaels, R.M., and Weiler, N.S. (1974) Transportation Needs of the Mobility Limited. Transportation Center, Northwestern University, Evanston, Illinois.

Miller, J.R. (1970) Professional Decision-Making: A Procedure for Evaluating Complex Alternatives. Praeger Publishers, New York.

North Central Texas Council of Governments (1977) Functional Specifications for Small Vehicles for Use by the Handicapped and the Elderly. Arlington, Texas.

Paaswell, R.E. (1977) "Estimation of Demand for Transit Service Among the Transportation Disadvantaged." Transportation Research Record 660. pp. 38-49.

Paaswell. R.E., and Recker, W. (1978) Problems of the Carless. Praeger Publishers, New York.

Pagano. A.M., and McKnight, C.E. (1983) Quality of Service in Special Transportation for the Elderly and Handicapped. Prepared for U.S. Department of Transportation.

Recker, W.W., and Golob, T.F. (1976) "An Attitudinal Modal Choice Model." Transportation Research 10, pp. 299-310.

Recker, W.W., and Stevens, R.F. (1976) "Attitudinal Models of Mode Choice: the Multinomial Case for Selected Nonwork Trips." Transportation 4. pp. 355-375.

Tomazinis, A. (1975) Productivity, Efficiency, and Quality in Urban Transportation Systems. Lexington Books, Massachusetts.

Wachs, M. (1980) The Changing Lifestyles and Travel Patterns of the Elderly. University of California Press, Berkeley, California.

8
Workshop reports

WORKSHOP 1
WORKSHOP ON ACTIVITY ANALYSIS AND TRIP CHAINING

Eric I. Pas
Duke University

Three introductory papers were presented to the workshop. Two of these papers; namely, those by Clarke and by Recker and McNally, deal with the development and application of activity-based models in the analysis of alternative transportation and related policies. One of the important elements in these models is the explicit recognition of the chaining of trips for different purposes (generally at different locations) on the same excursion from home. The third paper presented to the workshop; namely, that by Houee, deals with trip chaining in the sense of the linking together of trips by one or more modes in order to reach a single destination. Such trips have long been referred to as 'linked trips', but the extent to which travel behavior is affected by the characteristics and number of the links in the chain has not been carefully studied. An interesting result of Houee's study is that the number of links in a chain does not in itself appear to inhibit people from using a chain. Rather, the quality of the chain is important, including waiting time and schedule reliability.

The workshop participants attempted unsuccessfully to integrate the two concepts of trip chaining dealt with in the background papers. The best resolution that could be found was to consider Houee's chains as 'modal chains' with the other chains as 'purpose (or activity) chains'. While the activity-based approach can probably be used to examine the modal chaining question, it was felt that simpler approaches are probably more appropriate. The bulk of the workshop discussion concentrated on the activity-based approach to travel demand analysis and this subject is the focus of the remainder of this report. This report first summarizes the state-of-the-art in activity analysis, and then presents the workshop's findings and identifies research needs.

STATE-OF-THE-ART IN ACTIVITY ANALYSIS

Activity analysis, also referred to as the activity-based approach, refers to a wide variety of concerns, concepts and methodologies that have emerged in the past decade. This area of travel demand analysis has its roots in the time geography paradigm developed at the University of Lund and has benefited considerably from the research carried out over the past decade at the Oxford University Transport Studies Unit.

The workshop participants found it useful to define activity analysis in terms of a number of themes that recur in what is generally accepted to be the literature dealing with this subject. These themes include: (1) analysis of the demand for activity participation (and the analysis of travel as a derived demand), (2) the scheduling of activities in time and space, (3) the constraints (spatio-temporal and interpersonal) on activity and travel choice, (4) the interactions between activity and travel decisions over the day (or week, or longer time period) as well as interactions between different individuals, and (5) the structure of the household and the roles played by the various household members.

In order to evaluate the state-of-the-art, it is important to differentiate activity-based modeling and activity-based analysis. The papers prepared for the workshop by Clarke and by Recker and McNally describe the state-of-the-art in activity-based simulation models. The CARLA and STARCHILD models described in these papers are very similar, with the latter providing a more general and flexible framework. Other researchers have developed econometric activity-based models which are concerned with activity participation and scheduling or with trip chaining. The state-of-the-art in activity-based analysis has progressed substantially in the past decade and much is known about activity patterns and their relationship to sociodemographic and location factors. The workshop discussions concentrated on modeling rather than on analysis because that was the orientation of the background papers.

The CARLA model, developed at the Oxford University Transport Studies Unit, has been used in the examination of an actual policy question; namely, a hypothesized reduction of bus service to a rural village in the Netherlands. In particular, the CARLA model was used to examine the impact of a number of alternative bus schedules some of which provided the same overall level of service, but differed on the exact timing of buses. The analysis yielded useful information about who would be impacted, how many people would be impacted, and the severity of the impacts. The major conclusion from this application was that the activity-based approach was useful in examining the impacts of a policy that could not be investigated using more traditional trip-based methodologies. The important characteristic of this application, which made it particularly suitable for CARLA, is the emphasis on details of the temporal dimension. It should be noted, however, that CARLA can currently be used only to predict how people will be affected, rather than how they will behave. On the other hand, the STARCHILD model incorporates a choice module which enables prediction of how people will behave in addition to providing information about how they will be affected. The STARCHILD model, developed at the University of California at Irvine, requires further development and testing before being applied to an actual policy analysis.

WORKSHOP FINDINGS

The workshop participants discussed and reached agreement on a number of issues. The major findings of the workshop are described below.

1. Activity analysis is a valid, useful and important approach in travel demand analysis.

Activity analysis is important from a policy perspective in that it can be used to address issues that cannot be examined with more conventional (i.e. trip-based) approaches. Those workshop participants who interact with policy-makers recognize and value the rich information which is provided by the activity-based approach. The approach is particularly suited in those cases in which the temporal dimension is important and/or where linkages (between trips and/or people) are likely to be significant. Thus, if these characteristics are not important in a particular situation, simpler approaches will probably be more appropriate than activity models. Therefore, travel demand analysis requires a range of tools, each with an appropriate domain of application.

2. The activity-based approach is not inherently 'data hungry'.

The activity-based approach is often incorrectly criticized as being very 'data hungry'. Considerable discussion was devoted to this issue and a number of important points emerged from the discussion. First, it is important to differentiate between the data needed for studies aimed at improving our understanding of travel behavior and the data needed for policy application of activity-based models. Second, most of the activity-based work up to the present time has been aimed at improved understanding of an extremely complex phenomenon. Thus, very detailed data were collected in many early studies because it was not known what data were necessary, but this situation is changing rapidly. Third, it must be recognized that activity diary data are useful for application in fields other than transportation and that such data have recently been collected in national surveys in some countries. Fourth, future research is likely to demonstrate how more readily available data sets (e.g. traditional travel surveys) may be used in conducting activity-based studies.

3. A prediction of who will be affected by a policy is useful, and possibly sufficient, information for policy-makers.

The implicit assumption in traditional travel demand analyses is that the output should be a prediction of how behavior will change if a particular policy is implemented. The workshop participants examined the somewhat radical idea that, in some cases, predicting how many people will be impacted by a particular policy (and identifying the impacted groups) is sufficient information for a policy maker. In particular, this approach is relevant in the case of restrictive policies which potentially require some individuals to change their existing behavior by reducing the available options. Further, the reliability of the predictions obtained from models may be enhanced if we limit the information we ask them to provide.

4. The activity-based approach is particularly appropriate for analyzing policies and trends of increasing interest to policy-makers.

Policy-makers are increasingly confronted with examining policies, and the implications of trends, for which the traditional trip-based methodologies are either deficient or entirely inappropriate. For example, the trip-based methodologies pay little or no attention to the temporal scheduling of the desired activity program. Thus, the traditional methodologies are inappropriate to the investigation of the transportation impacts of temporal management policies such as flextime or four-day work weeks. On the other hand, the activity-based approach is particularly well suited for analysis of such policies. Similarly, the effects of recent and continuing sociodemographic trends on travel and related behavior are best examined by the activity-based approach. These trends include increased labor force participation by women, smaller family size, etc.

5. A comprehensive, integrated theory does not exist although much is known about particular elements.

The workshop participants agreed that no comprehensive, integrated theory underlies the activity-based approach and that the theoretical frameworks in use are based on introspection rather than a formal behavioral theory. Some workshop participants considered the development of a comprehensive behavioral theory an important research need while others felt that this was not an important concern.

Despite the lack of a comprehensive behavioral theory, the activity-based research has contributed much to our understanding of travel behavior, and this knowledge will be invaluable in the development of a comprehensive theory.

6. The diffusion of the activity-based approach into practice is not unduly slow.

Some critics of the activity-based approach have noted that it has been very slow to diffuse into the practice of travel demand analysis and forecasting. The workshop participants noted that roughly a decade has passed since the origin of activity-based research and that during this period there has been only one identifiable application of a formal activity-based model in an actual policy analysis. While this might be considered disappointing, a number of factors must be taken into account. First, it took a number of years before the discrete choice random utility models were used in practice. Second, the activity-based approach began developing during the time when practitioners were just beginning to accept the use of discrete choice models. Thus, it was unlikely for practitioners to give careful consideration to the activity-based approach. Third, the activity-based approach recognizes and attempts to deal with the inherent complexity of travel and related behavior. Therefore, the development of practical forecasting techniques based on this approach will inevitably require considerable time and effort. This consideration is, at least partially, offset by the final finding of the workshop.

7. The activity-based approach does not require the development of entirely new models and techniques.

The activity-based approach is seen by many to be exactly that, an approach. Thus, the introduction of activity analysis does not necessarily require the development of completely new models and techniques. A number of recent developments illustrate the possibilities for extremely productive cross-fertilization between the activity-based approach and the on-going development of discrete choice random utility models. For example, recent efforts to extend discrete choice models to incorporate constraints on travel behavior have been influenced by the knowledge acquired through activity-based research. On the other hand, discrete choice models have been used to operationalize activity analysis models. Similarly, the considerable knowledge gained in particular areas of travel demand analysis during the past 15 years is likely to be invaluable in the further development and implementation of activity-based research. For example developments in model transferability and stated preference techniques will certainly contribute to the further development of activity analysis methods.

RESEARCH NEEDS

A number of research needs were identified and discussed by the workshop participants. These research areas are summarized below.

1. The Choice Process

A better understanding of the process used by individuals in selecting among travel and related alternatives is not a new research area in travel demand analysis. For example, the question of whether individuals employ optimizing or satisficing decision rules in travel choice behavior has been an issue since the early development of disaggregate behavioral models. This question still exists and is an important issue in the development of the activity-based approach.

Considerable evidence has been accumulated that suggests that linkages exist between travel-activity decisions on the same and different days. However, little is known about the decision process which leads to the observed behavior. Is a simultaneous choice process employed? If so, what is the planning horizon? How are adjustments made to cope with unforeseen circumstances during the planning horizon? These are considered important questions to be addressed by future research in this area.

2. Intrahousehold Interdependencies

There is considerable evidence in the literature (both direct and indirect) which suggests that the travel and related behavior of the members of the same household are interdependent. Research is needed to understand these linkages better and to formulate models which capture these interdependencies. In particular, little is known about the extent to which observed linkages are considered binding by the individuals. For example, many families do eat their evening meal together, but how strongly do they feel constrained to do so?

3. Perceived Distinct Patterns of Behavior

Recent research has shown that, from a statistical standpoint, the large number of possible daily travel-activity patterns may be described by relatively few 'representative pattern types'. However, nothing is known about how individuals perceive these pattern types. Do they perceive only a few distinct pattern types or do individuals distinguish between many pattern types?

4. <u>Utility of Activity Patterns</u>

Research is needed to identify the attributes of activity patterns considered by individuals and how these attributes are combined to form an overall rating (utility) for an activity pattern. For example, it appears that individuals value the flexibility provided by an activity pattern. How should such attributes be measured and how do we determine the value placed on such attributes?

5. <u>Models Employing "Cheap" Data</u>

In the context of policy analysis, it is likely that activity analyses can be conducted using more commonly available data, including traditional travel surveys as well as secondary data sources. Research is needed to identify which data are needed for activity analyses and how to integrate different sources of data.

6. <u>Alternative Methods of Data Collection</u>

The activity-based approach first introduced the idea of alternatives to the more traditional survey and questionnaire data collection formats. An interesting research opportunity exists in examining the data collection (and analysis) possibilities made available by recent advances in computer and communication technologies.

7. <u>Methods of Displaying Results of Policy Analysis</u>

The output of an activity-based analysis is considerably more complex than that from a more traditional analysis of a particular policy and contains a wealth of relevant information about how people might behave as a result of that policy. Thus, there is a need to examine what results should be provided to policy-makers and how these results can best be presented. The use of modern electronic technology should once again be investigated.

WORKSHOP 2

LONGITUDINAL SURVEYS AND ANALYSIS METHODS

A. van der Hoorn, Ministerie van Verkeer en Waterstaat, The Netherlands

The papers in this workshop were not presented collectively in the first session, but were presented and discussed consecutively during the three days of the workshop. The discussions which took place can be summarised under four headings: the policy context, issues in data collection, measurement of change, and modelling change.

a. The policy context

The research framework is largely determined by the policy context. The emphasis in the Dutch panel is on monitoring public transport use, which forms a relatively small part of total mobility. Tripdiaries must, therefore, be collected over a large number of person-days to have enough observations for analysis. This has a major influence upon costs and upon drop-out from the panel.
The panel in Australia has more limited data requirements. Trip-diaries are not collected. The objective is tot study fuel consumption in relation to vehicle holdings and vehicle use and to make long-term predictions for these. It is expected that a longitudinal panel can give better insight into decision structures of individuals over time and, thus, give better forecasts.

The Dutch panel, among other things, must give answers to short-term policy questions (effect of the Fare Plan 1984-1987), which causes heavy pressure on the fieldwork and analysis.
Another problem can be that the costs of the panel are shared by various government agencies each with its own research needs. This can cause the questionnaires to unduly expand.
The presentation of results of panel studies to policy makers was not considered to be more demanding than those of traditional studies. There is a large scope for e.g. the use of graphics.

b. Data collection

The Dutch panel has a relatively large non-response and drop-out. The main reasons are the high burden on respondents of the 7-day diary and the postal method used in wave 2. The organisational structure is also very important. It is essential to have good interviewer-household relationships and to spend much time and effort in convincing marginal respondents to remain in the panel. A good cross-section-survey interviewer (maximizing survey output) is not necessarily a good panel interviewer.

Panel attrition is not by definition harmful. If drop-out occurs in a random fashion, analysis results may be unbiased. In addition correction methods for bias (based on factors derived from attrition models) are available.

The attrition in the Dutch panel is likely to be non-random: older households making few trips and low-income households are more likely to drop out. Also in the Cardiff continuous panel (6 months) on shopping behaviour the later drop-outs had different characteristics from the

earlier ones. There seems to be a tendency for panels to become more mobile on average in the course of time: people making few trips are more likely to fall out.
In addition, drop-out is largest in the earlier stages of a panel: once less motivated people have fallen out the remaining households will continue to participate.

There is no well-established evidence yet on the relation between drop-out and time span between successive waves. The same applies to the influence of payment of respondents. The Cardiff panel members were paid 25 pounds per household, the Dutch panelists only received a small gift, the Australian panel members were not compensated in any way.

It is very important to conduct a short preliminary interview when approaching households and asking them to participate in a wave. If some key information about household characterictics is collected at that stage, non-response analysis will be much easier later on.

Besides non-response and drop-out there is also the problem of 'interviewee fatigue', reflected in a declining trip rate per diary-day. In the Dutch panel the main cause of this in an increasing number of days without any trips. Short trips and discretionary trips have the highest chance of not being reported. The relation between panel wave number and interviewee fatigue is a subject for further research.

c. Measuring change

The paper by Dix and Layzell argues for a dwelling-based instead of a household-based panel in some circumstances. The workshop agreed that this can be useful to monitor changes in a particular study-area. Such a panel, however, cannot answer all relevant questions, e.g. motivations for moving. A dwelling-based panel does not differ much from a rotating panel.

One should be very careful in changing the survey instrument during the lifetime of a panel, because of the influence of the questionnaire on measured mobility. The Dutch panel swifted from a household to a postal survey and then back to a household survey. It may be better to accept some additional drop-out instead of directing the survey instrument to a maximum response with alle the risks associated.

'Learning effects' of panel members during successive waves, attempts to influence the study results, and feedbacks from panel membership to household mobility are a subject for further research.

Seasonal effects and influence of the weather on mobility could be important, but a thorough analysis of the Dutch panel data is not yet possible because of lack of funds.

It was stipulated that monitoring 'no change' can also be important, because of the role of habit formation on household behaviour.

In the Australian panel information is collected both on the changes themselves and on the time they occurred. A problem is analysis, however, is that the values of explanatory variables ideally should then be available for all intermediate points of time between two survey waves.

Attitudinal questions are hardly included in panel questionnaires. Analysis of attitudinal changes over time seems to have a relatively low priority.

The paper by Hensher and Wrigley indicates in which circumstances a panel can be useful:

i) If parameters values are needed at distinct time point of during periods of time and the changes are treated as negligable, repeated cross sections (RCS) can be used.

ii) The same applies to the determination of average values of parameters across periods of time.

iii) If one is interested in measuring net change on an aggregate level (e.g. in public transport use) a RCS is also the relevant tool.

iv) A panel is needed when various components of change should be estimated:

- gross change, i.e. at elemental level between two points
- average change for a individual across several time points
- instability for an individual, e.g. variance of public transport use across 12 months.

v) A panel is also necessary to aggregate data for individuals over time.

vi) To measure frequency, timing and duration of events occuring within a given time period, both RCS (but with telescoping errors) and panels can be used.

vii) To cumulate samples over time, especially samples of rare populations (e.g. handicapped people), both RCS and panels are suitable.

d. Modelling change

A classification of models is given in the paper by Hensher and Wrigley. Critical issues are:

i) structural versus single equation models
ii) ordered versus unordered choices and discrete versus continuous choices
iii) state dependency
  - Markovian
  - occurrence
  - Duration
  - Lagged duration

iv) stationarity
v) heterogeneity
vi) discrete versus continuous time.

In estimation the options are wave-specific modelling, pooling of data, and consideration of the data as an explicit sequence.

Analysis methods for panels are now operational, also in the advanced fields.
The paper presents some of these for discrete choice in discrete time, but also for continuous choice methods are available, e.g. LISREL, COSAM, PLS and EQS.

The problem of 'initial conditions' for the Sydney panel was solved by retrospective questioning about vehicle holdings during the past 15 years.

The MIDCATS model in the paper by Dix and Layzell is based on micro-simulation and is intended for long-term forecasting.
It can also be a useful tool for simulation of household structure and behaviour in a particular study area and for giving insight in the effects of various panel sampling and panel replenishment strategies in that area.

The Cardiff panel was intended to set up a tool for short-term predictions of turnover, number of visitors and the like of shopping centers. It is only intended for situations without major change. The Sydney panel has more ambitious objectives. The long-term forecasting abilities of panel-based models cannot be assessed yet.

WORKSHOP 3

MODEL STRUCTURE AND APPLICABILITY

P. Goodwin, Transport Studies Unit, Oxford (chairman)
F. Moïsi, Ministère de l'Urbanisme du Logement et des Transports (rapporteur)

1. Summary of the workshop papers

The main arguments of the three papers on which the workshop discussion was based were as follows:

Swait and Ben-akiva have implemented a now well established proposition that 'choice' does not occur in a vaccum, but is modified by all sorts of pressures and constraints. One type of these constraints can be dealt with by the concept of 'captivity'. The simplest case of this is the non car-owner who is 'captive' to public transport, but there are many other examples related to more subtle constraints. The analysis demonstrated three essential points. First, the calibration requires more complex techniques (which can, however, be derived by the extension of the theory which has led to convential mode-choice models). Secondly, some results and diagnostics are adequately robust even when the data are unreliable. Thirdly, such analysis actually makes a difference to results and policy assessment -for example as judged by the estimated demand elasticities produced.

Verron and Martens'paper had the primary intention of investigating the use of subjective or attitudinal data on travellers' preference, when assessing reactions not just to simple policy measures but to combinations or 'packages' of policies, especially in relation to public transport. Some of the conclusions from this approach are broadly in line with conclusions that have been derived from more conventional work, for example the importance of reducing walking time, waiting time, and the number of transfers on a journey. Other conclusions go rather further, in particular in emphasising the importance of user comfort (eg protection from the weather), and in stressing the need for 'area-oriented' public transport services. While there factors are widely appreciated, conventional approaches have rarely managed to model them well.

Young surveyed a wide range of different contexts in which 'thresholds' have been used in travel choice models, revealing that the same word has in fact been used in a number of different ways, particularly in residential location, freight, and urban choice models. He argued that the 'Elimination by Aspects' technique may be considered as a probabilistic threshold model, and this can be usefully grouped with the logit model, using improved statistical tests. In some test exercises, the use of implied elasticities again proved a useful diagnostic in demonstrating the differences between threshold and multi-nomial logit models.

2. Overlapping concepts: captivity, thresholds and perception

The three papers were by no means on similar themes, and used very different approaches and methods. Nevertheless taken together they indicate some closely related difficulties in handling their subject.

First there is a close relationship between captivity and thresholds, because there must inevitably be some discontinuity or 'break' involved in becoming captive, or overcoming captivity. This applies equally to a larger class of behavioural phenomena, such as 'habits', constraints or the cost of acquiring information, all of which lead to asymmetry or discontinuity or both in demand relationships.

Secondly, captivity and thresholds simultaneously influence, and are influenced by, attitudes and perception, Therefore there is a real problem about whether preference factors should be an input to, or an output from, demand models. This is especially important when defining the choice set, which clearly must take account of what people think their choice set is. Yet to use a totally subjective definition would involve an unacceptable circularity, and necessitate some rules for knowing how and why the perceived choice set might change.

Thirdly, the formation of preferences and the existence of constraints are both necessarily time-dependent. A binding constraint in the short run, and preference structures at the point of a decision may be quite different from those obtaining after the same decision has been repeated many time and become routine.

3. Context and recommendations

In deciding on appropriate directions for development of the model structure , it is necessary to consider the changing context and the type of applications that will be necessary.

We are no longer in a period of constant , uniform growth with buoyant public finding for transport improvements. This means that models must be able to cope both with situations where incomes, car ownership, public expenditure etc are increasing, and also where they are decreasing. The question of the reversibility of demand relationships therefore becomes much more important, and cannot be taken for granted (as is done, in effect, in all conventional models). In addition we now know that some phenomena are considerably more complex than was thought, and that the track record of well established (and energetically championed) models has not lived up to all the claims made for them, in their ability to avoid biasses, make accurate forecasts and evaluate policies.

Therefore there remains a need to develop models which are more genuinely behavioural, give sound results, are practical to use, and which are based both on established quantitative techniques and on the less restrictive investigations which are possible using social research techniques.

The range of policy interests is now much broader than the traditional preoccupation with journey time and costs (though these remain important) and it is important to be able to address this wider range. The workshop in particular suggested the following policy themes of interest:

- public transport fares are now widely collected by some form of season ticket, cards, or strips which seperate the act of travel from the act of payment.
  Models using the concept of partial captivity can more adequately cope with this form of payment.

- concerning infrastructure development and management, there is increasing need for low cost measures and more efficient use of existing resources. There will be a need for cheaper, more subtle (and perhaps more worthwhile) expenditures than the very large scale traditional new road building. Such policies will include for example real time information technology and adjustment of service frequency and timing. For these model forms which are strong in subjective elements, the analysis of perception and timing constraints will be necessary.

- with the growing issues of highway reconstruction and traffic management, it is necessary to create bridges between transport infrastructure, traffic and land use policies, testing (for example) policies such as flexi-time working hours. Then it becomes important for models to be able to handle time-of-day choice, which in some contexts is more important than mode choice.

- it will also be increasingly useful to cross-fertilize models used for passenger transport and there used for freight which have till now tended to develop quite separately.

In all of these areas, it is considered useful to consider the time-scale or pace, of change. This requires a disequilibrium , dynamic approach in stead of the static, equilibrium methods. It is necessary to address such questions as: how long do thresholds last? What is the mechanism by which short term effects are transformed into long term ones?
In developing a new range of models with an explicit time dimension (based inevitably on longitudinal data) it is expected that considerable advances will be possible in understanding constraints, thresholds, subjective perceptions and the processes whereby each of these are established and changed.

WORKSHOP 4

REPORT OF THE WORKSHOP ON "TRANSFERABILITY AND COST-EFFECTIVENESS"

J. Vogelaar, Netherlnads Economic Institute, The Netherlands

1. INTRODUCTION

The concept of transferability has been widely discussed for several years. At the Eibsee-Conference in 1979, it was the subject of a plenary session in which a number of researchers expressed their confidence in the models' ability to prove transferability of behaviour over time and space. Now, after a period of time with increasing budget constraints, the emphasis has changed to the cost-effectiveness aspect of transferability research.

Section 2 of this report deals with the main conclusions of the papers and discussions. Section 3 summarizes conclusions and R&D recommendations.

2. PAPERS AND DISCUSSIONS

2.1 Applicability and Stability of Logit-Models in Sweden - Some Recent Findings With Policy Implications by Staffan Algers and Staffan Widlert

The results emerge from an ongoing research project concerning the stability of logit models applied in a number of Swedish cities. This is being done by comparing results from various studies, by estimating new models, and by re-estimating earlier models.

It has been shown that it is possible to achieve detailed information on preferences on a level of detail which is of great policy importance, such as having a seat in public transport and the existence of bicycle-lanes.

The models turned out to be reasonably stable across Swedish cities, particularly as far as the value of time in home-to-work travel is concerned. The results are very sensitive to the quality of the level of service data and further work will explore this aspect of the findings. Attempts will be made to compare the Swedish results with evidence from other countries.

2.2 Cost-Effective Methods for National-Level Demand Forecasting, by Andrew Daly and High Gunn

The planning of infrastructure investments for road and railway networks in the Netherlands requires the development of a long term forecasting system. The paper gives a summary outline of the system[1] and of the demand models for passenger transport that are being developed. A number of specific characteristics of the models are:

- the application of pivot-point analyses;
- the dynamic features of the car-ownership and license-holding models;
- the construction of artificial samples in order to deal with regional differences in socio-economic factors, e.g. lifestyles;
- the application of transferability results (see Section 2.3).

2.3 Model Transferability: The Potential for Increasing Cost-Effectiveness, by High F. Gunn and Heero Pol

The paper presents some results from the "Overdraagbaarheid" study, an experiment conducted recently in the Netherlands in which an entire regional travel demand model system was transferred from one region to another. The aim of the experiment was to explore the extent to which results obtained in the first study were replicated in the second. The authors conclude that the results are sufficiently encouraging to proceed to the stage of developing techniques to implement transfers of such model systems without the costly and time-consuming task of conducting additional transportation surveys. They

---

1 On the suggestion of the Project Leader, Dr. Toon van der Hoorn, the version of the paper "Cost-Effective Methods for National Level Demand Forecasting" that is finally published will be extended to include a fuller description of the Landelijk Model project aims and overall organisation, and to clarify some methodological points of relevance of the simultaneous operation of the N.S.E.S. system with other models in the larger project.

stress the importance of gaining cost-effectiveness in the adjustment of earlier results rather than proving a priori hypothesised stability in behaviour.

2.4 Temporal Stability of Transport Demand Models in a Colombian City, by Valerie Jean Kozel

Accurate forecasts require that the model coefficients are temporally stable and that accurate forecasts of explanatory variables can be made.

The author was in the fortunate position to evaluate the temporal stability of logit mode-choice models estimated for Bogota, Colombia, by having two separate travel surveys for the city -- one performed in 1972 and another in 1978. The results show that the influence of income on mode choice is strong and remarkably stable over time, but the level-of-service coefficient turned out to be rather unstable. This could be attributed to the drastic changes of the urban and economic environment that occurred between the two survey dates.

In general, it is difficult to draw strong policy implications from the empirical results since the models failed to predict well over what must be considered a short period with reference to typical transport planning applications.

4. CONCLUSIONS AND R&D RECOMMENDATIONS

The conclusions reached in the workshop discussions are of two types: fundamental observations with respect to the methodology of transferability research and a number of more technical recommendations.

Methodology

- The concept of transferability should not necessarily be associated with the specific type of models (logit) reported in the papers. Its applicability is more general; transferability is about improving models within a continuous learning process, in which one makes use of former experiences in a systematic way.

- With respect to earlier discussions (e.g. Eibsee Conference), the emphasis has changed from transferability to adaptability of models. This is the result of the tendency towards higher cost-effectiveness, dictated by tightened research budgets.

- A great deal of discussion was focused on the notion of "context." Context relates to political and constitutional constraints, cultural aspects, attitudes and even climate, which may vary substantially over time and place. For instance: why do so many Dutch use the bicycle, while in other countries that appear to be suitable for bicycling people do not? Can we measure the context variables? Can we model them and, if not, can we make adjustments for them in our transferability analyses? In other words, in how far can the "context" be made endogenous and where are the boundaries between model and context?

- The considerations on "context" are particularly relevant to the supply side. Our behavioral models are predominantly demand-oriented and much more emphasis should be put on the representation of supply factors, influencing demand. Moreover, it should be indicated which part of the supply side is under government control, i.e. policy-sensitive. It is recommended to put more effort into studies of the behavioural supply side models.

Technical Issues

- The type and quality of survey design has been found extremely important in transferability analysis.
- Choice-based enrichment can lead to substantial cost reduction.
- The techniques of transferring results from a rich data base to another situation with few data should be elaborated further.
- Proper validation is crucial in transferability analysis.
- Further exploration of methods of including transfer errors into forecasts is recommended.
- Exchange of data bases and modelling techniques between countries should be promoted.

WORKSHOP 5

CHAIRMAN'S REPORT: WORKSHOP ON STATED PREFERENCE METHODS

J. Louvière, University of Iowa, U.S.A.

Stated Preference methods involve theory and methods of application that are used to study how human decision processes vary in response to variation in sets of variables of research interest. These methods are based on theory in psychology, known variously as information integration theory (Anderson, 1981, 1982), social judgment theory (See Hammond, et al., 1977), or conjoint analysis (Green, 1974; Green and Srnivasan, 1978). All such methods share in common the fact that individuals are exposed to a set of scenarios or descriptions of some phenomenon of interest; for example, individuals might be asked to evaluate a number of different commuter train options described by fare, travel time, scheduling, and points of arrival and departure. These descriptions or scenarios are generated by some controlled, experimental design procedure, usually some variation of a factorial design. These designs are used to insure that the independent variables are truly independent so that efficient estimates of the effects of these variables on the individual's responses can be obtained.

Stated preference techniques have a number of advantages over traditional methods for studying traveler responses to new or improved transport alternatives. These advantages include the ability to study the effects of variables that may not now exist, or that have only limited variability. In addition, modes or routes that do not now exist can be analyzed and the natural correlations that exist among transport variables, such as travel time and travel cost can be removed. Because stated preference techniques make effieicnt use of data, separate models for each individual traveler can often be obtained, permitting detailed microanalysis, as well as providing a rich base for segmentation and target marketing of transport options. The charge of this workshop was to examine the state-of-the art in these techniques, and to reach conclusions and make recommendations regarding both scientific and practical research. This charge required the workshop to discuss what is known about such techniques, as well as what is not known; and to identify priority areas for both research and application. This report focuses upon these discussions, conclusions and recommendations.

The workshop had the advantage of having little disagreement over whether stated preference methods can be of value in transport research, and of having three excellent papers that provided a good review of relevant work in transport and empirical examples of applications. As a result, agreement on the direction of the workshop and an agenda for discussion were quickly reached. This permitted the workshop to cover considerable ground, the full extent of which is reflected in the summary of conclusions and recommendations discussed later in this report.

The workshop began its agenda by asking the authors of the three resource papers to briefly outline the main points of their papers, with an eye to using communalities as a point of departure. Although the papers ostensibly dealt with different topics, each raised a number of questions regarding unresolved problems in applications, most of which were considered by the worskhop. The three papers illustrate the diversity of problems to which stated preference methods can be applied: John Bates used stated preference methods to derive values of time for use in evaluating investment options, Mark Bradley and Piet Bovy used stated preference methods to tackle the difficult problem of defining route preferences, while Eric Kroes and Robert Sheldon compared and contrasted stated preference methods and more traditional attitudinal methods and made suggestions regarding useful areas of application for stated preference methods.

In addition to its resource papers, the workshop considered a wide variety of other applications, including applications already completed and possibilities not yet realized. Of particular note were presentations made by John Segal of British Rail, outlining the current program of research applications using stated preference techniques, by Eric Kroes and Robert Sheldon concerning their approach to developing respondent-specific data collection procedures accomplished with a personal interview format, and by Jordan Louviere, dealing with the development of fully competitive choice models that are compatible with current econometric choice models. (See, e.g., Louviere and Woodworth, 1983). These applications served to illustrate the wide variety of practical problems that can be addressed using stated preference methods, the variations possible in applying the methods and new developments in the state-of-the-art.

Following the discussion of the resource papers, the workshop settled on an overall agenda, focusing upon two general topics : 1) theory and methods, and 2) surrounding issues. Each general topic had several subtopics. For example, the theory and methods topics included nine general issues: i. orthogonality versus "natural" correlation; ii. repeated measures analysis problems; iii. the number and nature of levels that can be used with experimental factors; iv. the number of factors that can be investigated in practical research problems; v. how to handle non-linearities and non-additivities in utility specifications; vi. survey format possibilities; vii. using stated preference methods to tests deductions from or hypotheses about models and/or theories; viii. design of experiments that permit a researcher to obtain estimates of the effects of the variables of interest without greatly complicating the problem or the survey; ix. pros and cons of the use of objective compared with subjective measures of explanatory variables. The surrounding issues topic area had eight subtopics: i.comparison of predictions of behavior from stated preference models with real behavior and the general nature of the linkage to same; ii. external validity tests of stated preference methods; iii. extrapolating into the future; iv. determining appropriate for different types of problems; v. learning effects in stated preference surveys; vi. non-response bias in stated preference surveys; vii. respondent game-playing in stated preference surveys; and viii. practical application considerations.

Following agreement upon the preceeding agenda, the workshop discussed as many of these topics as possible, the goal being to summarize the state-of-the-art for each subtopic, and to make research recommendations about unresolved problems, as well as recommendations for the implementation of practical applications.

The orthogonality versus natural correlation issue involves the use of uncorrelated versus correlated independent variables, respectively. This controversy has existed over 50 years and is still unresolved, althought recent research in decision making suggests that orthogonality does not preclude high external validity, and supports the external validity of orthogonal designs. New developments in design theory reported by Jordan Louviere also make it possible to incorporate natural correlations into stated preference studies (such as the positive relationship between travel time and travel cost) while still retaining orthogonality properties. Such designs require differences in attribute levels to be orthogonal, but the attributes themselves need not be orthogonal.

The problem of analyzing repeated measures data was outlined, and it was suggested that researchers use experimental designs that make it possible to estimate the effects of all of the variables of interest for each respondent. When this becomes difficult or impossible, researchers should consider designs that allow some effects to be estimated for individuals, while permitting other effects (typically interaction effects) to be estimated for groups of individuals. This problem is similar to pooling cross-sectional and time-series data; and requires similar statistical reasoning to obtain the appropriate tests of the effects of interest.

How many factors and how many levels of these factors could be investigated in practical problems was also considered. Well-known technology in the design of fractional factorial experiments permits the use of as many as 15 variables or factors, with designs small enough to be useful in practical applications. New developments also make it possible to greatly increase the number of levels of quantitative factors, while not increasing design sizes in practical applications (See Louviere and Anderson, 1985). Indeed, a wide variety of designs and combinations of designs are already cataloged and available (e.g., Hahn and Shapiro, 1966, reprinted in Kocur, et al., 1982); these fractional designs can be adapted to most practical problems, making it possible to find one or more designs that are both small enought to administer in practical situations and are sufficient to estimate the effects of interest. The workshop also recommended that the levels of the variables used in stated preference studies be as realistic as possible. The use of photographs, mock-ups, models, real variables and the like are therefore encouraged because the results of stated preference studies depend upon replication of the decision situation of interest for the respondents so that they will respond as they would in the real world.

Non-linearities and non-additives in utility specifications were discussed, and it was concluded that if one believes such effects exist, the experimental design utilized should be able to detect them. The only barrier to estimating such effects is that usually this requires enlargement of the experimental design. It was noted, however, that most effects of interest in practical transport research should be monotonically related to utility. If true, small designs can efficiently estimate the effects because most of the variation in such effects is concentrated in the linear-by-linear interaction components, or in the linear components of non-linearities. So, it was concluded that both theory and practical designs were already available to solve this problem.

Survey formats for the administration of stated preference studies were considered, and it was concluded that a variety of formats such as personal interviews or mail surveys were possible, but telephone applications were probably infeasible. Learning effects were of major concern, but could be dealt with by giving respondents 3-5 practice treatments or profiles which used a mix of all of the levels of all of the factors. This allows the respondents to learn the context of the decision task, and to decide on their decision strategy. If practice trials are not given, respondents use the first few survey questions of interest to learn, which biases the resulting model estimates.

Problems of objective versus subjective measurement and the use of stated preference methods to test models or theory were given lower priority by the workshop, and no resolution was reached on them.

The workshop, however, did spend a considerable amount of time on external validity and the linkage of stated preference results to real behavior. It was concluded that existing evidence (e.g., Levin, et al., 1983) was inconclusive as to whether stated preference results were linked to real behavior in a general way. In particular, almost all available evidence deals with aggregates of people, and is therefore inconclusive, even if positive. The workshop recommends that priority be given to investigating the external validity of stated preference methods, so that generalizations can be made regarding practical applications. It was recommended that such tests concentrate on forecasts of changes in real systems that can be compared "before and after", or on "backcasts" to one or more changes in transport systems (See Louviere and Kocur, 1982 for an example of "backcasting").

Regarding what types of samples are appropriate for what problems, it was noted that samples of size 30 are generally sufficient to gain a preliminary idea of the effects of interest. Thus, pilot work with small samples using relatively larger designs or combinations of designs can be used in many studies. Otherwise, stated preference methods have sampling problems similar to those of traditional econometric demand methods. The major advantage of stated preference methods is that they obtain considerably more data from individuals, making use of the additional data in statistical estimation. Thus, smaller samples are generally required compared to more traditional transport planning methods.

Non-response bias and respondent game playing were considered, and it was concluded that respondents should be informed that attempts to influence policy instead of honest responses to questions would be detected. There has been insufficient research into non-response bias for the workshop to reach any general conclusions. However, the workshop felt that this should be a priority research item for the near future.

The workshop concluded that stated preference methods were appropriate for short-term forecasts, but that their use for longer-term forecasts should be discouraged. Instead, the workshop recommended regular updating of results, perhaps using panels established for that purpose. Because stated preference studies are less expensive and can be designed and administered more easily than more traditional data collection methods, it should be possible to frequently update results, perhaps every year or so.

Finally, the workshop participants developed a research agenda for the future, dividing their research recommendations into scientific and practical research questions. The scientific research recommendations are as follows:

1. External validity should be given high priority because it is important to know the degree of external validity both from a scientific perspective and a practical perspective. Policy makers who rely of the results of stated preference methods must be able to place confidence in them, and at present we can only conclude that there is no negative evidence.

2. Research should be undertaken to determine the minimum sample sizes required for efficient estimation of effects for various stated preference approaches.

3. Comparisons of the results obtained from verbal, pictoral

and actual levels of experimental factors should be undertaken to provide directions for using one or the other of these options.

4. Results obtained from mail surveys should be compared to those from personal interviews to provide insights into biases that may exist and to give advice on when to use what.

5. The effects (if any) of non-response bias for a variety of transport decisions should be studied to provide guidance for the control of this source of bias in results.

6. Incorporation of external factors, such as budget constraints, perception biases and context effects, into stated preference studied should be investigated. Little evidence currently exists to guide the incorporation of these factors into the results other than as segmentation or sampling criteria.

7. A solution should be found to the problem of analyzing repeated measures data from stated preference studies. The workshop believes that the solution to this problem must already exist, probably in the econometrics literature.

The recommendations for practice are as follows:

1. Stated preference techniques be used to define traveler priorities for particular modes, e.g., to determine the relative importance of travel times and costs for inter-city buses, and to determine traveler tradeoffs among these variables.

2. Stated preference methods be used to define hard-to-measure qualitative variables, and to establish their importance in traveler choices, e.g., measuring the "convenience" of modes, and developing a functional specification for "convenience" based on a set of measureable variables that vary within and between modes.

3. Stated preference approaches be applied to the measurement of user valuations for use in planning studies, e.g., establishing

dollar values for in-vehicle and out-of-vehicle time.

4. That stated preference techniques be used to include variables that are hard-to-measure or have limited variability in demand models, e.g., there is often limited variation in traveler perceptions of the cost of traveling by automobile, and in variables like gasoline and parking costs.

5. Stated preference methods be used to develop detailed market segments based on similarities in preferences, and that these segments be used for target marketing and analysis of new transport alternatives as well as transport improvements.

6. Stated preference approaches be used to evaluate traveler responses to new transport alternatives, and to forecast likely choices and/or changes in choices that might result from the implementation of various planning alternatives.

LIST OF REFERENCES

Anderson, N.H. (1981) Foundations of Information Integration Theory. New York: Academic Press.

Anderson, N.H. (1982) Methods of Information Integration Theory. New York: Academic Press.

Green, P.S. (1974) "On the Design of Choice Experiments Involving Multifactor Alternatives". Journal of Consumer Research. 1 (September), 61-68.

Green, P.E. and V. Srinivasan. (1978) "Conjoint Analysis in Consumer Research: Issues and Outlook". Journal of Consumer Research. 5 (September), 103-123.

Hahn, G.J. and S.S. Shapiro. (1966) "A Catalog and Computer Program for the Design and Analysis of Orthogonal Symmetric and Asymmetric Fractional Factorial Experiments". Technical Report No. 66-C-165. Schenectady, N.Y.: General Electric Research and Development Center, General Electric Corporation.

Hammond, K.R., Rohrbaugh, J., Mumpower, J. and L. Adelman. (1977) "Social Judgment Theory: Applications in Policy Formulations". In Kaplan, M.F. and S. Schwartz (Eds.), Human Judgment and Decision Processes in Applied Settings. New York: Academic Press, 1-29.

Kocur, G., Adler, T., Hyman, W. and B. Aunet (1982) "Guide to Forecasting Travel Demand with Direct Utility Assessment". Washington, D.C.: U.S. Department of Transportation, Urban Mass Transit Administration.

Levin, I.P., Louviere, J.J., Norman, K.L. and A. Schepanski (1983) "Validity Tests and Applications of Laboratory Studies of Information Integration". Organizational Behavior and Human Performance, 31 (April), 173-193.

Louviere, J.J. and D.A. Anderson. (1985) "New Developments in Second-Order Response Surface Designs and Their Application to Conjoint Judgment and Choice Problems". Paper to be presented to the Association for Consumer Research, Special Session on Advances in Consumer Choice Modeling, Las Vegas, Nevada, October, 1985.

Louviere, J.J. and Kocur, G. (1983) "The Magnitude of Individual Level Variations in Demand Coefficients: A Xenia, Ohio, Case Example". Transportation Research A. 17A (5), 363-374.

Louviere, J.J. and Woodworth, G.G. (1983) "Design and Analysis of Simulated Consumer Choice or Allocation Experiments: An Approach Based On Aggregate Data". Journal of Marketing Research. 10 (November), 350-367.

WORKSHOP 6

HOUSEHOLD STRUCTURE AND LIFE STYLE

Edward Weiner

Department of Transportation, Washington DC

Methods and Definitions

Research in the area of household structure and life style is directed at integrating new variables from social science fields into analysis techniques to improve the explanation of household decisionmaking and travel behavior. This has become important because of the large changes that have occurred, and continue to occur, in household structure and life style. These changes include increases in the divorced persons, single parent households, two-worker households, childless couples, retired persons and greater labor participation rate of women.

These new variables should not simply be added to the vector of current variables but be integrated in such a manner that they capture the more complex aspects of household decisionmaking and travel behavior. The result should be new household classification schemes which define homogeneous groups in terms of their travel behavior and patterns.

Conventional travel forecasting methods embody travel relationships that may no longer be valid for many groups in the population. The paper "Household Lifecycles: Predictor of Travel Expenditure," by L. Kostyniuk and R. Kitamura discusses the changes in household travel patterns over time for Detroit, Michigan and Rochester, New York.

The focus of conventional travel forecasting has been on workers and work trips. Too little attention has been given to other segments of the travel market. In fact, entire groups have not been included in conventional forecasting procedures particularly women and children. Travel patterns of non-traditional families in Lyons, Rotterdam, and Austin, Texas are discussed in the paper, "Employment, Childcare and Travel Behavior: France, The Netherlands, The United States," by C. Raux and S. Rosenbloom.

Consequently, new relationships will need to be formulated for many travel forecasting purposes. These new relationships may also change in the future and procedures need to provide for such changes.

Current forecasting procedures were developed for regional scale transportation planning directed and major transportation facilities. New methods are needed to focus on community and neighborhood level transportation problems. Some progress has been made in this regard. The paper, "Environment, Conditions of Life and Traffic," by J. Keldborg describes one such technique for Haderslev, Denmark. These approaches promise to be more helpful for involving the public in identifying transportation problems and solutions.

In regard to the definition of "life style" and its relationship to the concept of "life cycle", life style provides additional dimensions to life cycle which improves the explanation of household decisions and travel behavior. There is not sufficient information at this time to develop a single formulation for integrating life cycle and life style.

Currently, there are not any methods for forecasting life styles into the future. Consequently, the use of life styles in planning will need to be confined to short term problems for the time being.

Research Recommendations

In the area of household structure and life style, further research is needed to develop new classification schemes which capture relevant life style variables. These new classification schemes need to be sensitive to changes which have occurred in household structure and which contribute to the explanation of household decisions and travel behavior. The schemes should not be totally static but allow for incorporation of future changes that might occur.

More research is needed on evolving household structure particularly the behavior of atypical groups that have received little attention so far. These include women, one-parent households, children, retired persons and "yuppies" (young upwardly mobile persons).

Greater attention needs to be focused on longer distance travel. This is especially true for travel on the weekends for which there has been very few studies. There will be increased non-work travel in the future and a better understanding of this travel market will be even more important than it is today.

Increased consideration is being given to the deregulation of regional and local public transport. More reliance will be on the market place and less on regulation. This will likely result in greater numbers of transport providers in the market, particularly private sector providers. Less coordination of these services is likely to be another consequence. The tripmaker will be faced with a wider range of services, fares, and other attributes.

Reliability may also vary. Obtaining complete and accurate information on the various options may be a problem for tripmakers. Currently, little is known about these changes on travel decisions and trip patterns. There are a wide range of issues that are required to be researched on the effects of deregulation.

Research has be conducted on market segmentation techniques for public transport users. Some of these market have been analyzed especially these specialized travel requirements of the elderly and handicapped. With the growing trend toward pricing public transport closer to actual cost, there is interest in segments of the population that can and are willing to pay the full cost of high quality public transport.

Over the last decade, greater value has been placed on the "quality of life" by some groups in the population. This has certainly affected their life style. Research is needed to explore these life style changes and the resulting affect on travel behavior.

Preliminary evidence suggests that two-worker households tend to select locations that are accessible to many jobs and then move infrequently. Other evidence suggests that transportation has become less important in all household location decisions. More needs to be known about household decisionmaking in regard to residential location and mobility especially for two-worker households which are sharply increasing in numbers in most Western nations.

As the industrial mix of nations evolves from manufacturing to information type industries, transportation requirements are changing. Information based industries have different location behavior than manufacturing industries and will likely cause different household decisions on residential location and travel. Although some research has been conducted on these subjects, much more needs to known about the effects of these major shifts in economic structure.

Many of the changes in travel patterns in recent years have been due to changes in society. Changes are likely to continue. Further research is needed to identify these emerging trends in society and to determine their affect on household decisions and life style.

Policy Implication

Most of the changes in travel patterns over the last 20 years have been the result of changes in household structure and life style. It is evident, therefore, that these changes need to be monitored in in the future.

The development of transportation systems is still not well integrated with land development and use. Households are giving little consideration to transportation factors when making location choices. Transportation is considered afterwards. Households often do not understand the extent of their travel problems when making location decisions, especially where households locate in lower density suburbs. It is then expected that the transportation system can be modified to straighten the resulting problems.

There is poor coordination of public decisions made in different sectors of government. Transportation, housing, land use, industrial location and environmental decisions are often made with little consideration of the consequences in other sectors. Problems could have been prevented or reduced if early consultation had occurred when implementing public programs and policies.

Trends in household structure and lifestyle suggest decreased usage of public transit. For example, two-worker and single-parent households need highly reliable transportation for short trips to serve their complex, time constrained lifestyle. Public transport systems are not serving this market well or, in some instances, not at all. Moreover, rising incomes and dispersed living patterns are fostering increased automobile ownership and use. In addition, younger women are more likely to have a driver's license and drive than older women and, thereby, more frequently use an automobile for travel. Even older persons who have in the past relied heavily on public transport are healthier and have sufficient income so that many of them are continuing to use automobiles rather than switch to public transport.

These trends indicate that the need for public transport is for short trips within a community rather than for large regional scale projects of constructing new infrastructure particularly rail transit systems. Public transport services will need to be better targeted to the various market segments. These can be facilitated using studies of specific market segments particularly those for non-work trips. Travel behavioralists will need to better explain behavioral techniques and their application to the problems faced by transit operators.

There are groups in the population which rely predominantly on public transport for mobility. As public transport operators face financial difficulties and reduce service, the mobility of these groups will become constrained. These problems will need to be identified and addressed.

Recent evidence indicates that current travel forecasting techniques may embody demand relationships which are no longer accurate. Consequently, these techniques may no longer be forecasting correctly. They need to be reevaluated and updated where they are found invalid.

There is strong evidence that travel decisions are based on other factors in addition to travel times and costs. Household decisionmaking takes into account other variables such as reliability, schedule times, needs for chaining of trips, travel requirements of other household members, etc. These factors should also be considered in the selection of transportation improvements.

A large share of tripmaking is for short trips made generally within the same community. In certain areas, bicycles are a good means of providing this mobility. They can even provide a suitable alternative for automobile especially for these short trips. Greater consideration needs to be given to planning for the safe movements of bicycles in the transport system.

Decisionmakers need to be informed on the various aspects of societal change which affect the travel behavior of households. Changes in the economy, demography, attitudes, migration trends, social structure, life styles, for example, can be provided to decisionmakers as background so that they can make more informed decisions.

WORKSHOP 7
MEASUREMENT OF VALUATIONS

David T. Hartgen,
New York State Department of Transportation,
Albany, NY, U.S.A.

Carina van Knippenberg,
University of Groningen,
The Netherlands

Workshop 7 focused on structural and technical issues relating to the measurement of perceptions, attitudes, values, and related issues. The goals of the workshop were to (1) describe and codify the methods presently available for measuring these concepts, (2) relate methods to transportation policy applications, and (3) identify research needs for the future.

The workshop heard several papers describing attitude and evaluation measurement techniques. The paper by Forsyth and Smyth described applications of attitudinal research to strategies for improving the bus system of Tyne and Wear, England. The paper by McKnight, Pagano, and Paaswell described an application of attitude measurement to estimating demand for specialized (handicapped persons) services in the Chicago area. Benjamin's paper provided an overview of various attitude measurement techniques, and an assessment of appropriate applications and research needs. In addition, the workshop heard a brief presentation by Kroes and Sheldon describing attitude models and stated preference techniques. These papers formed a review of state-of-the-art in various attitude and user valuation techniques, and a sound underpinning for discussions.

The workshop extensively discussed the structure of attitudes and related concepts, and their applications to transportation. The primary output of these discussions was a conceptualisation of seven separate theories of human behavior with respect to transportation, and clarification of the subject matter of each of these theories. The results are shown schematically in Figure 1 and are summarized briefly as follows:

1. Fishbein and Ajzen. A structural theory relating attitudes and social norms to behavioral intent (and occasionally to behavior), based on the principle that attitudes are formed by the components value and belief, and social norms by the components motivation to comply and normative belief.

2. Attribute Adequacy. A variant of the above, in which attitudes are linked directly to behavior by correlating measures of satisfaction and importance on numerous model attributes with e.g. mode choice, within the context of broad social norms.

3. Information Integration. A theory relating elements of the physical environment directly to behavioral intent and behavior, bypassing the specific structure of internal evaluation. Most commonly described techniques are functional measurement, trade-off, and conjoint measurement.

4. Household Adaptation. A broad theory in which elements of the environment, household roles and responsibilities are assessed and allocated to individual family members for accomplishment. Behavior and behavioral intent are seen as outgrowth of household interactions.

5. Individual Adaptation. A microscopic theory, in which individual perceptions, emotions, values, suppressed feelings, etc. are assumed to influence individual choices of products and services. Similar to typical marketing ideas.

6. Cognitive Dissonance. The well-known theory of Festinger, in which attitudes and behavior are seen to interact and seek consistency.

7. Self-perception. A theory that states that verbalizations (reports) of attitudes depend on prior behavior and decision making.

The status of these theories in transportation is described briefly in Table 1. The first two, (Fishbein and Ajzen and attribute adequacy) have led to the development of advanced techniques with a large number of applications in transportation, particularly in urban mode choice. Integration and adaptation theories are considerably less well developed, and have a limited number of applications. Their primary drawback appears to be the complexity of the methodologies, and the significant requirements for interviewer understanding which limits sample sizes substantially. On the other hand, they are considerably richer in ability to understand non-traditional alternatives and to relate choices to background factors. The last two theories, cognitive dissonance and self-perception, have seen only limited transportation development, although they have been applied extensively in other areas.

As mentioned above, potential applications of several of these techniques in the transportation sector have been extensive (Table 2). The Fishbein and Ajzen model has been applied extensively to mode choice problems, and has the potential for application in numerous other system-level studies as well as car ownership choice and marketing. It is aspatial in the sense that it does not require an urban context. Similarly, attribute adequacy has found a large number of applications, particularly in urban choice. It is particularly easy to understand and use and forms a basic methodology for improving transit services.

Applications of information integration and adaptation theories are considerably less extensive. Information integration, relating choice to the characteristics of alternatives, essentially by-passes the attitude formation stage and models around the internal decision making process of individuals. While it has wide application potential, its usefulness is limited by the complexity of the design and the necessity to stay within a small number of clearly articulated choices. Since such choices are declining, its future applications may be limited to major downside changes in services level, of large fare increases.

The remaining theories have not seen significant applications in transportation, although a possible exception may be the HATS and CARLA versions of household adaptation concepts. However, these techniques have been applied primarily outside of the United States, and even there the applications are quite limited.

The workshop extensively discussed research needs for these areas. An overall conclusion was that all the theories lack a coherent synthetic underpinning in behavior, and that basic research is necessary to flesh out the structure of these theories in clearer terms. For some well-developed techniques, the basic problem appears to be marketing of the theories: transportation analysts have been reluctant to use some of these techniques because of failure to understand their capabilities, as well as inaccessability of the literature and write-ups. This is particularly true for the integration and adaptation methods. The workshop concluded that the greatest likely payoff would come from expansions of household adaptation and individual adaptation theories, since these appear to have the greatest conceptual potential as well as the lowest present application. An important improvement in the information integration method would be to develop a procedure capable of being operated over the telephone. In addition to these specific research suggestions, the workshop also indentified understanding of measurement effects, the reactive effects of measurement on the design itself, and the role of habit, as two additional important areas of research.

The workshop concluded, in summary, that numerous techniques for measurement of valuation are now available and need further application. The field lacks a broad understanding of travel behavior which would overarch the various, more detailed theories and application techniques. Research on the development of such an overarching theory should continue. In the interim, many procedures can be applied in practice, but need better selling and dissemination of research results to achieve greater use. For the immediate future, the straightforward methods of Fishbein and Ajzen and attribute adequacy are likely to remain the backbone of most attitudinal applications in transportation.

TABLE 1

STATUS OF THEORIES

Theory	Purpose	Tools	Behaving Unit	Content
1. Fishbein & Ajzen	Explain behav. by beliefs	Semantic Differential	°individual (within HH, society) °interviews °postal quest self-admin.	researcher selects attributes of different options from open-ended pilot max. 5-7 choices
2. Attribute Adequacy	Explain/predict behavior	Unidimen. scales satisfaction-importance	°individual °postal quest °flexible (large sample)	researcher selects items from open-ended pilot 50+ quests
3. Information Integration	Product preferences	ratings or rankings of combinations of factor levels	°individuals °laboratory studies small can be large if self administered °complex	°limit to size °researcher determines °4-6 variables tops
4. Household Adaptation	Explain-understand-predict non-traditional choices	°Situational approach (Brög) °HATS/CARLA °observation/unstructured	°household (individual) °small sample (20) can be larger	°constraints links °very qualitative
5. Individual Adaptation	Understand individual's motivations, product attitudes	°In depth interviews °MDS °product market studies	°individual (in context of family)	°free form interviews °emotions °exploratory unstructured
6. Cognitive Dissonance	Explain behavior & cognitions	Unidimen. scales	°individual large scale cross-section before-after laboratory studies smaller)	literature review or select items from in-depth interviews
7. Self-Perception	select influencing strategies, modify behavior	unidimensional scales	°individual before-after laboratory studies (small)	researcher selected

A = Attitudes
C = Cognitions
BI = Behavior Intention
B = Behavior

Analysis	Interpretation	Generalizability	Attit- Beh.-Links
°Regression °logit °descriptive statistics	°Behavior intention °face value °direct from results	from data if random sample	A → BI ↘ B
°ANOVA °std. stats °factor analysis °clusters °regression °logits	°direct from results °graphics	OK from random sample	A → B
°regression °ANOVA °logit °on individuals and groups	°Coefficients have "elasticity" meaning °simulation of market response °Mlogit	°Laboratory can cause problems	predicts behavior (strongly?)
°content analysis °situational analysis °simulation possible (CARLA)	°direct from results °(needs specific interviewers skills)	°depends on sample	A → B (household view)
°content analysis °MDS °some statistics	°subjective °specific skills	°difficult	A → Beh.
°t-tests °ANOVA °MANOVA of repeated measures °econometric	°direct from results °graphics	°Laboratory can cause problems °OK from random sample	C ⇄ B (both measured)
°t-test °ANOVA °(M)ANOVA of repeated measures	direct from results	°random sample needed °good design °laboratory can cause problems	B → C

TABLE 2

POLICY APPLICATIONS & RESEARCH NEEDS

			Transportation Applications					Other Applications			Research Needs	
Theories		Goal	Subject Focus	Context Focus	Costs	Benefits (Theoretical)	Benefits (Practice)	Business	Academic	Public Sector	Applied	Basic
Fishbein & Ajzen	P	explain	mode choice dest. type change	urban (rural)	M	H	?	product marketing	psych exp.ps.	environment	sell! "change" uses	link to system vars norms link habit?
	F	predict change	syst.car/ market	intercity aspatial all								
Attribute Adequacy	P	describe predict	system ch. mode choice	intercity urban	L	L	H	?	marketing	?	"loss leader"	X
	F	---		all								
Information Integration	P	explain predict	mode choice Rt./dest. /serv.	urban	H	H	H	new product design	←		big downside differences	?
	F	(big difference choices, few# attributes)		interurban							use by telephone simple version	
Household Adaptation	P	describe explain understanding	mode/dest.	urban rural non traditional solutions	H	H	?	clinical	group dynamics roles lifestyle	?	"sell"outside Europe	validation how/why? quant. stats
Individual Adaptation	P	change	cars	ind. focus	H	?	H	clinical product design advert. marketing motivation	← ←	"I love NY"	modes/etc. urban intercity	"theories" !!!

	F		transport mode	All (inter)							
Cognitive Dissonance	P F	explain predict & change of attitudes	mode choice service dest./mode choice	urban all	L	M	?	marketing ←	psych.	?	adv. campaign for transit marketing strategy	exp.→theory e.g. mode choice
Self Perception	F	explain	service provided dest./mode choice	all	L-M	?	?	---	psych.	---		exp.research theory

P = present
F = future

FIGURE 1
THEORIES OF BEHAVIOR

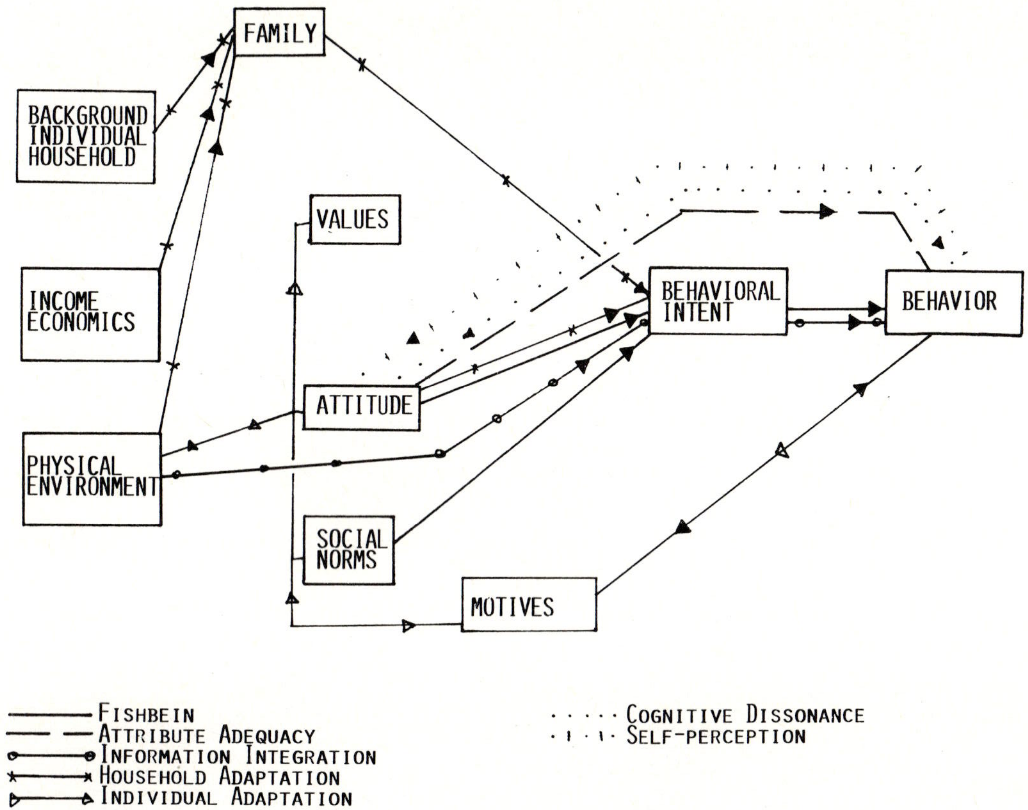